SIXTH EDITION

Fundamentals of Physics

PART 2

ENHANCED PROBLEMS VERSION

SIXTH EDITION

Fundamentals of Physics

PART 2

ENHANCED PROBLEMS VERSION

David Halliday
University of Pittsburgh

Robert Resnick
Rensselaer Polytechnic Institute

Jearl Walker
Cleveland State University

John Wiley & Sons, Inc.

ACQUISITIONS EDITOR   Stuart Johnson
DEVELOPMENTAL EDITOR   Ellen Ford
MARKETING MANAGER   Bob Smith
ASSOCIATE PRODUCTION DIRECTOR   Lucille Buonocore
SENIOR PRODUCTION EDITORS   Monique Calello, Elizabeth Swain
TEXT/COVER DESIGNER   Madelyn Lesure
COVER PHOTO   Tsuyoshi Nishiinoue/Orion Press
PHOTO MANAGER   Hilary Newman
DUMMY DESIGNER   Lee Goldstein
ILLUSTRATION EDITORS   Edward Starr and Anna Melhorn
ILLUSTRATION   Radiant/Precision Graphics
COPYEDITOR   Helen Walden
PROOFREADER   Lilian Brady
TECHNICAL PROOFREADER   Georgia Mederer
INDEXER   Dorothy M. Jahoda

This book was set in 10/12 Times Roman by Progressive Information Technologies and printed and bound by Von Hoffmann Press, Inc. The cover was printed by Von Hoffmann Press.

This book is printed on acid-free paper.

The paper in this book was manufactured by a mill whose forest management programs include sustained yield harvesting of its timberlands. Sustained yield harvesting principles ensure that the number of trees cut each year does not exceed the amount of new growth.

To order books or for customer service call 1-800-CALL-WILEY (225-5945).

ISBN 0-471-22859-1

Printed in the United States of America

10 9 8 7 6 5 4 3 2 1

BRIEF CONTENTS

TABLES

CONTENTS

CHAPTER 18

Waves—II   398

How does a bat detect a moth in total darkness?

CHAPTER 19

Temperature, Heat, and the First Law of Thermodynamics   425

What thermal protection do bees have against a hornet invasion?

CHAPTER 20

The Kinetic Theory of Gases   454

Why can a cold container of a carbonated drink form a fog when opened?

CHAPTER 21

Entropy and the Second Law of Thermodynamics   482

What in the world gives direction to time?

APPENDICES

ANSWERS TO CHECKPOINTS AND ODD-NUMBERED QUESTIONS, EXERCISES, AND PROBLEMS

INDEX

PREFACE

ENHANCED PROBLEMS VERSION

The Goal

The principle goal of this book is to provide instructors with a tool by which they can teach students how to effectively read scientific material and successfully reason through scientific questions. To sharpen this tool, this enhanced version of the sixth edition of *Fundamentals of Physics* contains over 1000 new, high-quality problems that require thought and reasoning rather than simplistic plugging of data into formulas.

The new problems are printed at the end of each chapter. They are not presorted in any way for students but, pedagogically, most of them can be grouped into three types:

➤ *Graphs as puzzles.* These are problems that give a graph and ask for a result that requires much more than just reading off a data point from the graph. Rather, the solution requires an understanding of the physical arrangement in a problem and the principles behind the associated equations. These problems are more like puzzles in that a student must decide what data are important, like Sherlock Holmes must decide what evidence is important. Here are three examples from over 130 similar problems:
- Chapter 9, Problem N14
- Chapter 22, Problem N3
- Chapter 30, Problem N6

➤ *Novel situations.* Here is one example:
- Chapter 5, problem N1: A true story of how Air Canada flight 143 ran out of fuel at an altitude of 7.9 km because the crew and airport personnel did not consider the units for the fuel. An important lesson for students who tend to "blow off" units.

➤ *Sorting-out problems.* These are problems that focus a student on a derivation, a commonly confusing point, or the limitation of an approximation. Here are three examples:
- Chapter 11, Problem N9
- Chapter 18, Problem N14
- Chapter 26, Problem N12

Sixth Edition

The sixth edition of *Fundamentals of Physics* contains a redesign and major rewrites of the widely used fifth edition, while maintaining many elements of the classic text first written by David Halliday and Robert Resnick. Nearly all the changes are based on suggestions from instructors and students using the fifth edition, from reviewers of the manuscripts for the sixth edition, and from research done on the process of learning. You can send suggestions, corrections, and positive or negative comments to John Wiley & Sons (http://www.wiley.com/college/halliday) or Jearl Walker (mail address: Physics Department, Cleveland State University, Cleveland OH 44115 USA; fax number: (USA) (216) 687-2424; or email address: physics@wiley.com). We may not be able to respond to all suggestions, but we keep and study each of them.

Design Changes

➤ *More open format.* Previous editions have been printed in a double-column format, which many students and instructors have found cluttered and distracting. In this edition, the narrative is presented in a single-column format with a wide margin for note-taking.

➤ *Streamlined presentation.* It is a common complaint of all texts that they cover too much material. As a response to this criticism, the sixth edition has been shortened in two ways.

1. Material regarding special relativity and quantum physics has been moved from the early chapters to the later chapters devoted to those subjects.

2. Only the essential sample problems have been retained in the book.

➤ *Vector notation.* Vectors are now presented with an overhead arrow (such as $\vec{F}$) instead of as a bold symbol (such as **F**).

➤ *Emphasis on metric units.* Except in Chapter 1 (in which various systems of units are employed) and certain problems involving baseball (in which English units are traditional), metric units are used almost exclusively.

➤ *Structured versus unstructured order of problems.* The homework problems in this book are still ordered approximately according to their difficulty and grouped under section titles corresponding to the narrative of the chapter. However, the new problems that are now at the end of each chapter are not ordered or grouped in any way.

➤ *Icons for additional help.* When worked-out solutions are provided either in print or electronically for certain of the odd-numbered homework problems, the statements for those problems include a trailing icon to alert both student and instructor as to where the solutions are located. An icon guide is provided here and at the beginning of each set of homework problems:

> ssm   Solution is in the Student Solutions Manual.
> www   Solution is available on the World Wide Web at:
>   http://www.wiley.com/college/halliday
> ilw   Solution is available on the Interactive LearningWare

These resources are described later in this preface.

Pedagogy Changes

➤ *Reasoning versus plug-and-chug.* A primary goal of this book is to teach students to reason through challenging situations, from basic principles to a solution. Although some plug-and-chug homework problems remain in this book, most homework problems emphasize reasoning.

➤ *Key Ideas in the sample problems.* The solutions to all the sample problems have been rewritten to begin with one or more Key Ideas based on basic principles.

➤ *Lengthened solutions to sample problems.* Most of the solutions to the sample problems are now longer because they build step by step from the beginning Key Ideas to an answer, often repeating some of the important reasoning of the narrative preceding the sample problems. For example, see Sample Problem 8-3 on page 148 and Sample Problem 10-2 on pages 200–201.

➤ *Use of vector-capable calculators.* When vector calculations in a sample problem can be performed directly on-screen with a vector-capable calculator, the solution of the sample problem indicates that fact but still carries through the traditional component analysis. When vector calculations cannot be performed directly on-screen, the solution explains why.

➤ *Problems with applied physics,* based on published research, have been added in many places, either as sample problems or homework problems. For example, see Sample Problem 11-6 on page 229, homework problem 64 on page 71, and homework problem 56 on page 214. For an example of homework problems that build with a continuing story, see problems 4, 32, and 48 (on pages 112, 114, and 115, respectively) in Chapter 6.

Content Changes

➤ *Chapter 5 on force and motion* contains clearer explanations of the gravitational force, weight, and normal force (pages 80–82).

➤ *Chapter 7 on kinetic energy and work* begins with a rough definition of energy. It then defines kinetic energy, work, and the work–kinetic energy theorem in ways that are more closely tied to Newton's second law than in the fifth edition, while keeping those definitions consistent with thermodynamics (pages 117–120).

➤ *Chapter 8 on the conservation of energy* avoids the much criticized definition of work done by a nonconservative force by explaining, instead, the energy transfers that occur due to a nonconservative force (page 153). (The wording still allows an instructor to superimpose a definition of work done by a nonconservative force.)

➤ *Chapter 10 on collisions* now presents the general situation of inelastic one-dimensional collisions (pages 198–200) before the special situation of elastic one-dimensional collisions (pages 202–204).

➤ *Chapters 16, 17, and 18 on SHM and waves* have been rewritten to better ease a student into these difficult subjects.

➤ *Chapter 21 on entropy* now presents a Carnot engine as the ideal heat engine with the greatest efficiency.

Chapter Features

➤ *Opening puzzlers.* A curious puzzling situation opens each chapter and is explained somewhere within the chapter, to entice a student to read the chapter.

➤ *Checkpoints* are stopping points that effectively ask the student, "Can you answer this question with some reasoning based on the narrative or sample problem that you just read?" If not, then the student should go back over that previous material before traveling deeper into the chapter. For example, see Checkpoint 3 on page 78 and Checkpoint 1 on page 101. **Answers to all checkpoints are in the back of the book.**

➤ *Sample problems* have been chosen to help the student organize the basic concepts of the narrative and to develop problem-solving skills. Each sample problem builds step by step from one or more Key Ideas to a solution.

➤ *Problem-solving tactics* contain helpful instructions to guide the beginning physics student as to how to solve problems and to avoid common errors.

➤ *Review & Summary* is a brief outline of the chapter contents that contains the essential concepts but which is not a substitute for reading the chapter.

➤ *Questions* are like the checkpoints and require reasoning and understanding rather than calculations. **Answers to the odd-number questions are in the back of the book.**

➤ *Exercises & Problems* are ordered approximately according to difficulty and grouped under section titles. **The odd-numbered ones are answered in the back of the book.** Worked-out solutions to the odd-numbered problems with trailing icons are available either in print or electronically. (See the icon guide at the beginning of the Exercises & Problems.) A problem number with a star indicates an especially challenging problem.

➤ *Additional Problems* appear at the end of the Exercises & Problems in certain chapters. They are not sorted according to section titles and many involve applied physics.

➤ *New Problems* appear at the ends of Chapters 1 through 38. They are not ordered or sorted in any way. Many use graphs where students must decide what data are important.

Versions of the Text

The Enhanced Problems Version of the sixth edition of *Fundamentals of Physics* is available in a number of different versions, to accommodate the individual needs of instructors and students. The Regular Edition consists of Chapters 1 through 38 (ISBN 0-471-22863-X). The Extended Edition contains seven additional chapters on quantum physics and cosmology (Chapters 1–45) (ISBN 0-471-22858-3). Both editions are available as single, hardcover books, or in the following alternative versions:

➤ *Volume 1—Chapters 1–21 (Mechanics/Thermodynamics), hardcover, 0-471-22861-3*

➤ *Volume 2—Chapters 22–45 (E&M and Modern Physics), hardcover, 0-471-22858-3*

➤ *Part 1—Chapters 1–12, paperback, 0-471-22860-5*

➤ *Part 2—Chapters 13–21, paperback, 0-471-22859-1*

➤ *Part 3—Chapters 22–33, paperback, 0-471-22857-5*

➤ *Part 4—Chapters 34–38, paperback, 0-471-22856-7*

➤ *Part 5—Chapters 39–45, paperback, 0-471-36038-4*

Supplements

The sixth edition of *Fundamentals of Physics* is supplemented by a comprehensive ancillary package carefully developed to help teachers teach and students learn.

Instructor's Supplements

➤ *Instructor's Manual* by J. RICHARD CHRISTMAN, U.S. Coast Guard Academy. This manual contains lecture notes outlining the most important topics of each chapter, demonstration experiments, laboratory and computer projects, film and video sources, answers to all Questions, Exercises & Problems, and Checkpoints, and a correlation guide to the Questions and Exercises & Problems in the previous edition.

➤ *Instructor's Solutions Manual* by JAMES WHITENTON, Southern Polytechnic University. This manual provides worked-out solutions for all the exercises and problems found at the end of each chapter within the text and in the Problem Supplement #1. *This supplement is available only to instructors.*

➤ *Test Bank* by J. RICHARD CHRISTMAN, U.S. Coast Guard Academy. More than 2200 multiple-choice questions are included in this manual. These items are also available in the Computerized Test Bank (see below).

➤ *Instructor's Resource CD.* This CD contains:
- All of the Instructor's Solutions Manual in both LaTex and PDF files.
- Computerized Test Bank in both IBM and Macintosh versions, with full editing features to help instructors customize tests.
- All text illustrations suitable for both classroom presentation and printing.

➤ *Transparencies.* More than 200 four-color illustrations from the text are provided in a form suitable for projection in the classroom.

➤ *On-line Course Management.*
- WebAssign, CAPA, and WebTest are on-line homework and quizzing programs that give instructors the ability to deliver and grade homework and quizzes over the Internet.
- Instructors will also have access to WebCT course materials. WebCT is a powerful Web site program that allows instructors to set up complete on-line courses with chat rooms, bulletin boards, quizzing, student tracking, etc. Please contact your local Wiley representative for more information.

Student's Supplements

➤ *A Student Companion* by J. RICHARD CHRISTMAN, U.S. Coast Guard Academy. This student study guide consists of a traditional print component and an accompanying Web site, which together provide a rich, interactive environment for review and study. The Student Companion Web site includes self-quizzes, simulation exercises, hints for solving end-of-chapter problems, the *Interactive LearningWare* program (see the next page), and links to other Web sites that offer physics tutorial help.

➤ *Student Solutions Manual* by J. RICHARD CHRISTMAN, U.S. Coast Guard Academy and EDWARD DERRINGH, Wentworth Institute. This manual provides students with complete worked-out solutions to 30 percent of the exercises and problems found at the end of each chapter within the text. These problems are indicated with an `ssm` icon in the text.

➤ *Interactive LearningWare.* This software guides students through solutions to 200 of the end-of-chapter problems. The solutions process is developed interactively, with appropriate feedback and access to error-specific help for the most common mistakes. These problems are indicated with an `itw` icon in the text.

➤ *CD-Physics, 3.0.* This CD-ROM based version of *Fundamentals of Physics,* Sixth Edition, contains the complete, extended version of the text, *A Student's Companion,* the *Student's Solutions Manual,* the *Interactive LearningWare,* and numerous simulations all connected with extensive hyperlinking.

➤ *Take Note!* This bound notebook lets students take notes directly onto large, black-and-white versions of textbook illustrations. All of the illustrations from the transparency set are included. In-class time spent copying illustrations is substantially reduced by this supplement.

➤ *Physics Web Site.* This Web site, **http://www.wiley.com/ college/halliday**, was developed specifically for *Fundamentals of Physics,* Sixth Edition, and is designed to further assist students in the study of physics and offers additional physics resources. The site also includes solutions to selected end-of-chapter problems. These problems are identified with a `www` icon in the text.

ACKNOWLEDGMENTS

A textbook contains far more contributions to the elucidation of a subject than those made by the authors alone. J. Richard Christman, of the U.S. Coast Guard Academy, has once again created many fine supplements for us; his knowledge of our book and his recommendations to students and faculty are invaluable. James Tanner, of Georgia Institute of Technology, and Gary Lewis, of Kennesaw State College, have provided us with innovative software, closely tied to the text exercises and problems. James Whitenton, of Southern Polytechnic State University, and Jerry Shi, of Pasadena City College, performed the Herculean task of working out solutions for every one of the Exercises & Problems in the text. We thank John Merrill, of Brigham Young University, and Edward Derringh, of the Wentworth Institute of Technology, for their many contributions in the past. We also thank George W. Hukle of Oxnard, California, for his check of the "old problems" and Renee M. Goertzen of Cleveland State University for her check of the "new problems."

At John Wiley publishers, we have been fortunate to receive strong coordination and support from our former editor, Cliff Mills. Cliff guided our efforts and encouraged us along the way. When Cliff moved on to other responsibilities at Wiley, we were ably guided to completion by his successor, Stuart Johnson. Ellen Ford has coordinated the developmental editing and multilayered preproduction process. Sue Lyons, our marketing manager, has been tireless in her efforts on behalf of this edition. Joan Kalkut has built a fine supporting package of ancillary materials. Thomas Hempstead managed the reviews of manuscript and the multiple administrative duties admirably.

We thank Lucille Buonocore and Pam Kennedy, our production directors, and Monique Calello and Elizabeth Swain, our production editors, for pulling all the pieces together and guiding us through the complex production process. We also thank Maddy Lesure, for her design; Helen Walden for her copyediting; Edward Starr and Anna Melhorn, for managing the illustration program; Georgia Mederer, Katrina Avery, and Lilian Brady, for their proofreading; and all other members of the production team.

Hilary Newman and her team of photo researchers were inspired in their search for unusual and interesting photographs that communicate physics principles beautifully. We also owe a debt of gratitude for the line art to the late John Balbalis, whose careful hand and understanding of physics can still be seen in every diagram.

We especially thank Edward Millman for his developmental work on the manuscript. With us, he has read every word, asking many questions from the point of view of a student. Many of his questions and suggested changes have added to the clarity of this volume.

We owe a particular debt of gratitude to the numerous students who used the previous editions of *Fundamentals of Physics* and took the time to fill out the response cards and return them to us. As the ultimate consumers of this text, students are extremely important to us. By sharing their opinions with us, your students help us ensure that we are providing the best possible product and the most value for their textbook dollars. We encourage the users of this book to contact us with their thoughts and concerns so that we can continue to improve this text in the years to come.

Finally, our external reviewers have been outstanding and we acknowledge here our debt to each member of that team:

Edward Adelson
Ohio State University

Mark Arnett
Kirkwood Community College

Arun Bansil
Northeastern University

J. Richard Christman
U.S. Coast Guard Academy

Robert N. Davie, Jr.
St. Petersburg Junior College

Cheryl K. Dellai
Glendale Community College

Eric R. Dietz
California State University at Chico

N. John DiNardo
Drexel University

Harold B. Hart
Western Illinois University

Rebecca Hartzler
Edmonds Community College

Joey Huston
Michigan State University

Shawn Jackson
University of Tulsa

Hector Jimenez
University of Puerto Rico

Sudhakar B. Joshi
York University

Leonard M. Kahn
University of Rhode Island

Yuichi Kubota
Cornell University

Priscilla Laws
Dickinson College

Edbertho Leal
Polytechnic University of Puerto Rico

Dale Long
Virginia Tech

Andreas Mandelis
University of Toronto

Paul Marquard
Caspar College

James Napolitano
Rensselaer Polytechnic Institute

Des Penny
Southern Utah University

Joe Redish
University of Maryland

Timothy M. Ritter
University of North Carolina at Pembroke

Gerardo A. Rodriguez
Skidmore College

John Rosendahl
University of California at Irvine

Michael Schatz
Georgia Institute of Technology

Michael G. Strauss
University of Oklahoma

Dan Styer
Oberlin College

Marshall Thomsen
Eastern Michigan University

Fred F. Tomblin
New Jersey Institute of Technology

B. R. Weinberger
Trinity College

William M. Whelan
Ryerson Polytechnic University

William Zimmerman, Jr.
University of Minnesota

Reviewers of the Fifth and Previous Editions

Maris A. Abolins
Michigan State University

Barbara Andereck
Ohio Wesleyan University

Albert Bartlett
University of Colorado

Michael E. Browne
University of Idaho

Timothy J. Burns
Leeward Community College

Joseph Buschi
Manhattan College

Philip A. Casabella
Rensselaer Polytechnic Institute

Randall Caton
Christopher Newport College

J. Richard Christman
U.S. Coast Guard Academy

Roger Clapp
University of South Florida

W. R. Conkie
Queen's University

Peter Crooker
University of Hawaii at Manoa

William P. Crummett
Montana College of Mineral Science and Technology

Eugene Dunnam
University of Florida

Robert Endorf
University of Cincinnati

F. Paul Esposito
University of Cincinnati

Jerry Finkelstein
San Jose State University

Alexander Firestone
Iowa State University

Alexander Gardner
Howard University

Andrew L. Gardner
Brigham Young University

John Gieniec
Central Missouri State University

John B. Gruber
San Jose State University

Ann Hanks
American River College

Samuel Harris
Purdue University

Emily Haught
Georgia Institute of Technology

Laurent Hodges
Iowa State University

John Hubisz
North Carolina State University

Joey Huston
Michigan State University

Darrell Huwe
Ohio University

Claude Kacser
University of Maryland

Leonard Kleinman
University of Texas at Austin

Earl Koller
Stevens Institute of Technology

Arthur Z. Kovacs
Rochester Institute of Technology

Kenneth Krane
Oregon State University

Sol Krasner
University of Illinois at Chicago

Peter Loly
University of Manitoba

Robert R. Marchini
Memphis State University

David Markowitz
University of Connecticut

Howard C. McAllister
University of Hawaii at Manoa

W. Scott McCullough
Oklahoma State University

James H. McGuire
Tulane University

David M. McKinstry
Eastern Washington University

Joe P. Meyer
Georgia Institute of Technology

Roy Middleton
University of Pennsylvania

Irvin A. Miller
Drexel University

Eugene Mosca
United States Naval Academy

Michael O'Shea
Kansas State University

Patrick Papin
San Diego State University

George Parker
North Carolina State University

Robert Pelcovits
Brown University

Oren P. Quist
South Dakota State University

Jonathan Reichart
SUNY—Buffalo

Manuel Schwartz
University of Louisville

Darrell Seeley
Milwaukee School of Engineering

Bruce Arne Sherwood
Carnegie Mellon University

John Spangler
St. Norbert College

Ross L. Spencer
Brigham Young University

Harold Stokes
Brigham Young University

Jay D. Strieb
Villanova University

David Toot
Alfred University

J. S. Turner
University of Texas at Austin

T. S. Venkataraman
Drexel University

Gianfranco Vidali
Syracuse University

Fred Wang
Prairie View A & M

Robert C. Webb
Texas A & M University

George Williams
University of Utah

David Wolfe
University of New Mexico

13 Equilibrium and Elasticity

Rock climbing may be the ultimate physics exam. Failure can mean death, and even "partial credit" can mean severe injury. For example, in a long chimney climb, in which your torso is pressed against one wall of a wide vertical fissure and your feet are pressed against the opposite wall, you need to rest occasionally or you will fall due to exhaustion. Here the exam consists of a single question: What can you do to relax your push on the walls in order to rest? If you relax without considering the physics, the walls will not hold you up.

What is the answer to this life-and-death, one-question exam?

The answer is in this chapter.

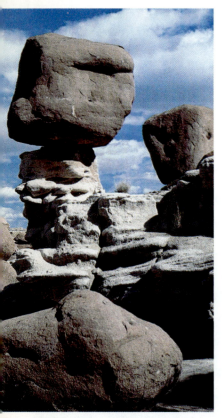

Fig. 13-1 A balancing rock near Petri-fied Forest National Park in Arizona. Although its perch seems precarious, the rock is in static equilibrium.

13-1 Equilibrium

Consider these objects: (1) a book resting on a table, (2) a hockey puck sliding across a frictionless surface with constant velocity, (3) the rotating blades of a ceiling fan, and (4) the wheel of a bicycle that is traveling along a straight path at constant speed. For each of these four objects:

1. The linear momentum $\vec{P}$ of its center of mass is constant.

2. Its angular momentum $\vec{L}$ about its center of mass, or about any other point, is also constant.

We say that such objects are in **equilibrium.** The two requirements for equilibrium are then

$$\vec{P} = \text{a constant} \quad \text{and} \quad \vec{L} = \text{a constant.} \tag{13-1}$$

Our concern in this chapter is with situations in which the constants in Eq. 13-1 are in fact zero; that is, we are concerned largely with objects that are not moving in any way—either in translation or in rotation—in the reference frame from which we observe them. Such objects are in **static equilibrium.** Of the four objects mentioned at the beginning of this section, only one—the book resting on the table—is in static equilibrium.

The balancing rock of Fig. 13-1 is another example of an object that, for the present at least, is in static equilibrium. It shares this property with countless other structures, such as cathedrals, houses, filing cabinets, and taco stands, that remain stationary over time.

As we discussed in Section 8-5, if a body returns to a state of static equilibrium after having been displaced from it by a force, the body is said to be in *stable* static equilibrium. A marble placed at the bottom of a hemispherical bowl is an example. However, if a small force can displace the body and end the equilibrium, the body is in *unstable* static equilibrium.

For example, suppose we balance a domino with the domino's center of mass vertically above the supporting edge, as in Fig. 13-2a. The torque about the supporting edge due to the gravitational force $\vec{F}_g$ on the domino is zero, because the line of action of $\vec{F}_g$ is through that edge. Thus, the domino is in equilibrium. Of course, even a slight force on it due to some chance disturbance ends the equilibrium. As the line of action of $\vec{F}_g$ moves to one side of the supporting edge (as in Fig. 13-2b), the torque due to $\vec{F}_g$ increases the rotation of the domino. Therefore, the domino in Fig. 13-2a is in unstable static equilibrium.

The domino in Fig. 13-2c is not quite as unstable. To topple this domino, a force would have to rotate it through and then beyond the balance position of Fig. 13-2a, in which the center of mass is above a supporting edge. A slight force will not topple this domino, but a vigorous flick of the finger against the domino certainly will. (If we arrange a chain of such upright dominos, a finger flick against the first can cause the whole chain to fall.)

Fig. 13-2 (a) A domino balanced on one edge, with its center of mass vertically above that edge. The gravitational force $\vec{F}_g$ on the domino is directed through the supporting edge. (b) If the domino is rotated even slightly from the balanced orientation, then $\vec{F}_g$ causes a torque that increases the rotation. (c) A domino upright on a narrow side is somewhat more stable than the domino in (a). (d) A square block is even more stable.

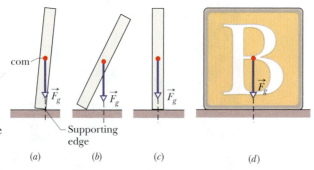

The child's square block in Fig. 13-2d is even more stable because its center of mass would have to be moved even farther to get it to pass above a supporting edge. A flick of the finger may not topple the block. (This is why you never see a chain of toppling square blocks.) The worker in Fig. 13-3 is like both the domino and the square block: Parallel to the beam, his stance is wide and he is stable; perpendicular to the beam, his stance is narrow and he is unstable (and at the mercy of a chance gust of wind).

The analysis of static equilibrium is very important in engineering practice. The design engineer must isolate and identify all the external forces and torques that may act on a structure and, by good design and wise choice of materials, ensure that the structure will remain stable under these loads. Such analysis is necessary to ensure, for example, that bridges do not collapse under their traffic and wind loads, and that the landing gear of aircraft will survive the shock of rough landings.

Fig. 13-3 A construction worker balanced above New York City is in static equilibrium but is more stable parallel to the beam than perpendicular to it.

13-2 The Requirements of Equilibrium

The translational motion of a body is governed by Newton's second law in its linear momentum form, given by Eq. 9-27 as

$$\vec{F}_{net} = \frac{d\vec{P}}{dt}. \tag{13-2}$$

If the body is in translational equilibrium—that is, if $\vec{P}$ is a constant—then $d\vec{P}/dt = 0$ and we must have

$$\vec{F}_{net} = 0 \qquad \text{(balance of forces).} \tag{13-3}$$

The rotational motion of a body is governed by Newton's second law in its angular momentum form, given by Eq. 12-29 as

$$\vec{\tau}_{net} = \frac{d\vec{L}}{dt}. \tag{13-4}$$

If the body is in rotational equilibrium—that is, if $\vec{L}$ is a constant—then $d\vec{L}/dt = 0$ and we must have

$$\vec{\tau}_{net} = 0 \qquad \text{(balance of torques).} \tag{13-5}$$

Thus, the two requirements for a body to be in equilibrium are as follows:

1. The vector sum of all the external forces that act on the body must be zero.

2. The vector sum of all the external torques that act on the body, measured about *any* possible point, must also be zero.

These requirements obviously hold for *static* equilibrium. They also hold for the more general equilibrium in which $\vec{P}$ and $\vec{L}$ are constant but not zero.

Equations 13-3 and 13-5, as vector equations, are each equivalent to three independent component equations, one for each direction of the coordinate axes:

Balance of forces	Balance of torques	
$F_{net,x} = 0$	$\tau_{net,x} = 0$	
$F_{net,y} = 0$	$\tau_{net,y} = 0$	(13-6)
$F_{net,z} = 0$	$\tau_{net,z} = 0$	

We shall simplify matters by considering only situations in which the forces that act on the body lie in the *xy* plane. This means that the only torques that can act on the body must tend to cause rotation around an axis parallel to the *z* axis. With this assumption, we eliminate one force equation and two torque equations from Eqs. 13-6, leaving

$$F_{\text{net},x} = 0 \qquad \text{(balance of forces)}, \qquad (13\text{-}7)$$

$$F_{\text{net},y} = 0 \qquad \text{(balance of forces)}, \qquad (13\text{-}8)$$

$$\tau_{\text{net},z} = 0 \qquad \text{(balance of torques)}. \qquad (13\text{-}9)$$

Here, $\tau_{\text{net},z}$ is the net torque that the external forces produce either about the *z* axis or about *any* axis parallel to it.

A hockey puck sliding at constant velocity over ice satisfies Eqs. 13-7, 13-8, and 13-9 and is thus in equilibrium *but not in static equilibrium*. For static equilibrium, the linear momentum $\vec{P}$ of the puck must be not only constant but also zero; the puck must be at rest on the ice. Thus, there is another requirement for static equilibrium:

> **3.** The linear momentum $\vec{P}$ of the body must be zero.

✔**CHECKPOINT 1:**   The figure gives six overhead views of a uniform rod on which two or more forces act perpendicularly to the rod. If the magnitudes of the forces are adjusted properly (but kept nonzero), in which situations can the rod be in static equilibrium?

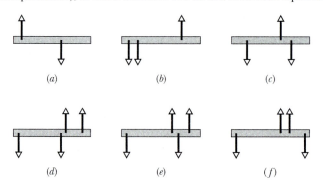

(a)   (b)   (c)

(d)   (e)   (f)

13-3 The Center of Gravity

The gravitational force on an extended body is the vector sum of the gravitational forces acting on the individual elements (the atoms) of the body. Instead of considering all those individual elements, we can say:

> The gravitational force $\vec{F}_g$ on a body effectively acts at a single point, called the **center of gravity** (cog) of the body.

Here the word "effectively" means that if the forces on the individual elements were somehow turned off and force $\vec{F}_g$ at the center of gravity were turned on, the net force and the net torque (about any point) acting on the body would not change.

Until now, we have assumed that the gravitational force $\vec{F}_g$ acts at the center of mass (com) of the body. This is equivalent to assuming that the center of gravity is

Fig. 13-4 (a) An element of mass m_i in an extended body. The gravitational force $\vec{F}_{gi}$ on it has moment arm x_i about the origin O of the coordinate system. (b) The gravitational force $\vec{F}_g$ on a body is said to act at the center of gravity (cog) of the body. Here it has moment arm x_{cog} about origin O.

at the center of mass. Recall that, for a body of mass M, the force $\vec{F}_g$ is equal to $M\vec{g}$, where $\vec{g}$ is the acceleration that the force would produce if the body were to fall freely. In the proof that follows, we show that

▶ If $\vec{g}$ is the same for all elements of a body, then the body's center of gravity (cog) is coincident with the body's center of mass (com).

This is approximately true for everyday objects because $\vec{g}$ varies only a little along Earth's surface and decreases in magnitude only slightly with altitude. Thus, for objects like a mouse or a moose, we have been justified in assuming that the gravitational force acts at the center of mass. After the following proof, we shall resume that assumption.

Proof

First, we consider the individual elements of the body. Figure 13-4a shows an extended body, of mass M, and one of its elements, of mass m_i. A gravitational force $\vec{F}_{gi}$ acts on each such element and is equal to $m_i\vec{g}_i$. The subscript on $\vec{g}_i$ means $\vec{g}_i$ is the gravitational acceleration *at the location of the element* (it can be different for other elements).

In Fig. 13-4a, each force $\vec{F}_{gi}$ produces a torque τ_i on the element about the origin O, with moment arm x_i. Using Eq. 11-33 ($\tau = r_\perp F$), we can write torque τ_i as

$$\tau_i = x_i F_{gi}. \tag{13-10}$$

The net torque on all the elements of the body is then

$$\tau_{net} = \sum \tau_i = \sum x_i F_{gi}. \tag{13-11}$$

Next, we consider the body as a whole. Figure 13-4b shows the gravitational force $\vec{F}_g$ acting at the body's center of gravity. This force produces a torque τ on the body about O, with moment arm x_{cog}. Again using Eq. 11-33, we can write this torque as

$$\tau = x_{cog} F_g. \tag{13-12}$$

The gravitational force $\vec{F}_g$ on the body is equal to the sum of the gravitational forces $\vec{F}_{gi}$ on all its elements, so we can substitute $\sum F_{gi}$ for F_g in Eq. 13-12 to write

$$\tau = x_{cog} \sum F_{gi}. \tag{13-13}$$

Now recall that the torque due to force $\vec{F}_g$ acting at the center of gravity is equal to the net torque due to all the forces $\vec{F}_{gi}$ acting on all the elements of the body. (That is how we defined the cog.) Therefore, τ in Eq. 13-13 is equal to τ_{net} in Eq. 13-11. Putting those two equations together, we can write

$$x_{cog} \sum F_{gi} = \sum x_i F_{gi}.$$

Substituting $m_i g_i$ for F_{gi} gives us

$$x_{cog} \sum m_i g_i = \sum x_i m_i g_i.$$

Now here is a key idea: If the accelerations g_i at all the locations of the elements are the same, we can cancel g_i from this equation to write

$$x_{cog} \sum m_i = \sum x_i m_i. \tag{13-14}$$

The sum $\sum m_i$ of the masses of all the elements is the mass M of the body. Therefore, we can rewrite Eq. 13-14 as

$$x_{cog} = \frac{1}{M} \sum x_i m_i. \tag{13-15}$$

The right side of this equation gives the coordinate x_{com} of the body's center of mass (Eq. 9-4). We now have what we sought to prove:

$$x_{cog} = x_{com}. \qquad (13\text{-}16)$$

13-4   Some Examples of Static Equilibrium

In this section we examine four sample problems involving static equilibrium. In each, we select a system of one or more objects to which we apply the equations of equilibrium (Eqs. 13-7, 13-8, and 13-9). The forces involved in the equilibrium are all in the xy plane, which means that the torques involved are parallel to the z axis. Thus, in applying Eq. 13-9, the balance of torques, we select an axis parallel to the z axis about which to calculate the torques. Although Eq. 13-9 is satisfied for *any* such choice of axis, you will see that certain choices simplify the application of Eq. 13-9 by eliminating one or more unknown force terms.

Sample Problem 13-1

In Fig. 13-5a, a uniform beam, of length L and mass $m = 1.8$ kg, is at rest with its ends on two scales. A uniform block, with mass $M = 2.7$ kg, is at rest on the beam, with its center a distance $L/4$ from the beam's left end. What do the scales read?

SOLUTION: The first steps in the solution of *any* problem about static equilibrium are these: Clearly define the system to be analyzed and then draw a free-body diagram of it, indicating all the forces on the system. Here, let us choose the system as the beam and block taken together. Then the forces on the system are shown in the free-body diagram of Fig. 13-5b. (Choosing the system takes experience and often there can be more than one good choice; see Problem Solving Tactic 1 below.)

The normal forces on the beam from the scales are $\vec{F}_l$ on the left and $\vec{F}_r$ on the right. The scale readings that we want are equal to the magnitudes of those forces. The gravitational force $\vec{F}_{g,beam}$ on the beam acts at the beam's center of mass and is equal to $m\vec{g}$. Similarly, the gravitational force $\vec{F}_{g,block}$ on the block acts at the block's center of mass and is equal to $M\vec{g}$. However, to simplify Fig. 13-5b, the block is represented by a dot within the boundary of the beam and the vector $\vec{F}_{g,block}$ is drawn with its tail on that dot. (This downward shift of vector $\vec{F}_{g,block}$ along its line of action does not alter the torque due to $\vec{F}_{g,block}$ about any axis perpendicular to the figure.)

The **Key Idea** here is that, because the system is in static equilibrium, we can apply the balance of forces equations (Eqs. 13-7 and 13-8) and the balance of torques equation (13-9) to it. The forces have no x components, so Eq. 13-7 ($F_{net,x} = 0$) provides no information. For the y components, Eq. 13-8 ($F_{net,y} = 0$) gives us

$$F_l + F_r - Mg - mg = 0. \qquad (13\text{-}17)$$

This equation contains two unknowns, the forces F_l and F_r, so we also need to use Eq. 13-9, the balance of torques equation.

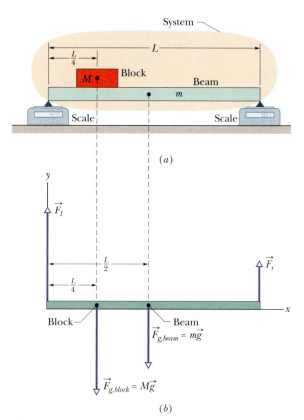

Fig. 13-5 Sample Problem 13-1. (a) A beam of mass m supports a block of mass M. (b) A free-body diagram, showing the forces that act on the system *beam* + *block*.

We can apply it to *any* rotation axis perpendicular to the plane of Fig. 13-5. Let us choose a rotation axis through the left end of the beam. We shall also use our general rule for assigning signs to torques: If a torque would cause an initially stationary body to rotate clockwise about the rotation axis, the torque is negative. If the rotation would be counterclockwise, the torque is positive. Finally, we shall write the torques in the form $r_\perp F$, where the moment arm $r_\perp$ is 0 for $\vec{F}_l$, $L/4$ for $M\vec{g}$, $L/2$ for $m\vec{g}$, and L for $\vec{F}_r$.

We now can write the balancing equation ($\tau_{net,z} = 0$) as

$$(0)(F_l) - (L/4)(Mg) - (L/2)(mg) + (L)(F_r) = 0,$$

which gives us

$$F_r = \tfrac{1}{4}Mg + \tfrac{1}{2}mg$$
$$= \tfrac{1}{4}(2.7\text{ kg})(9.8\text{ m/s}^2) + \tfrac{1}{2}(1.8\text{ kg})(9.8\text{ m/s}^2)$$
$$= 15.44\text{ N} \approx 15\text{ N}. \qquad \text{(Answer)}$$

Now, solving Eq. 13-17 for F_l and substituting this result, we find

$$F_l = (M + m)g - F_r$$
$$= (2.7\text{ kg} + 1.8\text{ kg})(9.8\text{ m/s}^2) - 15.44\text{ N}$$
$$= 28.66\text{ N} \approx 29\text{ N}. \qquad \text{(Answer)}$$

Notice the strategy in the solution: When we wrote an equation for the balance of force components, we got stuck with two unknowns. If we had written an equation for the balance of torques around some *arbitrary* axis, we would have again gotten stuck with those two unknowns. However, because we chose the axis to pass through the point of application of one of the unknown forces, here $\vec{F}_l$, we did not get stuck. Our choice neatly eliminated that force from the torque equation, allowing us to solve for the other unknown force magnitude F_r. Then we returned to the equation for the balance of force components to find the remaining unknown force magnitude.

✔ CHECKPOINT 3: The figure gives an overhead view of a uniform rod in static equilibrium. (a) Can you find the magnitudes of unknown forces $\vec{F}_1$ and $\vec{F}_2$ by balancing the forces? (b) If you wish to find the magnitude of force $\vec{F}_2$ by using a single equation, where should you place a rotational axis? (c) The magnitude of $\vec{F}_2$ turns out to be 65 N. What then is the magnitude of $\vec{F}_1$?

Sample Problem 13-2

In Fig. 13-6a, a ladder of length $L = 12$ m and mass $m = 45$ kg leans against a slick (frictionless) wall. Its upper end is at height $h = 9.3$ m above the pavement on which the lower end rests (the pavement is not frictionless). The ladder's center of mass is $L/3$ from the lower end. A firefighter of mass $M = 72$ kg climbs the ladder until her center of mass is $L/2$ from the lower end. What then are the magnitudes of the forces on the ladder from the wall and the pavement?

SOLUTION: First, we choose our system as being the firefighter and ladder, together, and then we draw the free-body diagram of Fig. 13-6b. The firefighter is represented with a dot within the boundary of the ladder. The gravitational force on her is represented with its equivalent $M\vec{g}$, and that vector has been shifted along its line of action, so that its tail is on the dot. (The shift does not alter a torque due to $M\vec{g}$ about any axis perpendicular to the figure.)

The only force on the ladder from the wall is the horizontal force $\vec{F}_w$ (there cannot be a frictional force along a frictionless wall). The force $\vec{F}_p$ on the ladder from the pavement has a horizontal component $\vec{F}_{px}$ that is a static frictional force and a vertical component $\vec{F}_{py}$ that is a normal force.

A **Key Idea** here is that the system is in static equilibrium, so the balancing equations (Eqs. 13-7 through 13-9) apply to it. Let us start with Eq. 13-9 ($\tau_{net,z} = 0$). To choose an axis about which to calculate the torques, note that we have unknown forces ($\vec{F}_w$ and $\vec{F}_p$) at the two ends of the ladder. To eliminate, say, $\vec{F}_p$ from the calculation, we place the axis at point O, perpendicular to the figure. We also place the origin of an xy coordinate system at O. We can find torques about O with any of Eqs. 11-31 through 11-33, but Eq. 11-33 ($\tau = r_\perp F$) is easiest to use here.

Fig. 13-6 Sample Problem 13-2 (*a*) A firefighter climbs halfway up a ladder that is leaning against a frictionless wall. The pavement beneath the ladder is not frictionless. (*b*) A free-body diagram, showing the forces that act on the firefighter–ladder system. The origin O of a coordinate system is placed at the point of application of the unknown force $\vec{F}_p$ (whose vector components $\vec{F}_{px}$ and $\vec{F}_{py}$ are shown).

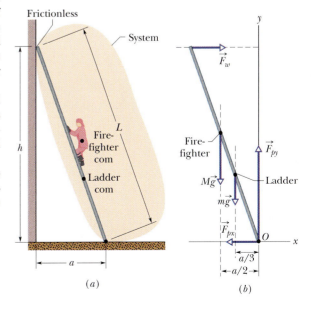

To find the moment arm $r_\perp$ of $\vec{F}_w$, we draw a line of action through that vector (Fig. 13-6b). Then $r_\perp$ is the perpendicular distance between O and the line of action. In Fig. 13-6b, it extends along the y axis and is equal to the height h. We similarly draw lines of action for $M\vec{g}$ and $m\vec{g}$ and see that their moment arms extend along the x axis. For the distance a shown in Fig. 13-6a, the moment arms are $a/2$ (the firefighter is halfway up the ladder) and $a/3$ (the ladder's center of mass is one-third of the way up the ladder), respectively. The moment arms for $\vec{F}_{px}$ and $\vec{F}_{py}$ are zero.

Now, with torques written in the form $r_\perp F$, the balancing equation $\tau_{net,z} = 0$ becomes

$$-(h)(F_w) + (a/2)(Mg) + (a/3)(mg)$$
$$+ (0)(F_{px}) + (0)(F_{py}) = 0. \quad (13\text{-}18)$$

(Recall our rule: A positive torque corresponds to counterclockwise rotation and a negative torque corresponds to clockwise rotation.)

Using the Pythagorean theorem, we find that

$$a = \sqrt{L^2 - h^2} = 7.58 \text{ m}.$$

Then Eq. 13-18 gives us

$$F_w = \frac{ga(M/2 + m/3)}{h}$$
$$= \frac{(9.8 \text{ m/s}^2)(7.58 \text{ m})(72/2 \text{ kg} + 45/3 \text{ kg})}{9.3 \text{ m}}$$
$$= 407 \text{ N} \approx 410 \text{ N}. \quad \text{(Answer)}$$

Now we need to use the force balancing equations. The equation $F_{net,x} = 0$ gives us

$$F_w - F_{px} = 0, \quad (13\text{-}19)$$

so

$$F_{px} = F_w = 410 \text{ N}. \quad \text{(Answer)}$$

The equation $F_{net,y} = 0$ gives us

$$F_{py} - Mg - mg = 0, \quad (13\text{-}20)$$

so

$$F_{py} = (M + m)g = (72 \text{ kg} + 45 \text{ kg})(9.8 \text{ m/s}^2)$$
$$= 1146.6 \text{ N} \approx 1100 \text{ N}. \quad \text{(Answer)}$$

Sample Problem 13-3

Figure 13-7a shows a safe, of mass $M = 430$ kg, hanging by a rope from a boom with dimensions $a = 1.9$ m and $b = 2.5$ m. The boom consists of a hinged beam and a horizontal cable that connects the beam to a wall. The uniform beam has a mass m of 85 kg; the mass of the cable and rope are negligible.

(a) What is the tension T_c in the cable? In other words, what is the magnitude of the force $\vec{T}_c$ on the beam from the cable?

SOLUTION: The system here is the beam alone, and the forces on it are shown in the free-body diagram of Fig. 13-7b. The force from the cable is $\vec{T}_c$. The gravitational force on the beam acts at the beam's center of mass (at the beam's center) and is represented by its equivalent $m\vec{g}$. The vertical component of the force on the beam from the hinge is $\vec{F}_v$, and the horizontal component of the force from the hinge is $\vec{F}_h$. The force from the rope supporting the safe is $\vec{T}_r$. Because beam, rope, and safe are stationary, the magnitude of $\vec{T}_r$ is equal to the weight of the safe: $T_r = Mg$. We place the origin O of an xy coordinate system at the hinge.

One **Key Idea** here is that our system is in static equilibrium, so the balancing equations apply to it. Let us start with Eq. 13-9 ($\tau_{net,z} = 0$). Note that we are asked for the magnitude of force $\vec{T}_c$ and not of forces $\vec{F}_h$ and $\vec{F}_v$ acting at the hinge, at point O. Thus, a second **Key Idea** is that, to eliminate $\vec{F}_h$ and $\vec{F}_v$ from the torque calculation, we should calculate torques about an axis that is perpendicular to the figure at point O. Then $\vec{F}_h$ and $\vec{F}_v$ will have moment arms of zero. The lines of action for $\vec{T}_c$, $\vec{T}_r$, and $m\vec{g}$ are dashed in Fig. 13-7b. The corresponding moment arms are a, b, and $b/2$.

Writing torques in the form of $r_\perp F$ and using our rule about signs for torques, the balancing equation $\tau_{net,z} = 0$ becomes

$$(a)(T_c) - (b)(T_r) - (\tfrac{1}{2}b)(mg) = 0.$$

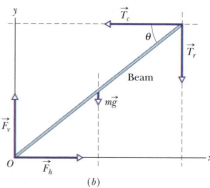

Fig. 13-7 Sample Problem 13-3. (a) A heavy safe is hung from a boom consisting of a horizontal steel cable and a uniform beam. (b) A free-body diagram for the beam.

Substituting Mg for T_r and solving for T_c, we find that

$$T_c = \frac{gb(M + \frac{1}{2}m)}{a}$$

$$= \frac{(9.8 \text{ m/s}^2)(2.5 \text{ m})(430 \text{ kg} + 85/2 \text{ kg})}{1.9 \text{ m}}$$

$$= 6093 \text{ N} \approx 6100 \text{ N}. \qquad \text{(Answer)}$$

(b) Find the magnitude F of the net force on the beam from the hinge.

SOLUTION: Now we want F_h and F_v so we can combine them to get F. Because we know T_c, our Key Idea here is to apply the force balancing equations to the beam. For the horizontal balance, we write $F_{\text{net},x} = 0$ as

$$F_h - T_c = 0,$$

and so

$$F_h = T_c = 6093 \text{ N}.$$

For the vertical balance, we write $F_{\text{net},y} = 0$ as

$$F_v - mg - T_r = 0.$$

Substituting Mg for T_r and solving for F_v, we find that

$$F_v = (m + M)g = (85 \text{ kg} + 430 \text{ kg})(9.8 \text{ m/s}^2)$$

$$= 5047 \text{ N}.$$

From the Pythagorean theorem, we now have

$$F = \sqrt{F_h^2 + F_v^2}$$

$$= \sqrt{(6093 \text{ N})^2 + (5047 \text{ N})^2} \approx 7900 \text{ N}. \qquad \text{(Answer)}$$

Note that F is substantially greater than either the combined weights of the safe and the beam, 5000 N, or the tension in the horizontal wire, 6100 N.

✔CHECKPOINT 4:   In the figure, a stationary 5 kg rod AC is held against a wall by a rope and friction between rod and wall. The uniform rod is 1 m long, and angle $\theta = 30°$. (a) If you are to find the magnitude of the force $\vec{T}$ on the rod from the rope with a single equation, at what labeled point should a rotational axis be placed? With that choice of axis and counterclockwise torques positive, what is the sign of (b) the torque τ_w due to the rod's weight and (c) the torque τ_r due to the pull on the rod by the rope? (d) Is the magnitude of τ_r greater than, less than, or equal to the magnitude of τ_w?

Sample Problem 13-4

In Fig. 13-8, a rock climber with mass $m = 55$ kg rests during a "chimney climb," pressing only with her shoulders and feet against the walls of a fissure of width $w = 1.0$ m. Her center of mass is a horizontal distance $d = 0.20$ m from the wall against which her shoulders are pressed. The coefficient of static friction between her shoes and the wall is $\mu_1 = 1.1$, and between her shoulders and the wall it is $\mu_2 = 0.70$. To rest, the climber wants to minimize her horizontal push on the walls. The minimum occurs when her feet and her shoulders are both on the verge of sliding.

Fig. 13-8 Sample Problem 13-4. The forces on a climber resting in a rock chimney. The push of the climber on the chimney walls results in rise to the normal forces $\vec{N}$ and the static frictional forces $\vec{f}_1$ and $\vec{f}_2$.

(a) What is that minimum horizontal push on the walls?

SOLUTION: Our system is the climber, and Fig. 13-8 shows the forces that act on her. The only horizontal forces are the normal forces $\vec{N}$ on her from the walls, at her feet and shoulders. The static frictional forces on her are $\vec{f}_1$ and $\vec{f}_2$, directed upward. The gravitational force $\vec{F}_g = m\vec{g}$ acts at her center of mass.

A Key Idea is that, because the system is in static equilibrium, we can apply the force balancing equations (Eqs. 13-7 and 13-8) to it. The equation $F_{\text{net},x} = 0$ tells us that the two normal forces on her must be equal in magnitude and opposite in direction. We seek the magnitude N of these two forces, which is also the magnitude of her push against either wall.

The balancing equation $F_{\text{net},y} = 0$ gives us

$$f_1 + f_2 - mg = 0. \qquad (13\text{-}21)$$

We want the climber to be on the verge of sliding at both her feet and her shoulders. That means we want the static frictional forces there to be at their maximum values. Those maximum values are, from Eq. 6-1 ($f_{s,\text{max}} = \mu_s N$),

$$f_1 = \mu_1 N \quad \text{and} \quad f_2 = \mu_2 N. \qquad (13\text{-}22)$$

Substituting these expressions into Eq. 13-21 and solving for N give us

$$N = \frac{mg}{\mu_1 + \mu_2} = \frac{(55 \text{ kg})(9.8 \text{ m/s}^2)}{1.1 + 0.70} = 299 \text{ N} \approx 300 \text{ N}.$$

Thus, her minimum horizontal push must be about 300 N.

(b) For that push, what must be the vertical distance h between her feet and her shoulders if she is to be stable?

SOLUTION: A Key Idea here is that the climber will be stable if the torque balancing equation ($\tau_{net,z} = 0$) applies to her. This means that the forces on her must not produce a net torque about *any* rotation axis. Another Key Idea is that we are free to choose a rotation axis that helps simplify the calculation. We shall write the torques in the form $r_\perp F$, where $r_\perp$ is the moment arm of force F. In Fig. 13-8, we choose a rotation axis at her shoulders, perpendicular to the figure's plane. Then the moment arms of the forces acting there ($\vec{N}$ and $\vec{f}_2$) are zero. Frictional force $\vec{f}_1$, the normal force $\vec{N}$ at her feet, and the gravitational force $\vec{F}_g = m\vec{g}$ have the corresponding moment arms w, h, and d.

Recalling our rule about the signs of torques and the corresponding directions, we can now write $\tau_{net,z} = 0$ as

$$-(w)(f_1) + (h)(N) + (d)(mg) + (0)(f_2) + (0)(N) = 0.   \quad (13\text{-}23)$$

(Note how the choice of rotation axis neatly eliminates f_2 from the calculation.) Next, solving Eq. 13-23 for h, setting $f_1 = \mu_1 N$, and substituting $N = 299$ N and other known values, we find that

$$h = \frac{f_1 w - mgd}{N} = \frac{\mu_1 N w - mgd}{N} = \mu_1 w - \frac{mgd}{N}$$

$$= (1.1)(1.0\ \text{m}) - \frac{(55\ \text{kg})(9.8\ \text{m/s}^2)(0.20\ \text{m})}{299\ \text{N}}$$

$$= 0.739\ \text{m} \approx 0.74\ \text{m}.   \qquad \text{(Answer)}$$

We would find the same required value of h if we wrote the torques about any other rotation axis perpendicular to the page, such as one at her feet.

If h is more than *or* less than 0.74 m, she must exert a force greater than 299 N on the walls to be stable. Here, then, is the advantage of knowing the physics before you climb a chimney. When you need to rest, you will avoid the (dire) error of novice climbers who place their feet too high or too low. Instead, you will know that there is a "best" distance between shoulders and feet, requiring the least push, and giving you a good chance to rest.

PROBLEM-SOLVING TACTICS

Tactic 1: *Static Equilibrium Problems*
Here is a list of steps for solving static equilibrium problems:

1. Draw a *sketch* of the problem.

2. Select the *system* to which you will apply the laws of equilibrium, drawing a closed curve around it on your sketch to fix it clearly in your mind. In some situations you can select a single object as the system; it is the object you wish to be in equilibrium (such as the rock climber in Sample Problem 13-4). In other situations, you might include additional objects in the system if their inclusion simplifies the calculations for equilibrium. For example, suppose in Sample Problem 13-2 you select only the ladder as the system. Then in Fig. 13-6*b* you will have to account for additional unknown forces exerted on the ladder by the hands and feet of the firefighter. These additional unknowns complicate the equilibrium calculations. The system of Fig. 13-6 was chosen to include the firefighter so that those unknown forces are *internal* to the system and thus need not be found in order to solve Sample Problem 13-2.

3. Draw a *free-body diagram* of the system. Show all the forces that act on the system, labeling them and making sure that their points of application and lines of action are correctly shown.

4. Draw in the *x* and *y axes* of a coordinate system. Choose them so that at least one axis is parallel to one or more unknown force. Resolve into components the forces that do not lie along one of the axes. In all our sample problems it made sense to choose the *x* axis horizontal and the *y* axis vertical.

5. Write the two *balance of forces equations,* using symbols throughout.

6. Choose one or more rotation axes perpendicular to the plane of the figure and write the *balance of torques equation* for each axis. If you choose an axis that passes through the line of action of an unknown force, the equation will be simplified because that force will not appear in it.

7. *Solve* your equations *algebraically* for the unknowns. Some students feel more confident in substituting numbers with units in the independent equations at this stage, especially if the algebra is particularly involved. However, experienced problem solvers prefer the algebraic approach, which reveals the dependence of solutions on the various variables.

8. Finally, *substitute numbers* with units in your algebraic solutions, obtaining numerical values for the unknowns.

9. Look at your answer—does it make sense? Is it obviously too large or too small? Is the sign correct? Are the units appropriate?

13-5   Indeterminate Structures

For the problems of this chapter, we have only three independent equations at our disposal, usually two balance of forces equations and one balance of torques equation about a given rotation axis. Thus, if a problem has more than three unknowns, we cannot solve it.

It is easy to find such problems. In Sample Problem 13-2, for example, we could have assumed that there is friction between the wall and the top of the ladder. Then

Fig. 13-9 The table is an indeterminate structure. The four forces on the table legs are different in magnitude and cannot be found from the laws of static equilibrium alone.

there would have been a vertical frictional force acting where the ladder touches the wall, making a total of four unknown forces. With only three equations, we could not have solved this problem.

Consider also an unsymmetrically loaded car. What are the forces—all different—on the four tires? Again, we cannot find them because we have only three independent equations with which to work. Similarly, we can solve an equilibrium problem for a table with three legs but not for one with four legs. Problems like these, in which there are more unknowns than equations, are called **indeterminate.**

Yet solutions to indeterminate problems exist in the real world. If you rest the tires of the car on four platform scales, each scale will register a definite reading, the sum of the readings being the weight of the car. What is eluding us in our efforts to find the individual forces by solving equations?

The problem is that we have assumed—without making a great point of it—that the bodies to which we apply the equations of static equilibrium are perfectly rigid. By this we mean that they do not deform when forces are applied to them. Strictly, there are no such bodies. The tires of the car, for example, deform easily under load until the car settles into a position of static equilibrium.

We have all had experience with a wobbly restaurant table, which we usually level by putting folded paper under one of the legs. If a big enough elephant sat on such a table, however, you may be sure that if the table did not collapse, it would deform just like the tires of a car. Its legs would all touch the floor, the forces acting upward on the table legs would all assume definite (and different) values as in Fig. 13-9, and the table would no longer wobble. How do we find the values of those forces acting on the legs?

To solve such indeterminate equilibrium problems, we must supplement equilibrium equations with some knowledge of *elasticity,* the branch of physics and engineering that describes how real bodies deform when forces are applied to them. The next section provides an introduction to this subject.

CHECKPOINT 5: A horizontal uniform bar of weight 10 N is to hang from a ceiling by two wires that exert upward forces $\vec{F}_1$ and $\vec{F}_2$ on the bar. The figure shows four arrangements for the wires. Which arrangements, if any, are indeterminate (so that we cannot solve for numerical values of $\vec{F}_1$ and $\vec{F}_2$)?

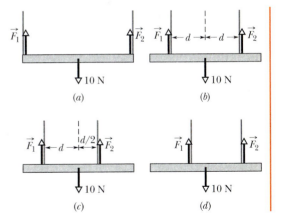

13-6 Elasticity

When a large number of atoms come together to form a metallic solid, such as an iron nail, they settle into equilibrium positions in a three-dimensional *lattice,* a repetitive arrangement in which each atom has a well-defined equilibrium distance from its nearest neighbors. The atoms are held together by interatomic forces that are modeled as tiny springs in Fig. 13-10. The lattice is remarkably rigid, which is another way of saying that the "interatomic springs" are extremely stiff. It is for this

Fig. 13-10 The atoms of a metallic solid are distributed on a repetitive three-dimensional lattice. The springs represent interatomic forces.

reason that we perceive many ordinary objects such as metal ladders, tables, and spoons as perfectly rigid. Of course, some ordinary objects, such as garden hoses or rubber gloves, do not strike us as rigid at all. The atoms that make up these objects *do not* form a rigid lattice like that of Fig. 13-10 but are aligned in long, flexible molecular chains, each chain being only loosely bound to its neighbors.

All real "rigid" bodies are to some extent **elastic,** which means that we can change their dimensions slightly by pulling, pushing, twisting, or compressing them. To get a feeling for the orders of magnitude involved, consider a vertical steel rod 1 m long and 1 cm in diameter. If you hang a subcompact car from the end of such a rod, the rod will stretch, but only by about 0.5 mm, or 0.05%. Furthermore, the rod will return to its original length when the car is removed.

If you hang two cars from the rod, the rod will be permanently stretched and will not recover its original length when you remove the load. If you hang three cars from the rod, the rod will break. Just before rupture, the elongation of the rod will be less than 0.2%. Although deformations of this size seem small, they are important in engineering practice. (Whether a wing under load will stay on an airplane is obviously important.)

Figure 13-11 shows three ways in which a solid might change its dimensions when forces act on it. In Fig. 13-11a, a cylinder is stretched. In Fig. 13-11b, a cylinder is deformed by a force perpendicular to its axis, much as we might deform a pack of cards or a book. In Fig. 13-11c, a solid object, placed in a fluid under high pressure, is compressed uniformly on all sides. What the three deformation types have in common is that a **stress,** or deforming force per unit area, produces a **strain,** or unit deformation. In Fig. 13-11, *tensile stress* (associated with stretching) is illustrated in (a), *shearing stress* in (b), and *hydraulic stress* in (c).

The stresses and the strains take different forms in the three situations of Fig. 13-11, but—over the range of engineering usefulness—stress and strain are proportional to each other. The constant of proportionality is called a **modulus of elasticity,** so that

$$\text{stress} = \text{modulus} \times \text{strain}. \qquad (13\text{-}24)$$

In a standard test of tensile properties, the tensile stress on a test cylinder (like that in Fig. 13-12) is slowly increased from zero to the point at which the cylinder fractures, and the strain is carefully measured and plotted. The result is a graph of stress versus strain like that in Fig. 13-13. For a substantial range of applied stresses, the stress–strain relation is linear, and the specimen recovers its original dimensions when the stress is removed; it is here that Eq. 13-24 applies. If the stress is increased beyond the **yield strength** S_y of the specimen, the specimen becomes permanently deformed. If the stress continues to increase, the specimen eventually ruptures, at a stress called the **ultimate strength** S_u.

Fig. 13-11 (a) A cylinder subject to *tensile stress* stretches by an amount ΔL. (b) A cylinder subject to *shearing stress* deforms by an amount Δx, somewhat like a pack of playing cards would. (c) A solid sphere subject to uniform *hydraulic stress* from a fluid shrinks in volume by an amount ΔV. All the deformations shown are greatly exaggerated.

Fig. 13-12 A test specimen, used to determine a stress–strain curve such as that of Fig. 13-13. The change ΔL that occurs in a certain length L is measured in a tensile stress–strain test.

Tension and Compression

For simple tension or compression, the stress on an object is defined as F/A, where F is the magnitude of the force applied perpendicularly to the area A on the object. The strain, or unit deformation, is then the dimensionless quantity $\Delta L/L$, the fractional (or sometimes percentage) change in the length of the specimen. If the specimen is a long rod and the stress does not exceed the yield strength, then not only the entire rod but also every section of it experiences the same strain when a given stress is applied. Because the strain is dimensionless, the modulus in Eq. 13-24 has the same dimensions as the stress, namely, force per unit area.

The modulus for tensile and compressive stresses is called the **Young's modulus** and is represented in engineering practice by the symbol E. Equation 13-24 becomes

$$\frac{F}{A} = E \frac{\Delta L}{L}. \tag{13-25}$$

The strain $\Delta L/L$ in a specimen can often be measured conveniently with a *strain gauge* (Fig. 13-14). This simple and useful device, which can be attached directly to operating machinery with an adhesive, is based on the principle that its electrical properties are dependent on the strain it undergoes.

Although the Young's modulus for an object may be almost the same for tension and compression, the object's ultimate strength may well be different for the two types of stress. Concrete, for example, is very strong in compression but is so weak in tension that it is almost never used in that manner. Table 13-1 shows the Young's modulus and other elastic properties for some materials of engineering interest.

Fig. 13-13 A stress–strain curve for a steel test specimen such as that of Fig. 13-12. The specimen deforms permanently when the stress is equal to the *yield strength* of the material. It ruptures when the stress is equal to the *ultimate strength* of the material.

Shearing

In the case of shearing, the stress is also a force per unit area, but the force vector lies in the plane of the area rather than perpendicular to it. The strain is the dimensionless ratio $\Delta x/L$, with the quantities defined as shown in Fig. 13-11b. The corresponding modulus, which is given the symbol G in engineering practice, is called the **shear modulus.** For shearing, Eq. 13-24 is written as

$$\frac{F}{A} = G \frac{\Delta x}{L}. \tag{13-26}$$

Shearing stresses play a critical role in the buckling of shafts that rotate under load and in bone fractures caused by bending.

Fig. 13-14 A strain gauge of overall dimensions 9.8 mm by 4.6 mm. The gauge is fastened with adhesive to the object whose strain is to be measured; it experiences the same strain as the object. The electrical resistance of the gauge varies with the strain, permitting strains up to 3% to be measured. Courtesy Measurements Groups, Inc., Raleigh, NC, USA.

TABLE 13-1   Some Elastic Properties of Selected Materials of Engineering Interest

Material	Density ρ (kg/m³)	Young's Modulus E (10^9 N/m²)	Ultimate Strength S_u (10^6 N/m²)	Yield Strength S_y (10^6 N/m²)
Steel[a]	7860	200	400	250
Aluminum	2710	70	110	95
Glass	2190	65	50[b]	—
Concrete[c]	2320	30	40[b]	—
Wood[d]	525	13	50[b]	—
Bone	1900	9[b]	170[b]	—
Polystyrene	1050	3	48	—

[a]Structural steel (ASTM-A36).   [b]In compression.   [c]High strength.   [d]Douglas fir.

Hydraulic Stress

In Fig. 13-11c, the stress is the fluid pressure p on the object, which, as you will see in Chapter 15, is a force per unit area. The strain is $\Delta V/V$, where V is the original volume of the specimen and ΔV is the absolute value of the change in volume. The corresponding modulus, with symbol B, is called the **bulk modulus** of the material. The object is said to be under *hydraulic compression,* and the pressure can be called the *hydraulic stress.* For this situation, we write Eq. 13-24 as

$$p = B \frac{\Delta V}{V}. \tag{13-27}$$

The bulk modulus is 2.2×10^9 N/m² for water and 16×10^{10} N/m² for steel. The pressure at the bottom of the Pacific Ocean, at its average depth of about 4000 m, is 4.0×10^7 N/m². The fractional compression $\Delta V/V$ of a volume of water due to this pressure is 1.8%; that for a steel object is only about 0.025%. In general, solids — with their rigid atomic lattices — are less compressible than liquids, in which the atoms or molecules are less tightly coupled to their neighbors.

Sample Problem 13-5

A structural steel rod has a radius R of 9.5 mm and a length L of 81 cm. A 62 kN force $\vec{F}$ stretches it along its length. What are the stress on the rod and the elongation and strain of the rod?

SOLUTION: The first Key Idea here has to do with what is meant by the second sentence in the problem statement. We assume the rod is held stationary by, say, a clamp or vise at one end. Then force $\vec{F}$ is applied at the other end, parallel to the length of the rod and thus perpendicular to the end face there. Therefore, the situation is like that in Fig. 13-11a.

The next Key Idea is that we assume the force is applied uniformly across the end face and thus over an area $A = \pi R^2$. Then the stress on the rod is given by the left side of Eq. 13-25:

$$\text{stress} = \frac{F}{A} = \frac{F}{\pi R^2} = \frac{6.2 \times 10^4 \text{ N}}{(\pi)(9.5 \times 10^{-3} \text{ m})^2}$$

$$= 2.2 \times 10^8 \text{ N/m}^2. \qquad \text{(Answer)}$$

The yield strength for structural steel is 2.5×10^8 N/m², so this rod is dangerously close to its yield strength.

Another Key Idea is that the elongation of the rod depends on the stress, the original length L, and the type of material in the rod. The last determines which value we use for Young's modulus E (from Table 13-1). Using the value for steel, Eq. 13-25 gives us

$$\Delta L = \frac{(F/A)L}{E} = \frac{(2.2 \times 10^8 \text{ N/m}^2)(0.81 \text{ m})}{2.0 \times 10^{11} \text{ N/m}^2}$$

$$= 8.9 \times 10^{-4} \text{ m} = 0.89 \text{ mm.} \qquad \text{(Answer)}$$

The last Key Idea we need here is that strain is the ratio of the change in length to the original length, so we have

$$\frac{\Delta L}{L} = \frac{8.9 \times 10^{-4} \text{ m}}{0.81 \text{ m}}$$

$$= 1.1 \times 10^{-3} = 0.11\%. \qquad \text{(Answer)}$$

Sample Problem 13-6

A table has three legs that are 1.00 m in length and a fourth leg that is longer by $d = 0.50$ mm, so that the table wobbles slightly. A heavy steel cylinder with mass $M = 290$ kg is placed upright on the table (with a mass much less than M) so that all four legs are compressed and the table no longer wobbles. The legs are wooden cylinders with cross-sectional area $A = 1.0$ cm². The Young's modulus E for the wood is 1.3×10^{10} N/m². Assume that the tabletop remains level and that the legs do not buckle. What are the magnitudes of the forces on the legs from the floor?

SOLUTION: We take the table plus steel cylinder as our system. The situation is like that in Fig. 13-9, except now we have a steel cylinder on the table. One Key Idea is that if the tabletop remains level, the legs must be compressed in the following ways: Each of the short legs must be compressed by the same amount (call it ΔL_3) and thus by the same force of magnitude F_3. The single long leg

must be compressed by a larger amount ΔL_4 and thus by a force with a larger magnitude F_4. In other words, for a level tabletop, we must have

$$\Delta L_4 = \Delta L_3 + d. \tag{13-28}$$

A second Key Idea is that, from Eq. 13-25, we can relate a change in length to the force causing the change with $\Delta L = FL/AE$, where L is the original length of a leg. We can use this relation to replace ΔL_4 and ΔL_3 in Eq. 13-28. However, note that we can approximate the original length L as being the same for all four legs. The replacements then give us

$$\frac{F_4 L}{AE} = \frac{F_3 L}{AE} + d. \tag{13-29}$$

We cannot solve this equation because it has two unknowns, F_4 and F_3.

To get a second equation containing F_4 and F_3, we can use a vertical y axis and then write the balance of vertical forces ($F_{net,y} = 0$) as

$$3F_3 + F_4 - Mg = 0, \qquad (13\text{-}30)$$

where Mg is equal to the magnitude of the gravitational force on the system. (*Three* legs have force $\vec{F}_3$ on them.) To solve the simultaneous equations 13-29 and 13-30 for, say, F_3, we first use Eq. 13-30 to find that $F_4 = Mg - 3F_3$. Substituting that into Eq. 13-29 then yields, after some algebra,

$$F_3 = \frac{Mg}{4} - \frac{dAE}{4L}$$

$$= \frac{(290 \text{ kg})(9.8 \text{ m/s}^2)}{4}$$

$$- \frac{(5.0 \times 10^{-4} \text{ m})(10^{-4} \text{ m}^2)(1.3 \times 10^{10} \text{ N/m}^2)}{(4)(1.00 \text{ m})}$$

$$= 548 \text{ N} \approx 550 \text{ N}. \qquad \text{(Answer)}$$

From Eq. 13-30 we then find

$$F_4 = Mg - 3F_3 = (290 \text{ kg})(9.8 \text{ m/s}^2) - 3(548 \text{ N})$$

$$\approx 1200 \text{ N}. \qquad \text{(Answer)}$$

You can show that to reach their equilibrium configuration, the three short legs are each compressed by 0.42 mm and the single long leg by 0.92 mm.

CHECKPOINT 6: The figure shows a horizontal block that is suspended by two wires, A and B, which are identical except for their original lengths. The center of mass of the block is closer to wire B than to wire A. (a) Measuring torques about the block's center of mass, state whether the magnitude of the torque due to wire A is greater than, less than, or equal to the magnitude of the torque due to wire B. (b) Which wire exerts more force on the block? (c) If the wires are now equal in length, which one was originally shorter?

REVIEW & SUMMARY

Static Equilibrium   A rigid body at rest is said to be in **static equilibrium.** For such a body, the vector sum of the external forces acting on it is zero:

$$\vec{F}_{net} = 0 \qquad \text{(balance of forces).} \qquad (13\text{-}3)$$

If all the forces lie in the xy plane, this vector equation is equivalent to two component equations:

$$F_{net,x} = 0 \quad \text{and} \quad F_{net,y} = 0 \qquad \text{(balance of forces).} \qquad (13\text{-}7, 13\text{-}8)$$

Static equilibrium also implies that the vector sum of the external torques acting on the body about *any* point is zero, or

$$\vec{\tau}_{net} = 0 \qquad \text{(balance of torques).} \qquad (13\text{-}5)$$

If the forces lie in the xy plane, all torque vectors are parallel to the z axis, and Eq. 13-5 is equivalent to the single component equation

$$\tau_{net,z} = 0 \qquad \text{(balance of torques).} \qquad (13\text{-}9)$$

Center of Gravity   The gravitational force acts individually on each element of a body. The net effect of all individual actions may be found by imagining an equivalent total gravitational force $\vec{F}_g$ acting at a specific point called the **center of gravity.** If the gravitational acceleration $\vec{g}$ is the same for all the elements of the body, the center of gravity is at the center of mass.

Elastic Moduli   Three **elastic moduli** are used to describe the elastic behavior (deformations) of objects as they respond to forces that act on them. The **strain** (fractional change in length) is linearly related to the applied **stress** (force per unit area) by the proper modulus, according to the general relation

$$\text{stress} = \text{modulus} \times \text{strain.} \qquad (13\text{-}24)$$

Tension and Compression   When an object is under tension or compression, Eq. 13-24 is written as

$$\frac{F}{A} = E \frac{\Delta L}{L}, \qquad (13\text{-}25)$$

where $\Delta L/L$ is the tensile or compressive strain of the object, F is the magnitude of the applied force $\vec{F}$ causing the strain, A is the cross-sectional area over which $\vec{F}$ is applied (perpendicular to A, as in Fig. 13-11a), and E is the **Young's modulus** for the object. The stress is F/A.

Shearing   When an object is under a shearing stress, Eq. 13-24 is written as

$$\frac{F}{A} = G \frac{\Delta x}{L}, \qquad (13\text{-}26)$$

where $\Delta x/L$ is the shearing strain of the object, Δx is the displacement of one end of the object in the direction of the applied force $\vec{F}$ (as in Fig. 13-11b), and G is the **shear modulus** of the object. The stress is F/A.

Hydraulic Stress   When an object undergoes *hydraulic compression* due to a stress exerted by a surrounding fluid, Eq. 13-24 is written as

$$p = B \frac{\Delta V}{V}, \qquad (13\text{-}27)$$

where p is the pressure (*hydraulic stress*) on the object due to the fluid, $\Delta V/V$ (the strain) is the absolute value of the fractional change in the object's volume due to that pressure, and B is the **bulk modulus** of the object.

QUESTIONS

1. Figure 13-15 shows an over-head view of a uniform stick on which four forces act. Suppose we choose a rotational axis through point O, calculate the torques about that axis due to the

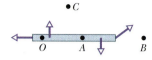
Fig. 13-15 Question 1.

forces, and find that these torques balance. Will the torques balance if, instead, the rotational axis is chosen to be at (a) point A, (b) point B, or (c) point C? (d) Suppose, instead, that we find that the torques about point O do not balance. Is there another point about which the torques will balance?

2. In Fig. 13-16, a rigid beam is attached to two posts that are fastened to a floor. A small but heavy safe is placed at the six positions indicated, in turn. Assume that the mass of the beam is neg-

Fig. 13-16 Question 2.

ligible compared to that of the safe. (a) Rank the positions according to the force on post A due to the safe, greatest compression first, greatest tension last, and indicate where, if anywhere, the force is zero. (b) Now rank them according to the force on post B.

3. Figure 13-17 shows four overhead views of rotating uniform disks that are sliding across a frictionless floor. Three forces, of magnitude F, $2F$, or $3F$, act on each disk, either at the rim, at the center, or halfway between rim and center. The force vectors rotate along with the disks, and in the "snapshots" of Fig. 13-17, point left or right. Which disks are in equilibrium?

(a)   (b)   (c)   (d)
Fig. 13-17 Question 3.

4. Figure 13-18 shows overhead views of two structures on which three forces act. The directions of the forces are as indicated. If the magnitudes of the forces are adjusted properly (but kept nonzero), which structure can be in static equilibrium?

5. Figure 13-19 shows a mobile of toy penguins hanging from a

(a)   (b)
Fig. 13-18 Question 4.

Fig. 13-19 Question 5.

ceiling. Each crossbar is horizontal, has negligible mass, and extends three times as far to the right of the wire supporting it as to the left. Penguin 1 has mass $m_1 = 48$ kg. What are the masses of the other penguins?

6. A ladder leans against a frictionless wall but is prevented from falling because of friction between it and the ground. Suppose you shift the base of the ladder toward the wall. Determine whether the following become larger, smaller, or stay the same (in magnitude): (a) the normal force on the ladder from the ground, (b) the force on the ladder from the wall, (c) the static frictional force on the ladder from the ground, and (d) the maximum value $f_{s,\max}$ of the static frictional force.

7. Three piñatas hang from the (stationary) assembly of massless pulleys and cords seen in Fig. 13-20. One long cord runs from the ceiling at the right to the lower pulley at the left. Several shorter cords suspend pulleys from the ceiling or piñatas from the pulleys. The weights (in newtons) of two piñatas are given. (a) What is the weight of the third piñata? (*Hint:* When a cord loops half-

Fig. 13-20 Question 7.

way around a pulley, it pulls on the pulley with a net force that is twice the tension in the cord.) (b) What is the tension in the short cord labeled with T?

8. (a) In Checkpoint 4, to express τ_r in terms of T, should you use $\sin\theta$ or $\cos\theta$? (b) If angle θ is decreased (by shortening the rope but still keeping the rod horizontal), does the torque τ_r required for equilibrium become larger, smaller, or stay the same? (c) Does the corresponding force magnitude T become larger, smaller, or stay the same?

9. The table gives the areas of three surfaces and the magnitude of a force that is applied perpendicular to the surface and uniformly across it. Rank the surfaces according to the stress on them, greatest first.

	Area	Force
Surface A	$0.5A_0$	$2F_0$
Surface B	$2A_0$	$4F_0$
Surface C	$3A_0$	$6F_0$

10. Four cylindrical rods are stretched as in Fig. 13-11a. The force magnitudes, the areas of the end faces, the changes in length, and the initial lengths are given here. Rank the rods according to their Young's moduli, greatest first.

Rod	Force	Area	Length Change	Initial Length
1	F	A	ΔL	L
2	$2F$	$2A$	$2\Delta L$	L
3	F	$2A$	$2\Delta L$	$2L$
4	$2F$	A	ΔL	$2L$

EXERCISES & PROBLEMS

ssm Solution is in the Student Solutions Manual.
www Solution is available on the World Wide Web at:
         http://www.wiley.com/college/hrw
ilw Solution is available on the Interactive LearningWare.

SEC. 13-4 Some Examples of Static Equilibrium

1E. A physics Brady Bunch, whose weights in newtons are indicated in Fig. 13-21, is balanced on a seesaw. What is the number of the person who causes the largest torque, about the rotation axis at *fulcrum f*, directed (a) out of the page and (b) into the page?

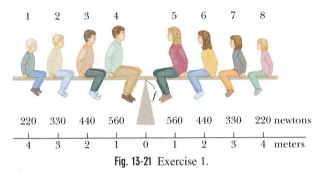

Fig. 13-21 Exercise 1.

2E. The leaning Tower of Pisa (Fig. 13-22) is 55 m high and 7.0 m in diameter. The top of the tower is displaced 4.5 m from the vertical. Treat the tower as a uniform, circular cylinder. (a) What additional displacement, measured at the top, would bring the tower to the verge of toppling? (b) What angle would the tower then make with the vertical?

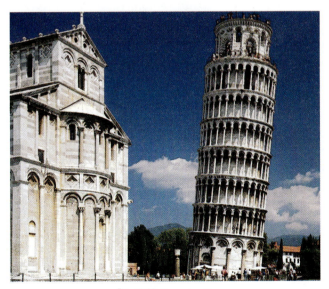

Fig. 13-22 Exercise 2.

3E. A particle is acted on by forces given, in newtons, by $\vec{F}_1 = 10\hat{i} - 4\hat{j}$ and $\vec{F}_2 = 17\hat{i} + 2\hat{j}$. (a) What force $\vec{F}_3$ balances these forces? (b) What direction does $\vec{F}_3$ have relative to the *x* axis? ssm

4E. A bow is drawn at its midpoint until the tension in the string is equal to the force exerted by the archer. What is the angle between the two halves of the string?

5E. A rope of negligible mass is stretched horizontally between two supports that are 3.44 m apart. When an object of weight 3160 N is hung at the center of the rope, the rope is observed to sag by 35.0 cm. What is the tension in the rope? ilw

6E. A scaffold of mass 60 kg and length 5.0 m is supported in a horizontal position by a vertical cable at each end. A window washer of mass 80 kg stands at a point 1.5 m from one end. What is the tension in (a) the nearer cable and (b) the farther cable?

7E. In Fig. 13-23, a uniform sphere of mass *m* and radius *r* is held in place by a massless rope attached to a frictionless wall a distance *L* above the center of the sphere. Find (a) the tension in the rope and (b) the force on the sphere from the wall. ssm

Fig. 13-23 Exercise 7.

8E. An automobile with a mass of 1360 kg has 3.05 m between the front and rear axles. Its center of gravity is located 1.78 m behind the front axle. With the automobile on level ground, determine the magnitude of the force from the ground on (a) each front wheel (assuming equal forces on the front wheels) and (b) each rear wheel (assuming equal forces on the rear wheels).

9E. A diver of weight 580 N stands at the end of a 4.5 m diving board of negligible mass (Fig. 13-24). The board is attached to two pedestals 1.5 m apart. What are the magnitude and direction of the force on the board from (a) the left pedestal and (b) the right pedestal? (c) Which pedestal is being stretched, and (d) which compressed? ssm

Fig. 13-24 Exercise 9.

10E. In Fig. 13-25, a man is trying to get his car out of mud on the shoulder of a road. He ties one end of a rope tightly around the front bumper and the other end tightly around a utility pole 18 m away. He then pushes sideways on the rope at its midpoint with a force of 550 N, displacing the center of the rope 0.30 m from its previous position, and the car barely moves. What is the magnitude of the force on the car from the rope? (The rope stretches somewhat.)

Fig. 13-25 Exercise 10.

11E. A meter stick balances horizontally on a knife-edge at the 50.0 cm mark. With two 5.0 g coins stacked over the 12.0 cm mark, the stick is found to balance at the 45.5 cm mark. What is the mass of the meter stick? **ssm**

12E. A uniform cubical crate is 0.750 m on each side and weighs 500 N. It rests on a floor with one edge against a very small, fixed obstruction. At what least height above the floor must a horizontal force of magnitude 350 N be applied to the crate to tip it?

13E. A 75 kg window cleaner uses a 10 kg ladder that is 5.0 m long. He places one end on the ground 2.5 m from a wall, rests the upper end against a cracked window, and climbs the ladder. He is 3.0 m up along the ladder when the window breaks. Neglecting friction between the ladder and window and assuming that the base of the ladder does not slip, find (a) the magnitude of the force on the window from the ladder just before the window breaks and (b) the magnitude and direction of the force on the ladder from the ground just before the window breaks. **ssm**

14E. Figure 13-26 shows the anatomical structures in the lower leg and foot that are involved in standing tiptoe with the heel raised off the floor so the foot effectively contacts the floor at only one point, shown as P in the figure. Calculate, in terms of a person's weight W, the forces on the foot from (a) the calf muscle (at A) and (b) the lower-leg bones (at B) when the person stands tiptoe on one foot. Assume that $a = 5.0$ cm and $b = 15$ cm.

Calf muscle

Lower leg bones

Fig. 13-26 Exercise 14.

$\leftarrow a \rightarrow \leftarrow \quad b \quad \rightarrow$

15P. In Fig. 13-27, an 817 kg construction bucket is suspended by a cable A that is attached at O to two other cables B and C, making angles of $51.0°$ and $66.0°$ with the horizontal. Find the tensions in (a) cable A, (b) cable B, and (c) cable C. (*Hint:* To avoid solving two equations in two unknowns, position the axes as shown in the figure.) **ssm**

Fig. 13-27 Problem 15.

16P. The system in Fig. 13-28 is in equilibrium, with the string in the center exactly horizontal. Find (a) tension T_1, (b) tension T_2, (c) tension T_3, and (d) angle θ.

40 N   50 N

Fig. 13-28 Problem 16.

17P. The force $\vec{F}$ in Fig. 13-29 keeps the 6.40 kg block and the pulleys in equilibrium. The pulleys have negligible mass and friction. Calculate the tension T in the upper cable. (*Hint:* When a cable wraps halfway around a pulley as here, the magnitude of its net force on the pulley is twice the tension in the cable.)

ssm   ilw

18P. A 15 kg block is being lifted by the pulley system shown in Fig. 13-30. The upper arm is vertical, whereas the forearm makes an angle of 30° with the horizontal. What are the forces on the forearm from (a) the triceps muscle and (b) the upper-arm bone (the humerus)? The forearm and hand together have a mass of 2.0 kg with a center of mass 15 cm (measured along the arm)

Fig. 13-29 Problem 17.

from the point where the forearm and upper-arm bones are in contact. The triceps muscle pulls vertically upward at a point 2.5 cm behind that contact point.

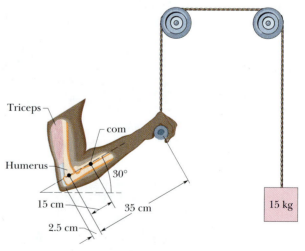

Triceps

com

Humerus

30°

15 cm

35 cm

15 kg

2.5 cm

Fig. 13-30 Problem 18.

19P. Forces $\vec{F}_1, \vec{F}_2$, and $\vec{F}_3$ act on the structure of Fig. 13-31, shown in an overhead view. We wish to put the structure in equilibrium by applying a fourth force, at a point such as P. The fourth force has vector components $\vec{F}_h$ and $\vec{F}_v$. We are given that $a = 2.0$ m,

$b = 3.0$ m, $c = 1.0$ m, $F_1 = 20$ N, $F_2 = 10$ N, and $F_3 = 5.0$ N. Find (a) F_h, (b) F_v, and (c) d. ilw

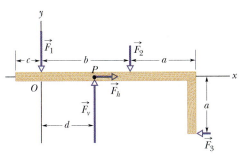

Fig. 13-31 Problem 19.

20P. In Fig. 13-32, a 50.0 kg uniform square sign, 2.00 m on a side, is hung from a 3.00 m horizontal rod of negligible mass. A cable is attached to the end of the rod and to a point on the wall 4.00 m above the point where the rod is hinged to the wall. (a) What is the tension in the cable? What are the magnitudes and directions of the (b) horizontal and (c) vertical components of the force on the rod from the wall?

Fig. 13-32 Problem 20.

21P. In Fig. 13-33, what magnitude of force $\vec{F}$ applied horizontally at the axle of the wheel is necessary to raise the wheel over an obstacle of height h? The wheel's radius is r and its mass is m. ssm www

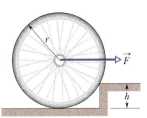

Fig. 13-33 Problem 21.

22P. In Fig. 13-34, a 55 kg rock climber is in a lie-back climb along a fissure, with hands pulling on one side of the fissure and feet pressed against the opposite side. The fissure has width $w = 0.20$ m, and the center of mass of the climber is a horizontal distance $d = 0.40$ m from the fissure. The coefficient of static friction between hands and rock is $\mu_1 = 0.40$, and between boots and rock it is $\mu_2 = 1.2$. (a) What is the least horizontal pull by the hands and push by the feet that will keep the climber stable? (b) For the horizontal pull of (a), what must be the vertical distance h between hands and feet? (c) If the climber encounters wet rock, so that μ_1 and μ_2 are reduced, what happens to the answers to (a) and (b), respectively?

Fig. 13-34 Problem 22.

23P. In Fig. 13-35, one end of a uniform beam that weighs 222 N is attached to a wall with a hinge. The other end is supported by a wire. (a) Find the tension in the wire. What are the (b) horizontal and (c) vertical components of the force of the hinge on the beam? ssm

24P. Four bricks of length L, identical and uniform, are stacked on top of one another (Fig. 13-36) in such a way that part of each extends beyond the one beneath. Find, in terms of L, the maximum values of (a) a_1, (b) a_2, (c) a_3, (d) a_4, and (e) h, such that the stack is in equilibrium.

Fig. 13-35 Problem 23.

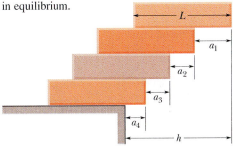

Fig. 13-36 Problem 24.

25P. The system in Fig. 13-37 is in equilibrium. A concrete block of mass 225 kg hangs from the end of the uniform strut whose mass is 45.0 kg. Find (a) the tension T in the cable and the (b) horizontal and (c) vertical force components on the strut from the hinge. ilw

Fig. 13-37 Problem 25.

26P. A door 2.1 m high and 0.91 m wide has a mass of 27 kg. A hinge 0.30 m from the top and another 0.30 m from the bottom each support half the door's mass. Assume that the center of gravity is at the geometrical center of the door, and determine the (a) vertical and (b) horizontal components of the force from each hinge on the door.

27P. A nonuniform bar is suspended at rest in a horizontal position by two massless cords as shown in Fig. 13-38. One cord makes the angle $\theta = 36.9°$ with the vertical; the other makes the angle $\phi = 53.1°$ with the vertical. If the length L of the bar is 6.10 m, compute the distance x from the left-hand end of the bar to its center of mass. ssm

Fig. 13-38 Problem 27.

28P. In Fig. 13-39, a thin horizontal bar AB of negligible weight and length L is hinged to a vertical wall at A and supported at B

by a thin wire BC that makes an angle θ with the horizontal. A load of weight W can be moved anywhere along the bar; its position is defined by the distance x from the wall to its center of mass. As a function of x, find (a) the tension in the wire, and the (b) horizontal and (c) vertical components of the force on the bar from the hinge at A.

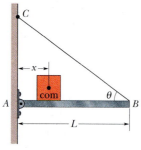

Fig. 13-39 Problems 28 and 30.

29P. In Fig. 13-40, a uniform plank, with a length L of 6.10 m and a weight of 445 N, rests on the ground and against a frictionless roller at the top of a wall of height $h = 3.05$ m. The plank remains in equilibrium for any value of $\theta \geq 70°$ but slips if $\theta < 70°$. Find the coefficient of static friction between the plank and the ground. ssm

30P. In Fig. 13-39, suppose the length L of the uniform bar is 3.0 m and its weight is 200 N. Also, let the load's weight $W = 300$ N and the angle $\theta = 30°$. The wire can withstand a maximum tension of 500 N. (a) What is the maximum possible distance x before the wire breaks? With the load placed at this maximum x, what are the (b) horizontal and (c) vertical components of the force on the bar from the hinge at A?

31P. For the stepladder shown in Fig. 13-41, sides AC and CE are each 2.44 m long and hinged at C. Bar BD is a tie-rod 0.762 m long, halfway up. A man weighing 854 N climbs 1.80 m along the ladder. Assuming that the floor is frictionless and neglecting the mass of the ladder, find (a) the tension in the tie-rod and the magnitudes of the forces on the ladder from the floor at (b) A and (c) E. (*Hint:* It will help to isolate parts of the ladder in applying the equilibrium conditions.) ssm

32P. Two uniform beams, A and B, are attached to a wall with hinges and then loosely bolted together as in Fig. 13-42. Find the

Fig. 13-40 Problem 29.

Fig. 13-41 Problem 31.

Fig. 13-42 Problem 32.

x and y components of the force on (a) beam A due to its hinge, (b) beam A due to the bolt, (c) beam B due to its hinge, and (d) beam B due to the bolt.

33P. A cubical box is filled with sand and weighs 890 N. We wish to "roll" the box by pushing horizontally on one of the upper edges. (a) What minimum force is required? (b) What minimum coefficient of static friction between box and floor is required? (c) Is there a more efficient way to roll the box? If so, find the smallest possible force that would have to be applied directly to the box to roll it. (*Hint:* At the onset of tipping, where is the normal force located?) ssm

34P. Four bricks of length L, identical and uniform, are stacked on a table in two ways, as shown in Fig. 13-43 (compare with Problem 24). We seek to maximize the overhang distance h in both arrangements. Find the optimum distances a_1, a_2, b_1, and b_2, and calculate h for the two arrangements. (See "The Amateur Scientist," *Scientific American,* June 1985, pp. 133–134, for a discussion and an even better version of arrangement (b).)

Fig. 13-43 Problem 34.

35P. A crate, in the form of a cube with edge lengths of 1.2 m, contains a piece of machinery; the center of mass of the crate and its contents is located 0.30 m above the crate's geometrical center. The crate rests on a ramp that makes an angle θ with the horizontal. As θ is increased from zero, an angle will be reached at which the crate will either start to slide down the ramp or tip over. Which event will occur (a) when the coefficient of static friction between ramp and crate is 0.60 and (b) when it is 0.70? In each case, give the angle at which the event occurs. (*Hint:* At the onset of tipping, where is the normal force located?) ssm www

SEC. 13-6 Elasticity

36E. Figure 13-44 shows the stress–strain curve for quartzite. What

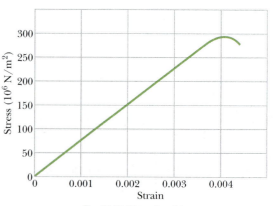

Fig. 13-44 Exercise 36.

are (a) the Young's modulus and (b) the approximate yield strength for this material?

37E. A horizontal aluminum rod 4.8 cm in diameter projects 5.3 cm from a wall. A 1200 kg object is suspended from the end of the rod. The shear modulus of aluminum is 3.0×10^{10} N/m². Neglecting the rod's mass, find (a) the shear stress on the rod and (b) the vertical deflection of the end of the rod. ssm ilw

38P. In Fig. 13-45, a lead brick rests horizontally on cylinders A and B. The areas of the top faces of the cylinders are related by $A_A = 2A_B$; the Young's moduli of the cylinders are related by $E_A = 2E_B$. The cylinders had identical lengths before the brick was placed on them. What fraction of the brick's mass is supported (a) by cylinder A and (b) by cylinder B? The horizontal distances between the center of mass of the brick and the centerlines of the cylinders are d_A for cylinder A and d_B for cylinder B. (c) What is the ratio d_A/d_B?

Fig. 13-45 Problem 38.

39P. In Fig. 13-46, a 103 kg uniform log hangs by two steel wires, A and B, both of radius 1.20 mm. Initially, wire A was 2.50 m long and 2.00 mm shorter than wire B. The log is now horizontal. What are the magnitudes of the forces on it from (a) wire A and (b) wire B? (c) What is the ratio d_A/d_B? ssm www

Fig. 13-46 Problem 39.

40P. A tunnel 150 m long, 7.2 m high, and 5.8 m wide (with a flat roof) is to be constructed 60 m beneath the ground. (See Fig. 13-47.) The tunnel roof is to be supported entirely by square steel columns, each with a cross-sectional area of 960 cm². The density of the ground material is 2.8 g/cm³. (a) What is the total mass of the material that the columns must support? (b) How many columns are needed to keep the compressive stress on each column at one-half its ultimate strength?

Fig. 13-47 Problem 40.

Additional Problems

41. Here is a way to move a heavy log through a tropical forest. Find a young tree in the general direction of travel; find a vine that hangs from the top of the tree down to ground level; pull the vine over to the log; wrap the vine around a limb on the log; pull hard enough on the vine to bend the tree over; and then tie off the vine on the limb. Repeat this procedure with several trees; eventually the net force of the vines on the log moves the log forward. Although tedious, this technique allowed workers to move heavy logs long before modern machinery was available. Figure 13-48 shows the essentials of the technique. There, a single vine is shown attached to a branch at one end of a uniform log of mass M. The coefficient of static friction between the log and the ground is 0.80. If the log is on the verge of sliding, with the left end raised slightly by the vine, what are (a) the angle θ and (b) the magnitude T of the force on the log from the vine?

Fig. 13-48 Problem 41.

42. You have been hired to build a large sand mound in an indoor playground and must be careful about the stress that the sand will put on the floor. Consulting research literature, you are surprised to find that the greatest stress occurs, not directly beneath the apex (top) of the mound, but at points that are a distance r_m from that central point (Fig. 13-49a). This outward displacement of the maximum stress is presumably due to the sand grains forming arches within the mound. For a mound of height $H = 3.00$ m and angle $\theta = 33°$, and with sand of density $\rho = 1800$ kg/m³, Fig. 13-49b gives the stress σ as a function of radius r from the central point of the mound's base. In that figure, $\sigma_0 = 40\,000$ N/m², $\sigma_m = 40\,024$ N/m², and $r_m = 1.82$ m.

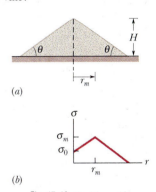

Fig. 13-49 Problem 42.

(a) What is the volume of sand contained in the mound for $r \leq r_m/2$? (Hint: The volume is that of a vertical cylinder plus a cone on top of the cylinder. The volume of the cone is $\pi R^2 h/3$, where R is the cone's radius and h is the cone's height.) (b) What is the weight W of that volume of sand? (c) Use Fig. 13-49b to write an expression for the stress σ on the floor as a function of radius r, for $r \leq r_m$. (d) On the floor, what is the area dA of a thin ring of radius r centered on the mound's central axis and with radial width dr? (e) What then is the magnitude dF of the downward force on the ring due to the sand? (f) What is the magnitude F of the net downward force on the floor due to all the sand contained in the mound for $r \leq r_m/2$? (Hint: Integrate the expression of (e) from $r = 0$ to $r = r_m/2$.) Now note the surprise: This force magnitude F on the floor is less than the weight W of the sand above the floor, as found in (b). (g) By what fraction is F reduced from W; that is, what is $(F - W)/W$?

NEW PROBLEMS

N1. Figure 13N-1*a* shows a horizontal uniform beam of mass m_b and length L that is supported on the left by a hinge with a wall and on the right by a cable at angle θ with the horizontal. A package of mass m_p is positioned on the beam at a distance x from the left end. The total mass is $m_b + m_p = 61.22$ kg. Figure 13N-1*b* gives the tension T in the cable as a function of the package's position given as a fraction x/L of the beam length. Evaluate (a) angle θ, (b) mass m_b, and (c) mass m_p.

(a)

(b)

Fig. 13N-1 Problem N1.

N2. In Fig. 13N-2*a*, a uniform 40 kg beam is centered over two rollers. Vertical lines across the beam mark off equal lengths. Two of the lines are centered over the rollers; a 10 kg package of tamale is centered over roller B. What are the magnitudes of the forces on the beam from (a) roller A and (b) roller B? The beam is then rolled to the left until the right-hand end is centered over roller B (Fig. 13N-2*b*). What now are the magnitudes of the forces on the beam from (c) roller A and (d) roller B? Next, the beam is rolled to the right. Assume that it has a length of 0.800 m. (e) What horizontal distance between the package and roller B puts the beam on the verge of losing contact with roller A?

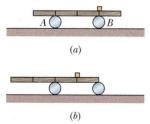

(a)

(b)

Fig. 13N-2 Problem N2.

N3. In an ice plant, 200 kg blocks of ice slide down a frictionless ramp that makes an angle of 10° with the horizontal. To keep the blocks of ice from moving too quickly, they are restrained by an attached cable that is parallel to the ramp. If the blocks are temporarily held at rest on the ramp by the cable, what is the tension in the cable?

N4. Figure 13N-3*a* shows a vertical uniform beam of length L that is hinged at its lower end. A horizontal force $\vec{F}_a$ is applied to the beam at a distance y from the lower end. The beam remains vertical because of a cable attached at the upper end, at angle θ with the horizontal. Figure 13N-3*b* gives the tension T in the cable as a function of the position of the applied force given as a fraction y/L of the beam length. Figure 13N-3*c* gives the magnitude F_h of the horizontal force on the beam from the hinge, also as a function of y/L. Evaluate (a) angle θ and (b) the magnitude of $\vec{F}_a$.

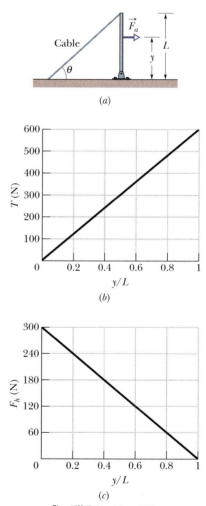

(a)

(b)

(c)

Fig. 13N-3 Problem N4.

N5. A 10 kg sphere is supported on a frictionless plane inclined at 45° from the horizontal as shown in Fig. 13N-4. Calculate the tension in the cable.

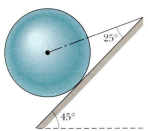

Fig. 13N-4 Problem N5.

N6. In Figure 13N-5, a package of mass m hangs from a short cord that is tied to the wall via cord 1 and to the ceiling via cord 2. Cord 1 is at an angle of 40° with the horizontal; cord 2 is at angle θ. (a) For what value of θ is the tension in cord 2 minimized? (b) In terms of mg, what is the minimum tension in cord 2?

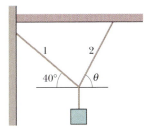

Fig. 13N-5 Problem N6.

N7. A makeshift swing is constructed by making a loop in one end of a rope and tying the other end to a tree limb. A child is sitting in the loop with the rope hanging vertically when an adult pulls on the child with a horizontal force and displaces the child to one side. Just before the child is released from rest, the rope makes an angle of 15° with the vertical and the tension in the rope is 280 N. (a) How much does the child weigh? (b) What is the magnitude of the (horizontal) force of the adult on the child just before the child is released? (c) If the maximum horizontal force that the adult can exert on the child is 93 N, what is the maximum angle with the vertical that the rope can make while the adult is pulling horizontally?

N8. In Fig. 13N-6, a uniform beam of weight 500 N and length 3.0 m is suspended horizontally. On the left it is hinged to a wall; on the right it is supported by a cable bolted to the wall at a distance D above the beam. The least tension that will snap the cable is 1200 N. (a) What value of D corresponds to that tension? (b) To prevent the cable from snapping, should D be greater or less than that value?

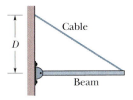

Fig. 13N-6 Problem N8.

N9. A uniform ladder whose length is 5.0 m and whose weight is 400 N leans against a frictionless vertical wall. The coefficient of static friction between the level ground and the foot of the ladder is 0.46. What is the greatest distance the foot of the ladder can be placed from the base of the wall without the ladder immediately slipping?

N10. *Center of gravity.* The gravitational acceleration g varies so little over the extent of most structures that the structure's cog effectively coincides with its com. Here is a fictitious example where the acceleration varies more significantly. Figure 13N-7 shows an array of six particles, each with mass m, that are fixed to the edge of a rigid structure of negligible mass. The distance between adjacent particles along the edge is 2.00 m. The following table gives the gravitational acceleration g at each particle's location. Using the coordinate system shown, find (a) the x coordinate x_{com} and (b) the y coordinate y_{com} of the center of mass of the six-particle system. Then find (c) the x coordinate x_{cog} and (d) the y coordinate y_{cog} of the center of gravity of the six-particle system. (*Hint:* See the equation above Eq. 13-14.)

Particle	g (m/s^2)	Particle	g (m/s^2)
1	8.00	4	7.40
2	7.80	5	7.60
3	7.60	6	7.80

Fig. 13N-7 Problem N10.

N11. A construction worker attempts to lift a uniform beam off the floor and raise it to a vertical position. The beam is 2.5 m long and weighs 500 N. At a certain instant the worker holds the beam momentarily at rest with one end 1.5 m off the floor, as shown in Fig. 13N-8, by exerting a force $\vec{P}$ on the beam, perpendicular to the beam. (a) What is the magnitude of the force exerted by the worker? (b) What is the magnitude of the (net) force of the floor on the beam? (c) What is the minimum value that the coefficient of static friction between the beam and the floor can have in order for the beam not to slip at this instant?

Fig. 13N-8 Problem N11.

N12. Figure 13N-9a shows a uniform ramp between two buildings that allows for motion between the buildings due to strong winds. At its left end, it is hinged to the building wall; at its right end, it has a roller that can roll along the building wall. There is no vertical force on the roller from the building, only a horizontal force with magnitude F_h. The horizontal distance between the buildings is $D = 4.00$ m. The rise of the ramp is $h = 0.490$ m. A man walks across the ramp from the left. Figure 13N-9b gives F_h as a function of the horizontal distance x of the man from the building at the left. What are the masses of (a) the ramp and (b) the man?

(a)

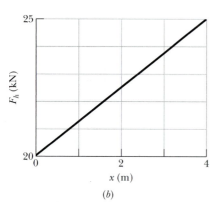

(b)

Fig. 13N-9 Problem N12.

N13. In Fig. 13N-10, a uniform diving board (mass = 40 kg) is 3.5 m long and is attached to two supports. When a diver stands on the end of the board, the support on the other end exerts a downward force of 1200 N on the board. Where on the board should the diver stand in order to reduce that force to zero?

Fig. 13N-10 Problem N13.

N14. In Fig. 13N-11, a horizontal scaffold, of length 2.00 m and uniform mass 50 kg, is suspended from a building by two cables. It has dozens of paint cans stacked on it at various points. The total mass of the paint cans is 75.0 kg. The tension in the cable at the right is 722 N. How far horizontally from that cable is the center of mass of the system of paint cans?

Fig. 13N-11 Problem N14.

N15. Figure 13N-12 shows a 300 kg horizontal cylinder. Three steel wires support the cylinder from a ceiling. Wires 1 and 3 are at the ends of the cylinder, and wire 2 is at the center. The wires each have a cross sectional area of 2.00×10^{-6} m². Initially (before the cylinder was put in place) wires 1 and 3 were 2.0000 m long and wire 2 was 6.00 mm longer than that. Now all three wires have been stretched. What are the tensions in (a) wire 1 and (b) wire 2?

Fig. 13N-12 Problem N15.

N16. In Fig. 13N-13, a uniform rod of mass m is hinged to a building at its lower end, while its upper end is held in place by a rope attached to the wall. If angle $\theta_1 = 60°$, what value must angle θ_2 have so that the tension in the rope is equal to $mg/2$?

Fig. 13N-13 Problem N16.

N17. Figure 13N-14 shows the stress versus strain plot for an aluminum wire that is stretched by a machine pulling in opposite directions at the two ends of the wire. The wire has an initial length of 0.800 m and an initial cross-sectional area of 2.00×10^{-6} m². How much work does the force from the machine do on the wire to produce a strain of 1.00×10^{-3}?

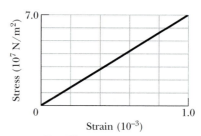

Fig. 13N-14 Problem N17.

N18. In Fig. 13N-15, horizontal scaffold 2, with uniform mass $m_2 = 30.0$ kg and length $L_2 = 2.00$ m, hangs from horizontal scaffold 1, with uniform mass $m_1 = 50.0$ kg. A 20.0 kg package of nails lies on scaffold 2, centered at distance $d = 0.500$ m from the left end. What is the tension T in the cable indicated?

Fig. 13N-15 Problem N18.

N19. A cylindrical aluminum rod, with an initial length of 0.8000 m and radius 1000.0 μm, is clamped in place at one end and then stretched by a machine pulling parallel to its length at its other end. Assuming that the rod's density (mass per unit volume) does not change, find the force magnitude that is required of the machine to decrease the radius to 999.9 μm. (The yield strength is not exceeded.)

N20. Figure 13N-16 shows a stationary arrangement of two crayon boxes and three cords. Box A has a mass of 11.0 kg and is on a ramp at angle $\theta = 30.0°$; box B has a mass of 7.00 kg and hangs on a cord. The cord connected to box A is parallel to the ramp, which is frictionless. What are (a) the tension and (b) the orientation of the upper cord?

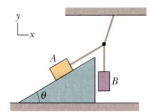

Fig. 13N-16 Problem N20.

N21*. A car on a horizontal road makes an emergency stop by applying the brakes so that all four wheels lock and skid along the road. The coefficient of kinetic friction between tires and road is 0.40. The separation between the front and rear axles is 4.2 m, and the center of mass of the car is located 1.8 m behind the front axle and 0.75 m above the road; see Fig. 13N-17. The car weighs 11 kN. Calculate (a) the braking deceleration of the car, (b) the normal force on each wheel, and (c) the braking force on each wheel. (*Hint:* Although the car is not in translational equilibrium, it *is* in rotational equilibrium.)

Fig. 13N-17 Problem N21.

14 Gravitation

The Milky Way galaxy is a disk-shaped collection of dust, planets, and billions of stars, including our Sun and solar system. The force that binds it or any other galaxy together is the same force that holds the Moon in orbit and you on Earth—the gravitational force. That force is also responsible for one of nature's strangest objects, the black hole, a star that has completely collapsed onto itself. The gravitational force near a black hole is so strong that not even light can escape it.

If that is the case, how can a black hole be detected?

The answer is in this chapter.

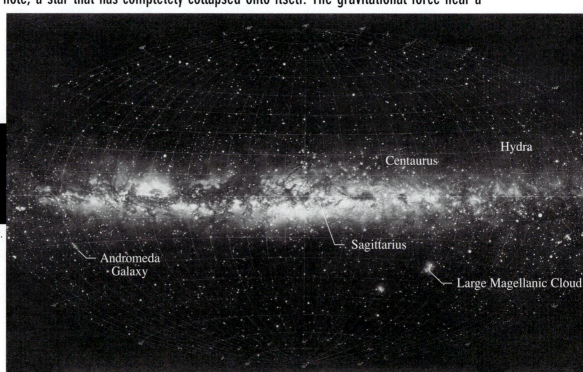

Hydra

Centaurus

Sagittarius

Andromeda
Galaxy

Large Magellanic Cloud

14-1 The World and the Gravitational Force

The drawing that opens this chapter shows our view of the Milky Way galaxy. We are near the edge of the disk of the galaxy, about 26 000 light-years (2.5×10^{20} m) from its center, which in the drawing lies in the star collection known as Sagittarius. Our galaxy is a member of the Local Group of galaxies, which includes the Andromeda galaxy (Fig. 14-1) at a distance of 2.3×10^6 light-years, and several closer dwarf galaxies, such as the Large Magellanic Cloud shown in the opening drawing.

The Local Group is part of the Local Supercluster of galaxies. Measurements taken during and since the 1980s suggest that the Local Supercluster and the supercluster consisting of the clusters Hydra and Centaurus are all moving toward an exceptionally massive region called the Great Attractor. This region appears to be about 300 million light-years away, on the opposite side of the Milky Way from us, past the clusters Hydra and Centaurus.

The force that binds together these progressively larger structures, from star to galaxy to supercluster, and may be drawing them all toward the Great Attractor, is the gravitational force. That force not only holds you on Earth but also reaches out across intergalactic space.

14-2 Newton's Law of Gravitation

Physicists like to study seemingly unrelated phenomena to show that a relationship can be found if they are examined closely enough. This search for unification has been going on for centuries. In 1665, the 23-year-old Isaac Newton made a basic contribution to physics when he showed that the force that holds the Moon in its orbit is the same force that makes an apple fall. We take this so much for granted now that it is not easy for us to comprehend the ancient belief that the motions of earthbound bodies and heavenly bodies were different in kind and were governed by different laws.

Newton concluded that not only does Earth attract an apple and the Moon but every body in the universe attracts every other body; this tendency of bodies to move toward each other is called **gravitation.** Newton's conclusion takes a little getting used to, because the familiar attraction of Earth for earthbound bodies is so great that it overwhelms the attraction that earthbound bodies have for each other. For example, Earth attracts an apple with a force magnitude of about 0.8 N. You also attract a nearby apple (and it attracts you), but the force of attraction has less magnitude than the weight of a speck of dust.

Quantitatively, Newton proposed a *force law* that we call **Newton's law of gravitation:** Every particle attracts any other particle with a **gravitational force** whose magnitude is given by

$$F = G \frac{m_1 m_2}{r^2} \qquad \text{(Newton's law of gravitation).} \qquad (14\text{-}1)$$

Here m_1 and m_2 are the masses of the particles, r is the distance between them, and G is the **gravitational constant,** with a value that is now known to be

$$G = 6.67 \times 10^{-11} \text{ N} \cdot \text{m}^2/\text{kg}^2$$
$$= 6.67 \times 10^{-11} \text{ m}^3/\text{kg} \cdot \text{s}^2. \qquad (14\text{-}2)$$

As Fig. 14-2 shows, a particle m_2 attracts a particle m_1 with a gravitational force $\vec{F}$ that is directed toward particle m_2, and particle m_1 attracts particle m_2 with a gravitational force $-\vec{F}$ that is directed toward m_1. The forces $\vec{F}$ and $-\vec{F}$ form a third-law

Fig. 14-1 The Andromeda galaxy. Located 2.3×10^6 light-years from us, and faintly visible to the naked eye, it is very similar to our home galaxy, the Milky Way.

Fig. 14-2 Two particles, of masses m_1 and m_2 and with separation r, attract each other according to Newton's law of gravitation, Eq. 14-1. The forces of attraction, $\vec{F}$ and $-\vec{F}$, are equal in magnitude and in opposite directions.

force pair; they are opposite in direction but equal in magnitude. They depend on the separation of the two particles, but not on their location: the particles could be in a deep cave or in deep space. Also, forces $\vec{F}$ and $-\vec{F}$ are not altered by the presence of other bodies, even if those bodies lie between the two particles we are considering.

The strength of the gravitational force—that is, how strongly two particles with given masses at a given separation attract each other—depends on the value of the gravitational constant G. If G—by some miracle—were suddenly multiplied by a factor of 10, you would be crushed to the floor by Earth's attraction. If G were divided by this factor, Earth's attraction would be weak enough that you could jump over a building.

Although Newton's law of gravitation applies strictly to particles, we can also apply it to real objects as long as the sizes of the objects are small compared to the distance between them. The Moon and Earth are far enough apart so that, to a good approximation, we can treat them both as particles—but what about an apple and Earth? From the point of view of the apple, the broad and level Earth, stretching out to the horizon beneath the apple, certainly does not look like a particle.

Newton solved the apple–Earth problem by proving an important theorem called the *shell theorem:*

> A uniform spherical shell of matter attracts a particle that is outside the shell as if all the shell's mass were concentrated at its center.

Fig. 14-3 The apple pulls up on Earth just as much as Earth pulls down on the apple.

Earth can be thought of as a nest of such shells, one within another, and each attracting a particle outside Earth's surface as if the mass of that shell were at the center of the shell. Thus, from the apple's point of view, Earth *does* behave like a particle, one that is located at the center of Earth and has a mass equal to that of Earth.

Suppose, as in Fig. 14-3, that Earth pulls down on an apple with a force of magnitude 0.80 N. The apple must then pull up on Earth with a force of magnitude 0.80 N, which we take to act at the center of Earth. Although the forces are matched in magnitude, they produce different accelerations when the apple is released. For the apple, the acceleration is about 9.8 m/s^2, the familiar acceleration of a falling body near Earth's surface. For Earth, the acceleration measured in a reference frame attached to the center of mass of the apple–Earth system is only about 1×10^{-25} m/s^2.

✔**CHECKPOINT 1:**   A particle is to be placed, in turn, outside four objects, each of mass m: (1) a large uniform solid sphere, (2) a large uniform spherical shell, (3) a small uniform solid sphere, and (4) a small uniform shell. In each situation, the distance between the particle and the center of the object is d. Rank the objects according to the magnitude of the gravitational force they exert on the particle, greatest first.

14-3   Gravitation and the Principle of Superposition

Given a group of particles, we find the net (or resultant) gravitational force on any one of them from the others by using the **principle of superposition.** This is a general principle that says a net effect is the sum of the individual effects. Here, the principle means that we first compute the gravitational force that acts on our selected particle due to each of the other particles, in turn. We then find the net force by adding these forces vectorially, as usual.

For n interacting particles, we can write the principle of superposition for gravitational forces as

$$\vec{F}_{1,\text{net}} = \vec{F}_{12} + \vec{F}_{13} + \vec{F}_{14} + \vec{F}_{15} + \cdots + \vec{F}_{1n}. \qquad (14\text{-}3)$$

Here $\vec{F}_{1,\text{net}}$ is the net force on particle 1 and, for example, $\vec{F}_{13}$ is the force on particle 1 from particle 3. We can express this equation more compactly as a vector sum:

$$\vec{F}_{1,\text{net}} = \sum_{i=2}^{n} \vec{F}_{1i}. \qquad (14\text{-}4)$$

What about the gravitational force on a particle from a real extended object? The force is found by dividing the object into parts small enough to treat as particles and then using Eq. 14-4 to find the vector sum of the forces on the particle from all the parts. In the limiting case, we can divide the extended object into differential parts of mass dm, each of which produces only a differential force $d\vec{F}$ on the particle. In this limit, the sum of Eq. 14-4 becomes an integral and we have

$$\vec{F}_1 = \int d\vec{F}, \qquad (14\text{-}5)$$

in which the integral is taken over the entire extended object and we drop the subscript "net." If the object is a uniform sphere or a spherical shell, we can avoid the integration of Eq. 14-5 by assuming that the object's mass is concentrated at the object's center and using Eq. 14-1.

Sample Problem 14-1

Figure 14-4a shows an arrangement of three particles, particle 1 having mass $m_1 = 6.0$ kg and particles 2 and 3 having mass $m_2 = m_3 = 4.0$ kg, and with distance $a = 2.0$ cm. What is the net gravitational force $\vec{F}_1$ that acts on particle 1 due to the other particles?

SOLUTION: One **Key Idea** here is that, because we have particles, the magnitude of the gravitational force on particle 1 due to either of the other particles is given by Eq. 14-1 ($F = Gm_1m_2/r^2$). Thus, the magnitude of the force $\vec{F}_{12}$ on particle 1 from particle 2 is

$$F_{12} = \frac{Gm_1m_2}{a^2}$$

$$= \frac{(6.67 \times 10^{-11}\ \text{m}^3/\text{kg} \cdot \text{s}^2)(6.0\ \text{kg})(4.0\ \text{kg})}{(0.020\ \text{m})^2}$$

$$= 4.00 \times 10^{-6}\ \text{N}.$$

Similarly, the magnitude of force $\vec{F}_{13}$ on particle 1 from particle 3 is

$$F_{13} = \frac{Gm_1m_3}{(2a)^2}$$

$$= \frac{(6.67 \times 10^{-11}\ \text{m}^3/\text{kg} \cdot \text{s}^2)(6.0\ \text{kg})(4.0\ \text{kg})}{(0.040\ \text{m})^2}$$

$$= 1.00 \times 10^{-6}\ \text{N}.$$

To determine the directions of $\vec{F}_{12}$ and $\vec{F}_{13}$, we use this **Key Idea**: Each force on particle 1 is directed toward the particle responsible for that force. Thus, $\vec{F}_{12}$ is directed in the positive direction of y (Fig. 14-4b) and has only the y component F_{12}. Similarly,

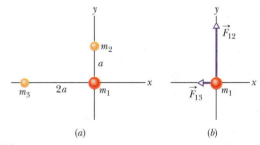

Fig. 14-4 Sample Problem 14-1. (a) An arrangement of three particles. (b) The forces acting on the particle of mass m_1 due to the other particles.

$\vec{F}_{13}$ is directed in the negative direction of x and has only the x component $-F_{13}$.

To find the net force $\vec{F}_{1,\text{net}}$ on particle 1, we first use this very important **Key Idea**: Because the forces are not directed along the same line, we *cannot* simply add or subtract their magnitudes or their components to get their net force. Instead, we must add them as vectors.

We can do so on a vector-capable calculator. However, here we note that $-F_{13}$ and F_{12} are actually the x and y components of $\vec{F}_{1,\text{net}}$. Therefore, we shall follow the guide of Eq. 3-6 to find first the magnitude and then the direction of $\vec{F}_{1,\text{net}}$. The magnitude is

$$F_{1,\text{net}} = \sqrt{(F_{12})^2 + (-F_{13})^2}$$

$$= \sqrt{(4.00 \times 10^{-6}\ \text{N})^2 + (-1.00 \times 10^{-6}\ \text{N})^2}$$

$$= 4.1 \times 10^{-6}\ \text{N}. \qquad \text{(Answer)}$$

Relative to the positive direction of the x axis, Eq. 3-6 gives the direction of $\vec{F}_{1,\text{net}}$ as

$$\theta = \tan^{-1}\frac{F_{12}}{-F_{13}} = \tan^{-1}\frac{4.00 \times 10^{-6}\ \text{N}}{-1.00 \times 10^{-6}\ \text{N}} = -76°.$$

Is this a reasonable direction? No, the direction of $\vec{F}_{1,\text{net}}$ must be between the directions of $\vec{F}_{12}$ and $\vec{F}_{13}$. Recall from Chapter 3 (Tactic 3) that a calculator displays only one of the two possible answers to a $\tan^{-1}$ function. We find the other answer by adding 180°. That gives us

$$-76° + 180° = 104°, \qquad \text{(Answer)}$$

which *is* a reasonable direction for $\vec{F}_{1,\text{net}}$.

CHECKPOINT 2:   The figure shows four arrangements of three particles of equal masses. (a) Rank the arrangements according to the magnitude of the net gravitational force on the particle labeled m, greatest first. (b) In arrangement 2, is the direction of the net force closer to the line of length d or to the line of length D?

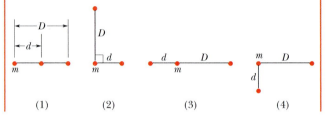

(1)   (2)   (3)   (4)

Sample Problem 14-2

Figure 14-5a shows an arrangement of five particles, with masses $m_1 = 8.0$ kg, $m_2 = m_3 = m_4 = m_5 = 2.0$ kg, and with $a = 2.0$ cm and $\theta = 30°$. What is the net gravitational force $\vec{F}_{1,\text{net}}$ on particle 1 due to the other particles?

SOLUTION: Our **Key Ideas** are the same as in Sample Problem 14-1. However, this problem has a lot of symmetry that can help simplify the solution.

For the magnitudes of the forces on particle 1, first note that particles 2 and 4 have equal masses and equal distances of $r = 2a$ from particle 1. Thus, from Eq. 14-1, we find

$$F_{12} = F_{14} = \frac{Gm_1m_2}{(2a)^2}. \qquad (14\text{-}6)$$

Similarly, since particles 3 and 5 have equal masses and are both a distance $r = a$ from particle 1, we find

$$F_{13} = F_{15} = \frac{Gm_1m_3}{a^2}. \qquad (14\text{-}7)$$

We can now substitute known data into these two equations to evaluate the magnitudes of the forces. Then we can indicate the directions of the forces on the free-body diagram of Fig. 14-5b, and find the net force in either of two basic ways: We could resolve the vectors into x and y components, find the net x component and the net y component, and then vectorially combine those net components. We could also add the vectors directly on a vector-capable calculator.

Instead, however, we shall make further use of the symmetry of the problem. First, we note that $\vec{F}_{12}$ and $\vec{F}_{14}$ are equal in magnitude but opposite in direction; thus, those forces *cancel*. Inspec-

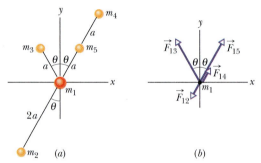

Fig. 14-5 Sample Problem 14-2. (*a*) An arrangement of five particles. (*b*) The forces acting on the particle of mass m_1 due to the other four particles.

tion of Fig. 14-5b and Eq. 14-7 reveals that the x components of $\vec{F}_{13}$ and $\vec{F}_{15}$ also *cancel*, and that their y components are identical in magnitude and both act in the positive direction of y. Thus, $\vec{F}_{1,\text{net}}$ acts in that same direction, and its magnitude is twice the y component of $\vec{F}_{13}$:

$$F_{1,\text{net}} = 2F_{13}\cos\theta = 2\frac{Gm_1m_3}{a^2}\cos\theta$$

$$= 2\frac{(6.67 \times 10^{-11}\ \text{m}^3/\text{kg}\cdot\text{s}^2)(8.0\ \text{kg})(2.0\ \text{kg})}{(0.020\ \text{m})^2}\cos 30°$$

$$= 4.6 \times 10^{-6}\ \text{N}. \qquad \text{(Answer)}$$

Note that the presence of particle 5 along the line between particles 1 and 4 does not alter the gravitational force on particle 1 from particle 4.

CHECKPOINT 3:   In the figure here, what is the direction of the net gravitational force on the particle of mass m_1 due to the other particles, each of mass m, that are arranged symmetrically relative to the y axis?

Tactic 1: *Drawing Gravitational Force Vectors*

When you are given a diagram of particles, such as Fig. 14-4*a*, and asked to find the net gravitational force on one of them, you should usually draw a free-body diagram showing only the particle of concern and the forces on *it alone*, as in Fig. 14-4*b*. If, instead, you choose to superimpose the force vectors on the given diagram, be sure to draw the vectors with either their tails (preferably) or their heads on the particle experiencing those forces. If you draw the vectors elsewhere, you invite confusion—and confusion is guar-

anteed if you draw the vectors on the particles *causing* the forces on the particle of concern.

Tactic 2: *Simplifying a Sum of Forces with Symmetry*

In Sample Problem 14-2 we used the symmetry of the situation: By realizing that particles 2 and 4 are positioned symmetrically about particle 1, and thus that $\vec{F}_{12}$ and $\vec{F}_{14}$ cancel, we avoided calculating either force. By realizing that the *x* components of $\vec{F}_{13}$ and $\vec{F}_{15}$ cancel and that their *y* components are identical and add, we saved even more effort.

14-4 Gravitation Near Earth's Surface

Let us assume that Earth is a uniform sphere of mass M. The magnitude of the gravitational force from Earth on a particle of mass m, located outside Earth a distance r from Earth's center, is then given by Eq. 14-1 as

$$F = G\frac{Mm}{r^2}. \tag{14-8}$$

If the particle is released, it will fall toward the center of Earth, as a result of the gravitational force $\vec{F}$, with an acceleration we shall call the **gravitational acceleration** $\vec{a}_g$. Newton's second law tells us that magnitudes F and a_g are related by

$$F = ma_g. \tag{14-9}$$

Now, substituting F from Eq. 14-8 into Eq. 14-9 and solving for a_g, we find

$$a_g = \frac{GM}{r^2}. \tag{14-10}$$

Fig. 14-6 The density of Earth as a function of distance from the center. The limits of the solid inner core, the largely liquid outer core, and the solid mantle are shown, but the crust of Earth is too thin to show clearly on this plot.

Table 14-1 shows values of a_g computed for various altitudes above Earth's surface.

Since Section 5-6, we have assumed that Earth is an inertial frame by neglecting its actual rotation. This simplification has allowed us to assume that the actual free-fall acceleration g of a particle is the same as the gravitational acceleration (which we now call a_g). Furthermore, we assumed that g has the constant value of 9.8 m/s² over Earth's surface. However, the g we would measure differs from the a_g we would calculate with Eq. 14-10 for three reasons: (1) Earth is not uniform, (2) it is not a perfect sphere, and (3) it rotates. Moreover, because g differs from a_g, the measured weight mg of the particle differs from the magnitude of the gravitational force on the particle as given by Eq. 14-8 for the same three reasons. Let us now examine those reasons.

1. **Earth is not uniform.** The density (mass per unit volume) of Earth varies radially as shown in Fig. 14-6, and the density of the crust (or outer section) of Earth varies from region to region over Earth's surface. Thus, g varies from region to region over the surface.

TABLE 14-1   Variation of a_g with Altitude

Altitude (km)	a_g (m/s²)	Altitude Example
0	9.83	Mean Earth surface
8.8	9.80	Mt. Everest
36.6	9.71	Highest manned balloon
400	8.70	Space shuttle orbit
35 700	0.225	Communications satellite

(a)

(b)

Fig. 14-7 (a) A crate lies on a scale at Earth's equator, as seen along Earth's rotation axis from above the north pole. (b) A free-body diagram for the crate, with a radially outward r axis. The gravitational force on the crate is represented with its equivalent $m\vec{a}_g$. The normal force on the crate from the scale is $\vec{N}$. Because of Earth's rotation, the crate has a centripetal acceleration $\vec{a}$ that is directed toward Earth's center.

2. ***Earth is not a sphere.*** Earth is approximately an ellipsoid, flattened at the poles and bulging at the equator. Its equatorial radius is greater than its polar radius by 21 km. Thus, a point at the poles is closer to the dense core of Earth than is a point on the equator. This is one reason the free-fall acceleration g increases as one proceeds, at sea level, from the equator toward either pole.

3. ***Earth is rotating.*** The rotation axis runs through the north and south poles of Earth. An object located on Earth's surface anywhere except at those poles must rotate in a circle about the rotation axis and thus must have a centripetal acceleration directed toward the center of the circle. This centripetal acceleration requires a centripetal net force that is also directed toward that center.

To see how Earth's rotation causes g to differ from a_g, let us analyze a simple situation in which a crate of mass m is on a scale at the equator. Figure 14-7a shows this situation as viewed from a point in space above the north pole.

Figure 14-7b, a free-body diagram for the crate, shows the two forces on the crate, both acting along a radial axis r that extends from Earth's center. The normal force $\vec{N}$ on the crate from the scale is directed outward, in the positive direction of axis r. The gravitational force, represented with its equivalent $m\vec{a}_g$, is directed inward. Because the crate travels in a circle about the center of Earth as Earth turns, it has a centripetal acceleration $\vec{a}$ directed inward. From Eq. 11-23, we know this acceleration is equal to $\omega^2 R$, where ω is Earth's angular speed and R is the circle's radius (approximately Earth's radius). Thus, we can write Newton's second law for the r axis ($F_{\text{net},r} = ma_r$) as

$$N - ma_g = m(-\omega^2 R). \qquad (14\text{-}11)$$

The magnitude N of the normal force is equal to the weight mg read on the scale. With mg substituted for N, Eq. 14-11 gives us

$$mg = ma_g - m(\omega^2 R), \qquad (14\text{-}12)$$

which says

$$\left(\begin{array}{c}\text{measured}\\\text{weight}\end{array}\right) = \left(\begin{array}{c}\text{magnitude of}\\\text{gravitational force}\end{array}\right) - \left(\begin{array}{c}\text{mass times}\\\text{centripetal acceleration}\end{array}\right).$$

Thus, the measured weight is actually less than the magnitude of the gravitational force on the crate, because of Earth's rotation.

To find a corresponding expression for g and a_g, we cancel m from Eq. 14-12 to write

$$g = a_g - \omega^2 R, \qquad (14\text{-}13)$$

which says

$$\left(\begin{array}{c}\text{free-fall}\\\text{acceleration}\end{array}\right) = \left(\begin{array}{c}\text{gravitational}\\\text{acceleration}\end{array}\right) - \left(\begin{array}{c}\text{centripetal}\\\text{acceleration}\end{array}\right).$$

Thus, the measured free-fall acceleration is actually less than the gravitational acceleration, because of Earth's rotation.

The difference between accelerations g and a_g is equal to $\omega^2 R$ and is greatest on the equator (for one reason, the radius of the circle traveled by the crate is greatest there). To find the difference, we can use Eq. 11-5 ($\omega = \Delta\theta/\Delta t$) and Earth's radius $R = 6.37 \times 10^6$ m. For one rotation of Earth, θ is 2π rad and the time period Δt is about 24 h. Using these values (and converting hours to seconds), we find that g is less than a_g by only about 0.034 m/s² (compared to 9.8 m/s²). Therefore, neglecting the difference in accelerations g and a_g is often justified. Similarly, neglecting the difference between weight and the magnitude of the gravitational acceleration is also often justified.

Sample Problem 14-3

(a) An astronaut whose height h is 1.70 m floats "feet down" in an orbiting space shuttle at a distance $r = 6.77 \times 10^6$ m from the center of Earth. What is the difference between the gravitational acceleration at her feet and that at her head?

SOLUTION: One Key Idea here is that we can approximate Earth as a uniform sphere of mass M_E. Then, from Eq. 14-10, the gravitational acceleration at any distance r from the center of Earth is

$$a_g = \frac{GM_E}{r^2}. \qquad (14\text{-}14)$$

We might simply apply Eq. 14-14 twice, first with, say, $r = 6.77 \times 10^6$ m for the feet and then with $r = 6.77 \times 10^6$ m + 1.70 m for the head. However, a calculator may give us the same value for a_g twice, and thus a difference of zero, because h is so small compared to r. A second Key Idea helps here: Because we have a differential change dr in r between the astronaut's feet and head, let us differentiate Eq. 14-14 with respect to r. That gives us

$$da_g = -2\frac{GM_E}{r^3}\, dr, \qquad (14\text{-}15)$$

where da_g is the differential change in the gravitational acceleration due to the differential change dr in r. For the astronaut, $dr = h$ and $r = 6.77 \times 10^6$ m. Substituting data into Eq. 14-15, we find

$$da_g = -2\frac{(6.67 \times 10^{-11}\ \text{m}^3/\text{kg}\cdot\text{s}^2)(5.98 \times 10^{24}\ \text{kg})}{(6.77 \times 10^6\ \text{m})^3}(1.70\ \text{m})$$

$$= -4.37 \times 10^{-6}\ \text{m/s}^2. \qquad \text{(Answer)}$$

This result means that the gravitational acceleration of the astronaut's feet toward Earth is slightly greater than the gravitational acceleration of her head toward Earth. This difference in acceleration tends to stretch her body, but the difference is so small that the stretching is unnoticeable.

(b) If the astronaut is now "feet down" at the same orbital radius r of 6.77×10^6 m about a black hole of mass $M_h = 1.99 \times 10^{31}$ kg (which is 10 times our Sun's mass), what is now the difference between the gravitational acceleration at her feet and that at her head? The black hole has a surface (called its *event horizon*) of radius $R_h = 2.95 \times 10^4$ m. Nothing, not even light, can escape from that surface or anywhere inside it. Note that the astronaut is (wisely) well outside the surface (at $r = 229R_h$).

SOLUTION: The Key Idea here is that we again have a differential change dr in r between the astronaut's feet and head, so we can again use Eq. 14-15. However, now we substitute $M_h = 1.99 \times 10^{31}$ kg for M_E. We find

$$da_g = -2\frac{(6.67 \times 10^{-11}\ \text{m}^3/\text{kg}\cdot\text{s}^2)(1.99 \times 10^{31}\ \text{kg})}{(6.77 \times 10^6\ \text{m})^3}(1.70\ \text{m})$$

$$= -14.5\ \text{m/s}^2. \qquad \text{(Answer)}$$

This means that the gravitational acceleration of the astronaut's feet toward the black hole is noticeably larger than that of her head. The resulting tendency to stretch her body would be bearable but quite painful. If she drifted closer to the black hole, the stretching tendency would increase drastically.

14-5 Gravitation Inside Earth

Newton's shell theorem can also be applied to a situation in which a particle is located *inside* a uniform shell, to show the following:

> A uniform shell of matter exerts no *net* gravitational force on a particle located inside it.

Caution: This statement does *not* mean that the gravitational forces on the particle from the various elements of the shell magically disappear. Rather, it means that the *sum* of the force vectors on the particle from all the elements is zero.

If the density of Earth were uniform, the gravitational force acting on a particle would be a maximum at Earth's surface and would decrease as the particle moved outward. If the particle were to move inward, perhaps down a deep mine shaft, the gravitational force would change for two reasons. (1) It would tend to increase because the particle would be moving closer to the center of Earth. (2) It would tend to decrease because the thickening shell of material lying outside the particle's radial position would not exert any net force on the particle.

For a uniform Earth, the second influence would prevail and the force on the particle would steadily decrease to zero as the particle approached the center of Earth. However, for the real (nonuniform) Earth, the force on the particle actually increases as the particle begins to descend. The force reaches a maximum at a certain depth; only then does it begin to decrease as the particle descends farther.

Sample Problem 14-4

In *Pole to Pole,* an early science fiction story by George Griffith, three explorers attempt to travel by capsule through a naturally formed (and, of course, fictional) tunnel directly from the south pole to the north pole (Fig. 14-8). According to the story, as the capsule approaches Earth's center, the gravitational force on the explorers becomes alarmingly large and then, exactly at the center, it suddenly but only momentarily disappears. Then the capsule travels through the second half of the tunnel, to the north pole.

Check Griffith's description by finding the gravitational force on the capsule of mass m when it reaches a distance r from Earth's center. Assume that Earth is a sphere of uniform density ρ (mass per unit volume).

SOLUTION: Newton's shell theorem gives us three **Key Ideas** here:

1. When the capsule is at a radius r from Earth's center, the portion of Earth that lies outside a sphere of radius r does *not* produce a net gravitational force on the capsule.

2. The portion that lies inside that sphere *does* produce a net gravitational force on the capsule.

Fig. 14-8 Sample Problem 14-4. A capsule of mass m falls from rest through a tunnel that connects Earth's south and north poles. When the capsule is at distance r from Earth's center, the portion of Earth's mass that is contained in a sphere of that radius is M_{ins}.

3. We can treat the mass M_{ins} of that inside portion of Earth as being the mass of a particle located at Earth's center.

All three ideas tell us that we can write Eq. 14-1, for the magnitude of the gravitational force on the capsule, as

$$F = \frac{GmM_{ins}}{r^2}. \tag{14-16}$$

To write the mass M_{ins} in terms of the radius r, we note that the volume V_{ins} containing this mass is $\frac{4}{3}\pi r^3$. Also, its density is Earth's density ρ. Thus, we have

$$M_{ins} = \rho V_{ins} = \rho \frac{4\pi r^3}{3}. \tag{14-17}$$

Then, after substituting this expression into Eq. 14-16 and canceling, we have

$$F = \frac{4\pi Gm\rho}{3} r. \qquad \text{(Answer)} \quad (14\text{-}18)$$

This equation tells us that the force magnitude F depends linearly on the capsule's distance r from Earth's center. Thus, as r decreases, F also decreases (opposite of Griffith's description), until it is zero at Earth's center. At least Griffith got zero-at-the-center correct.

Equation 14-18 can also be written in terms of the force vector $\vec{F}$ and the capsule's position vector $\vec{r}$ along a radial axis extending from Earth's center. Let K represent the collection of constants $4\pi Gm\rho/3$. Then, Eq. 14-18 becomes

$$\vec{F} = -K\vec{r}, \tag{14-19}$$

in which we have inserted a minus sign to indicate that $\vec{F}$ and $\vec{r}$ have opposite directions. Equation 14-19 has the form of Hooke's law (Eq. 7-20). Thus, under the idealized conditions of the story, the capsule would oscillate like a block on a spring, with the center of the oscillation at Earth's center. After the capsule had fallen from the south pole to Earth's center, it would travel from the center to the north pole (as Griffith said) and then back again.

14-6 Gravitational Potential Energy

In Section 8-3, we discussed the gravitational potential energy of a particle–Earth system. We were careful to keep the particle near Earth's surface, so that we could regard the gravitational force as constant. We then chose some reference configuration of the system as having a gravitational potential energy of zero. Often, in this configuration the particle was on Earth's surface. For particles not on Earth's surface, the gravitational potential energy decreased when the separation between the particle and Earth decreased.

Here, we broaden our view and consider the gravitational potential energy U of two particles, of masses m and M, separated by a distance r. We again choose a reference configuration with U equal to zero. However, to simplify the equations, the separation distance r in the reference configuration is now large enough to be approximated as *infinite*. As before, the gravitational potential energy decreases when

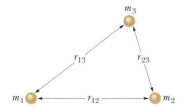

Fig. 14-9 Three particles form a system. (The separation for each pair of particles is labeled with a double subscript to indicate the particles.) The gravitational potential energy *of the system* is the sum of the gravitational potential energies of all three pairs of particles.

the separation decreases. Since $U = 0$ for $r = \infty$, the potential energy is negative for any finite separation and becomes progressively more negative as the particles move closer together.

With these facts in mind and as we shall justify next, we take the gravitational potential energy of the two-particle system to be

$$U = -\frac{GMm}{r} \qquad \text{(gravitational potential energy).}$$ (14-20)

Note that $U(r)$ approaches zero as r approaches infinity and that for any finite value of r, the value of $U(r)$ is negative.

The potential energy given by Eq. 14-20 is a property of the system of two particles rather than of either particle alone. There is no way to divide this energy and say that so much belongs to one particle and so much to the other. However, if $M \gg m$, as is true for Earth (mass M) and a baseball (mass m), we often speak of "the potential energy of the baseball." We can get away with this because, when a baseball moves in the vicinity of Earth, changes in the potential energy of the baseball–Earth system appear almost entirely as changes in the kinetic energy of the baseball, since changes in the kinetic energy of Earth are too small to be measured. Similarly, in Section 14-8 we shall speak of "the potential energy of an artificial satellite" orbiting Earth, because the satellite's mass is so much smaller than Earth's mass. When we speak of the potential energy of bodies of comparable mass, however, we have to be careful to treat them as a system.

If our system contains more than two particles, we consider each pair of particles in turn, calculate the gravitational potential energy of that pair with Eq. 14-20 as if the other particles were not there, and then algebraically sum the results. Applying Eq. 14-20 to each of the three pairs of Fig. 14-9, for example, gives the potential energy of the system as

$$U = -\left(\frac{Gm_1m_2}{r_{12}} + \frac{Gm_1m_3}{r_{13}} + \frac{Gm_2m_3}{r_{23}}\right).$$ (14-21)

Proof of Equation 14-20

Let us shoot a baseball directly away from Earth along the path in Fig. 14-10. We want to find an expression for the gravitational potential energy U of the ball at point P along its path, at radial distance R from Earth's center. To do so, we first find the work W done on the ball by the gravitational force as the ball travels from point P to a great (infinite) distance from Earth. Because the gravitational force $\vec{F}(r)$ is a variable force (its magnitude depends on r), we must use the techniques of Section 7-6 to find the work. In vector notation, we can write

$$W = \int_R^\infty \vec{F}(r) \cdot d\vec{r}.$$ (14-22)

The integral contains the scalar (or dot) product of the force $\vec{F}(r)$ and the differential displacement vector $d\vec{r}$ along the ball's path. We can expand that product as

$$\vec{F}(r) \cdot d\vec{r} = F(r) \, dr \cos \phi,$$ (14-23)

where ϕ is the angle between the directions of $\vec{F}(r)$ and $d\vec{r}$. When we substitute $180°$ for ϕ and Eq. 14-1 for $F(r)$, Eq. 14-23 becomes

$$\vec{F}(r) \cdot d\vec{r} = -\frac{GMm}{r^2} \, dr,$$

where M is Earth's mass and m is the mass of the ball.

Fig. 14-10 A baseball is shot directly away from Earth, through point P at radial distance R from Earth's center. The gravitational force $\vec{F}$ on the ball and a differential displacement vector $d\vec{r}$ are shown, both directed along a radial r axis.

Substituting this into Eq. 14-22 and integrating gives us

$$W = -GMm \int_R^\infty \frac{1}{r^2}\, dr = \left[\frac{GMm}{r} \right]_R^\infty$$

$$= 0 - \frac{GMm}{R} = -\frac{GMm}{R}. \qquad (14\text{-}24)$$

W in Eq. 14-24 is the work required to move the ball from point P (at distance R) to infinity. Equation 8-1 ($\Delta U = -W$) tells us that we can also write that work in terms of potential energies as

$$U_\infty - U = -W.$$

The potential energy U_∞ at infinity is zero, and U is the potential energy at P. Thus, with Eq. 14-24 substituted for W, the previous equation becomes

$$U = W = -\frac{GMm}{R}.$$

Switching R to r gives us Eq. 14-20, which we set out to prove.

Path Independence

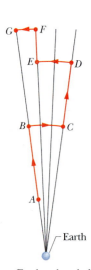

Fig. 14-11 Near Earth, a baseball is moved from point A to point G along a path consisting of radial lengths and circular arcs.

In Fig. 14-11, we move a baseball from point A to point G along a path consisting of three radial lengths and three circular arcs (centered on Earth). We are interested in the total work W done by Earth's gravitational force $\vec{F}$ on the ball as it moves from A to G. The work done along each circular arc is zero, because the direction of $\vec{F}$ is perpendicular to the arc at every point. Thus, the only works done by $\vec{F}$ are along the three radial lengths, and the total work W is the sum of those works.

Now, suppose we mentally shrink the arcs to zero. We would then be moving the ball directly from A to G along a single radial length. Does that change W? No. Because no work was done along the arcs, eliminating them does not change the work. The path taken from A to G now is clearly different, but the work done by $\vec{F}$ is the same.

We discussed such a result in a general way in Section 8-2. Here is the point: The gravitational force is a conservative force. Thus, the work done by the gravitational force on a particle moving from an initial point i to a final point f is independent of the actual path taken between the points. From Eq. 8-1, the change ΔU in the gravitational potential energy from point i to point f is given by

$$\Delta U = U_f - U_i = -W. \qquad (14\text{-}25)$$

Since the work W done by a conservative force is independent of the actual path taken, the change ΔU in gravitational potential energy is *also independent* of the actual path taken.

Potential Energy and Force

In the proof of Eq. 14-20, we derived the potential energy function $U(r)$ from the force function $\vec{F}(r)$. We should be able to go the other way—that is, to start from the potential energy function and derive the force function. Guided by Eq. 8-20, we can write

$$F = -\frac{dU}{dr} = -\frac{d}{dr}\left(-\frac{GMm}{r} \right)$$

$$= -\frac{GMm}{r^2}. \qquad (14\text{-}26)$$

This is Newton's law of gravitation (Eq. 14-1). The minus sign indicates that the force on mass m points radially inward, toward mass M.

Escape Speed

If you fire a projectile upward, usually it will slow, stop momentarily, and return to Earth. There is, however, a certain minimum initial speed that will cause it to move upward forever, theoretically coming to rest only at infinity. This initial speed is called the (Earth) **escape speed.**

Consider a projectile of mass m, leaving the surface of a planet (or some other astronomical body or system) with escape speed v. It has a kinetic energy K given by $\frac{1}{2}mv^2$ and a potential energy U given by Eq. 14-20:

$$U = -\frac{GMm}{R},$$

in which M is the mass of the planet, and R is its radius.

When the projectile reaches infinity, it stops and thus has no kinetic energy. It also has no potential energy because this is our zero-potential-energy configuration. Its total energy at infinity is therefore zero. From the principle of conservation of energy, its total energy at the planet's surface must also have been zero, so

$$K + U = \frac{1}{2}mv^2 + \left(-\frac{GMm}{R}\right) = 0.$$

This yields

$$v = \sqrt{\frac{2GM}{R}}. \tag{14-27}$$

The escape speed v does not depend on the direction in which a projectile is fired from a planet. However, attaining that speed is easier if the projectile is fired in the direction the launch site is moving as the planet rotates about its axis. For example, rockets are launched eastward at Cape Canaveral to take advantage of the Cape's eastward speed of 1500 km/h due to Earth's rotation.

Equation 14-27 can be applied to find the escape speed of a projectile from any astronomical body, provided we substitute the mass of the body for M and the radius of the body for R. Table 14-2 shows escape speeds from some astronomical bodies.

TABLE 14-2 Some Escape Speeds

Body	Mass (kg)	Radius (m)	Escape Speed (km/s)
Ceres[a]	1.17×10^{21}	3.8×10^5	0.64
Earth's moon[a]	7.36×10^{22}	1.74×10^6	2.38
Earth	5.98×10^{24}	6.37×10^6	11.2
Jupiter	1.90×10^{27}	7.15×10^7	59.5
Sun	1.99×10^{30}	6.96×10^8	618
Sirius B[b]	2×10^{30}	1×10^7	5200
Neutron star[c]	2×10^{30}	1×10^4	2×10^5

[a] The most massive of the asteroids.

[b] A *white dwarf* (a star in a final stage of evolution) that is a companion of the bright star Sirius.

[c] The collapsed core of a star that remains after that star has exploded in a *supernova* event.

Sample Problem 14-5

An asteroid, headed directly toward Earth, has a speed of 12 km/s relative to the planet when it is at a distance of 10 Earth radii from Earth's center. Neglecting the effects of Earth's atmosphere on the asteroid, find the asteroid's speed v_f when it reaches Earth's surface.

SOLUTION: One Key Idea is that, because we are to neglect the effects of the atmosphere on the asteroid, the mechanical energy of the asteroid–Earth system is conserved during the fall. Thus, the final mechanical energy (when the asteroid reaches Earth's surface) is equal to the initial mechanical energy. We can write this as

$$K_f + U_f = K_i + U_i, \qquad (14\text{-}28)$$

where K is kinetic energy and U is gravitational potential energy.

A second Key Idea is that, if we assume the system is isolated, the system's linear momentum must be conserved during the fall. Therefore, the momentum change of the asteroid and that of Earth must be equal in magnitude and opposite in sign. However, because Earth's mass is so great relative to the asteroid's mass, the change in Earth's speed is negligible relative to the change in the asteroid's speed. So, the change in Earth's kinetic energy is also negligible. Thus, we can assume that the kinetic energies in Eq. 14-28 are those of the asteroid alone.

Let m represent the asteroid's mass and M represent Earth's mass (5.98×10^{24} kg). The asteroid is initially at the distance $10R_E$ and finally at the distance R_E, where R_E is Earth's radius (6.37×10^6 m). Substituting Eq. 14-20 for U and $\frac{1}{2}mv^2$ for K, we

rewrite Eq. 14-28 as

$$\tfrac{1}{2}mv_f^2 - \frac{GMm}{R_E} = \tfrac{1}{2}mv_i^2 - \frac{GMm}{10R_E}.$$

Rearranging and substituting known values, we find

$$v_f^2 = v_i^2 + \frac{2GM}{R_E}\left(1 - \frac{1}{10}\right)$$

$$= (12 \times 10^3 \text{ m/s})^2$$

$$+ \frac{2(6.67 \times 10^{-11} \text{ m}^3/\text{kg} \cdot \text{s}^2)(5.98 \times 10^{24} \text{ kg})}{6.37 \times 10^6 \text{ m}} 0.9$$

$$= 2.567 \times 10^8 \text{ m}^2/\text{s}^2,$$

and thus

$$v_f = 1.60 \times 10^4 \text{ m/s} = 16 \text{ km/s}. \qquad \text{(Answer)}$$

At this speed, the asteroid would not have to be particularly large to do considerable damage at impact. As an example, if it were only 5 m across, the impact could release about as much energy as the nuclear explosion at Hiroshima. Alarmingly, about 500 million asteroids of this size are near Earth's orbit, and in 1994 one of them apparently penetrated Earth's atmosphere and exploded at an altitude of 20 km near a remote South Pacific island (setting off nuclear-explosion warnings on six military satellites). The impact of an asteroid 500 m across (there may be a million of them near Earth's orbit) could end modern civilization and almost eliminate humans worldwide.

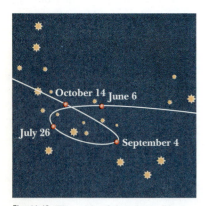

Fig. 14-12 The path of the planet Mars as it moved against a background of the constellation Capricorn during 1971. Its position on four selected days is marked. Both Mars and Earth are moving in orbits around the Sun so that we see the position of Mars relative to us; this sometimes results in an apparent loop in the path of Mars.

14-7   Planets and Satellites: Kepler's Laws

The motions of the planets, as they seemingly wander against the background of the stars, have been a puzzle since the dawn of history. The "loop-the-loop" motion of Mars, shown in Fig. 14-12, was particularly baffling. Johannes Kepler (1571–1630), after a lifetime of study, worked out the empirical laws that govern these motions. Tycho Brahe (1546–1601), the last of the great astronomers to make observations without the help of a telescope, compiled the extensive data from which Kepler was able to derive the three laws of planetary motion that now bear his name. Later, Newton (1642–1727) showed that his law of gravitation leads to Kepler's laws.

In this section we discuss each of Kepler's laws in turn. Although here we apply the laws to planets orbiting the Sun, they hold equally well for satellites, either natural or artificial, orbiting Earth or any other massive central body.

▶ **1. THE LAW OF ORBITS:** All planets move in elliptical orbits, with the Sun at one focus.

Figure 14-13 shows a planet of mass m moving in such an orbit around the Sun, whose mass is M. We assume that $M \gg m$, so that the center of mass of the planet–Sun system is approximately at the center of the Sun.

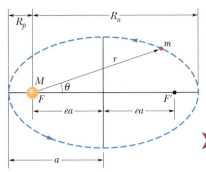

Fig. 14-13 A planet of mass m moving in an elliptical orbit around the Sun. The Sun, of mass M, is at one focus F of the ellipse. The other focus is F', which is located in empty space. Each focus is a distance ea from the ellipse's center, with e being the eccentricity of the ellipse. The semimajor axis a of the ellipse, the perihelion (nearest the Sun) distance R_p, and the aphelion (farthest from the Sun) distance R_a are also shown.

The orbit in Fig. 14-13 is described by giving its **semimajor axis** a and its **eccentricity** e, the latter defined so that ea is the distance from the center of the ellipse to either focus F or F'. *An eccentricity of zero corresponds to a circle,* in which the two foci merge to a single central point. The eccentricities of the planetary orbits are not large, so—sketched on paper—the orbits look circular. The eccentricity of the ellipse of Fig. 14-13, which has been exaggerated for clarity, is 0.74. The eccentricity of Earth's orbit is only 0.0167.

> **2. THE LAW OF AREAS:** A line that connects a planet to the Sun sweeps out equal areas in the plane of the planet's orbit in equal times; that is, the rate dA/dt at which it sweeps out area A is constant.

Qualitatively, this second law tells us that the planet will move most slowly when it is farthest from the Sun and most rapidly when it is nearest to the Sun. As it turns out, Kepler's second law is totally equivalent to the law of conservation of angular momentum. Let us prove it.

The area of the shaded wedge in Fig. 14-14a closely approximates the area swept out in time Δt by a line connecting the Sun and the planet, which are separated by a distance r. The area ΔA of the wedge is approximately the area of a triangle with base $r\,\Delta\theta$ and height r. Since the area of a triangle is one-half of the base times the height, $\Delta A \approx \frac{1}{2}r^2\,\Delta\theta$. This expression for ΔA becomes more exact as Δt (hence $\Delta\theta$) approaches zero. The instantaneous rate at which area is being swept out is then

$$\frac{dA}{dt} = \frac{1}{2}r^2\frac{d\theta}{dt} = \frac{1}{2}r^2\omega, \tag{14-29}$$

in which ω is the angular speed of the rotating line connecting Sun and planet.

Figure 14-14b shows the linear momentum $\vec{p}$ of the planet, along with its radial and perpendicular components. From Eq. 12-20 ($L = rp_\perp$), the magnitude of the angular momentum $\vec{L}$ of the planet about the Sun is given by the product of r and $p_\perp$, the component of $\vec{p}$ perpendicular to r. Here, for a planet of mass m,

$$L = rp_\perp = (r)(mv_\perp) = (r)(m\omega r)$$
$$= mr^2\omega, \tag{14-30}$$

where we have replaced $v_\perp$ with its equivalent ωr (Eq. 11-18). Eliminating $r^2\omega$ between Eqs. 14-29 and 14-30 leads to

$$\frac{dA}{dt} = \frac{L}{2m}. \tag{14-31}$$

If dA/dt is constant, as Kepler said it is, then Eq. 14-31 means that L must also be constant—angular momentum is conserved. Kepler's second law is indeed equivalent to the law of conservation of angular momentum.

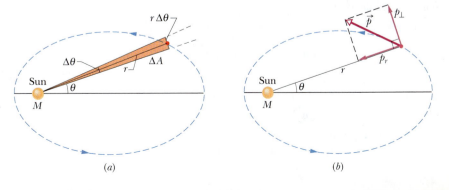

Fig. 14-14 (*a*) In time Δt, the line r connecting the planet to the Sun (of mass M) sweeps through an angle $\Delta\theta$, sweeping out an area ΔA (shaded). (*b*) The linear momentum $\vec{p}$ of the planet and its components.

Fig. 14-15 A planet of mass m moving around the Sun in a circular orbit of radius r.

> **3. THE LAW OF PERIODS:** The square of the period of any planet is proportional to the cube of the semimajor axis of its orbit.

To see this, consider the circular orbit of Fig. 14-15, with radius r (the radius of a circle is equivalent to the semimajor axis of an ellipse). Applying Newton's second law ($F = ma$) to the orbiting planet in Fig. 14-15 yields

$$\frac{GMm}{r^2} = (m)(\omega^2 r). \qquad (14\text{-}32)$$

Here we have substituted from Eq. 14-1 for the force magnitude F and used Eq. 11-23 to substitute $\omega^2 r$ for the centripetal acceleration. If we use Eq. 11-20 to replace ω with $2\pi/T$, where T is the period of the motion, we obtain Kepler's third law:

$$T^2 = \left(\frac{4\pi^2}{GM}\right)r^3 \qquad \text{(law of periods)}. \qquad (14\text{-}33)$$

The quantity in parentheses is a constant that depends only on the mass M of the central body about which the planet orbits.

Equation 14-33 holds also for elliptical orbits, provided we replace r with a, the semimajor axis of the ellipse. This law predicts that the ratio T^2/a^3 has essentially the same value for every planetary orbit around a given massive body. Table 14-3 shows how well it holds for the orbits of the planets of the solar system.

TABLE 14-3   Kepler's Law of Periods for the Solar System

Planet	Semimajor Axis a (10^{10} m)	Period T (y)	T^2/a^3 (10^{-34} y^2/m^3)
Mercury	5.79	0.241	2.99
Venus	10.8	0.615	3.00
Earth	15.0	1.00	2.96
Mars	22.8	1.88	2.98
Jupiter	77.8	11.9	3.01
Saturn	143	29.5	2.98
Uranus	287	84.0	2.98
Neptune	450	165	2.99
Pluto	590	248	2.99

✓**CHECKPOINT 5:** Satellite 1 is in a certain circular orbit about a planet, while satellite 2 is in a larger circular orbit. Which satellite has (a) the longer period and (b) the greater speed?

Sample Problem 14-6

Comet Halley orbits about the Sun with a period of 76 years and, in 1986, had a distance of closest approach to the Sun, its *perihelion distance* R_p, of 8.9×10^{10} m. Table 14-3 shows that this is between the orbits of Mercury and Venus.

(a) What is the comet's farthest distance from the Sun, its *aphelion distance* R_a?

SOLUTION: One Key Idea comes from Fig. 14-13, in which we see that $R_a + R_p = 2a$, where a is the semimajor axis of the orbit of comet

Halley. Thus, we can find R_a if we first find a. A second Key Idea is that we can relate a to the given period via the law of periods (Eq. 14-33) if we simply substitute the semimajor axis a for r. Doing so and then solving for a, we have

$$a = \left(\frac{GMT^2}{4\pi^2}\right)^{1/3}. \qquad (14\text{-}34)$$

If we substitute the mass M of the Sun, 1.99×10^{30} kg, and the period T of the comet, 76 years or 2.4×10^9 s, into Eq. 14-34, we find that $a = 2.7 \times 10^{12}$ m. Now we have

$$
\begin{aligned}
R_a &= 2a - R_p \\
&= (2)(2.7 \times 10^{12} \text{ m}) - 8.9 \times 10^{10} \text{ m} \\
&= 5.3 \times 10^{12} \text{ m}. \quad \text{(Answer)}
\end{aligned}
$$

Table 14-3 shows that this is a little less than the semimajor axis of the orbit of Pluto. Thus, the comet does not get farther from the Sun than Pluto.

(b) What is the eccentricity e of the orbit of comet Halley?

SOLUTION: The **Key Idea** here is that we can relate e, a, and R_p via Fig. 14-13. We see there that $ea = a - R_p$, or

$$
e = \frac{a - R_p}{a} = 1 - \frac{R_p}{a}
$$

$$
= 1 - \frac{8.9 \times 10^{10} \text{ m}}{2.7 \times 10^{12} \text{ m}} = 0.97. \quad \text{(Answer)}
$$

This cometary orbit, with an eccentricity approaching unity, is a long thin ellipse.

Sample Problem 14-7

Hunting a black hole. Observations of the light from a certain star indicate that it is part of a binary (two-star) system. This visible star has orbital speed $v = 270$ km/s, orbital period $T = 1.70$ days, and approximate mass $m_1 = 6M_s$, where M_s is the Sun's mass, 1.99×10^{30} kg. Assuming that the visible star and its companion star, which is dark and unseen, are both in circular orbits (see Fig. 14-16), determine the approximate mass m_2 of the dark star.

SOLUTION: Some of the **Key Ideas** in this challenging problem are as follows:

1. The two stars are in circular orbits, not about each other, but about the center of mass of this two-star system.

2. As with the two-particle systems of Section 9-2, the center of mass of the two-star system must lie along a line connecting the centers of the stars—that is, at point O in Fig. 14-16. The visible star orbits at radius r_1, the dark star at radius r_2.

3. The center of mass of the system is not even approximately at the center of a central, massive object (like the Sun). Therefore, Kepler's law of periods, Eq. 14-33, does *not* apply here and we cannot easily find mass m_2 with it.

4. The centripetal force causing each star to move in a circle is the gravitational force due to the other star. The magnitude of the force is Gm_1m_2/r^2, where r is the distance between the centers of the stars.

5. From Eq. 4-32, the centripetal acceleration a of the visible star is v^2/r_1.

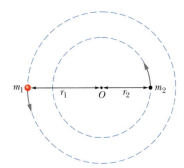

Fig. 14-16 Sample Problem 14-7. A visible star with mass m_1 and a dark, unseen star with mass m_2 orbit around the center of mass of the two-star system at O.

These ideas lead us to write Newton's second law ($F = ma$) for the visible star as

$$
\frac{Gm_1m_2}{r^2} = m_1 \frac{v^2}{r_1}. \quad (14\text{-}35)
$$

This equation contains the required mass m_2, but to find it we need expressions for r and r_1. (Note that m_1 cancels out.)

We start by locating the center of mass relative to the visible star, using Eq. 9-1. That star is at distance zero relative to itself, the center of mass is at distance r_1, and the dark star is at distance r. Equation 9-2 then becomes

$$
r_1 = \frac{m_1(0) + m_2 r}{m_1 + m_2}, \quad (14\text{-}36)
$$

which yields

$$
r = r_1 \frac{m_1 + m_2}{m_2}. \quad (14\text{-}37)
$$

To find an expression for r_1, we note that the visible star is moving in a circle of radius r_1, at speed v, and with period T. Thus, from Eq. 4-33, we have $v = 2\pi r_1/T$ or

$$
r_1 = \frac{vT}{2\pi}. \quad (14\text{-}38)
$$

Substituting this for r_1 in Eq. 14-37 results in

$$
r = \frac{vT}{2\pi} \frac{m_1 + m_2}{m_2}. \quad (14\text{-}39)
$$

Now we return to Eq. 14-35 and substitute for r with Eq. 14-39, for r_1 with Eq. 14-38, and for m_1 with the given $6M_s$. Rearranging the result and substituting known data then give us

$$
\frac{m_2^3}{(6M_s + m_2)^2} = \frac{v^3 T}{2\pi G}
$$

$$
= \frac{(2.7 \times 10^5 \text{ m/s})^3 (1.70 \text{ days})(86\,400 \text{ s/day})}{(2\pi)(6.67 \times 10^{-11} \text{ N} \cdot \text{m}^2/\text{kg}^2)}
$$

$$
= 6.90 \times 10^{30} \text{ kg},
$$

or

$$
\frac{m_2^3}{(6M_s + m_2)^2} = 3.47 M_s. \quad (14\text{-}40)
$$

We can solve this cubic equation for m_2 with a polynomial solver on a calculator. Instead, since we are working with approximate masses anyway, we can substitute integer multiples of M_s for m_2 until we find one that makes Eq. 14-40 nearly true. This occurs for

$$m_2 \approx 9M_s. \qquad \text{(Answer)}$$

The data here approximate those for the binary system LMC X-3 in the Large Magellanic Cloud (shown in the figure that begins this chapter). From other data, the dark object is known to be especially compact: It may be a star that collapsed under its own gravitational pull to become a neutron star or a black hole. Since a neutron star cannot have a mass larger than about $2M_s$, the result $m_2 \approx 9M_s$ strongly suggests that the dark object is a black hole.

Thus, we can detect the presence of a black hole provided it is part of a binary system with a visible star whose mass, orbital speed, and orbital period can be measured.

14-8   Satellites: Orbits and Energy

As a satellite orbits Earth on its elliptical path, both its speed, which fixes its kinetic energy K, and its distance from the center of Earth, which fixes its gravitational potential energy U, fluctuate with fixed periods. However, the mechanical energy E of the satellite remains constant. (Since the satellite's mass is so much smaller than Earth's mass, we assign U and E for the Earth–satellite system to the satellite alone.)

The potential energy of the system is given by Eq. 14-20 and is

$$U = -\frac{GMm}{r}$$

On February 7, 1984, at a height of 102 km above Hawaii and with a speed of about 29 000 km/h, Bruce McCandless stepped (untethered) into space from a space shuttle and became the first human satellite.

(with $U = 0$ for infinite separation). Here r is the radius of the orbit, assumed for the time being to be circular, and M and m are the masses of Earth and the satellite, respectively.

To find the kinetic energy of a satellite in a circular orbit, we write Newton's second law ($F = ma$) as

$$\frac{GMm}{r^2} = m\frac{v^2}{r}, \qquad (14\text{-}41)$$

where v^2/r is the centripetal acceleration of the satellite. Then, from Eq. 14-41, the kinetic energy is

$$K = \tfrac{1}{2}mv^2 = \frac{GMm}{2r}, \qquad (14\text{-}42)$$

which shows us that for a satellite in a circular orbit,

$$K = -\frac{U}{2} \qquad \text{(circular orbit).} \qquad (14\text{-}43)$$

The total mechanical energy of the orbiting satellite is

$$E = K + U = \frac{GMm}{2r} - \frac{GMm}{r}$$

or

$$E = -\frac{GMm}{2r} \qquad \text{(circular orbit).} \qquad (14\text{-}44)$$

This tells us that for a satellite in a circular orbit, the total energy E is the negative of the kinetic energy K:

$$E = -K \qquad \text{(circular orbit).} \qquad (14\text{-}45)$$

For a satellite in an elliptical orbit of semimajor axis a, we can substitute a for r in Eq. 14-44 to find the mechanical energy as

$$E = -\frac{GMm}{2a} \qquad \text{(elliptical orbit).} \qquad (14\text{-}46)$$

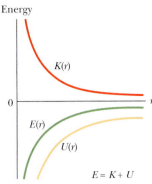

Fig. 14-17 Four orbits about an object of mass M. All four orbits have the same semimajor axis a and thus correspond to the same total mechanical energy E. Their eccentricities e are marked.

Fig. 14-18 The variation of kinetic energy K, potential energy U, and total energy E with radius r for a satellite in a circular orbit. For any value of r, the values of U and E are negative, the value of K is positive, and $E = -K$. As $r \to \infty$, all three energy curves approach a value of zero.

Equation 14-46 tells us that the total energy of an orbiting satellite depends only on the semimajor axis of its orbit and not on its eccentricity e. For example, four orbits with the same semimajor axis are shown in Fig. 14-17; the same satellite would have the same total mechanical energy E in all four orbits. Figure 14-18 shows the variation of K, U, and E with r for a satellite moving in a circular orbit about a massive central body.

✔**CHECKPOINT 6:** In the figure here, a space shuttle is initially in a circular orbit of radius r about Earth. At point P, the pilot briefly fires a forward-pointing thruster to decrease the shuttle's kinetic energy K and mechanical energy E. (a) Which of the dashed elliptical orbits shown in the figure will the shuttle then take? (b) Is the orbital period T of the shuttle (the time to return to P) then greater than, less than, or the same as in the circular orbit?

Sample Problem 14-8

A playful astronaut releases a bowling ball, of mass $m = 7.20$ kg, into circular orbit about Earth at an altitude h of 350 km.

(a) What is the mechanical energy E of the ball in its orbit?

SOLUTION: The **Key Idea** here is that we can get E from the orbital energy, given by Eq. 14-44 ($E = -GMm/2r$), if we first find the orbital radius r. That radius must be

$$r = R + h = 6370 \text{ km} + 350 \text{ km} = 6.72 \times 10^6 \text{ m},$$

in which R is the radius of Earth. Then, from Eq. 14-44, the mechanical energy is

$$E = -\frac{GMm}{2r}$$

$$= -\frac{(6.67 \times 10^{-11} \text{ N} \cdot \text{m}^2/\text{kg}^2)(5.98 \times 10^{24} \text{ kg})(7.20 \text{ kg})}{(2)(6.72 \times 10^6 \text{ m})}$$

$$= -2.14 \times 10^8 \text{ J} = -214 \text{ MJ}. \qquad \text{(Answer)}$$

(b) What is the mechanical energy E_0 of the ball on the launchpad at Cape Canaveral? From there to the orbit, what is the change ΔE in the ball's mechanical energy?

SOLUTION: The **Key Idea** here is that, on the launchpad, the ball is *not* in orbit and thus Eq. 14-44 does *not* apply. Instead, we must find $E_0 = K_0 + U_0$, where K_0 is the ball's kinetic energy and U_0 is the gravitational potential energy of the ball–Earth system. To find U_0, we use Eq. 14-20 to write

$$U_0 = -\frac{GMm}{R}$$

$$= -\frac{(6.67 \times 10^{-11} \text{ N} \cdot \text{m}^2/\text{kg}^2)(5.98 \times 10^{24} \text{ kg})(7.20 \text{ kg})}{6.37 \times 10^6 \text{ m}}$$

$$= -4.51 \times 10^8 \text{ J} = -451 \text{ MJ}.$$

The kinetic energy K_0 of the ball is due to the ball's motion with

Earth's rotation. You can show that K_0 is less than 1 MJ, which is negligible relative to U_0. Thus, the mechanical energy of the ball on the launchpad is

$$E_0 = K_0 + U_0 \approx 0 - 451 \text{ MJ} = -451 \text{ MJ.}\quad \text{(Answer)}$$

The *increase* in the mechanical energy of the ball from launch-

pad to orbit is

$$\Delta E = E - E_0 = (-214 \text{ MJ}) - (-451 \text{ MJ})$$
$$= 237 \text{ MJ.}\quad \text{(Answer)}$$

You can buy this amount of energy from your utility company for a few dollars. Obviously the high cost of placing objects into orbit is not due to the mechanical energy those objects require.

14-9 Einstein and Gravitation

Principle of Equivalence

Albert Einstein once said: "I was . . . in the patent office at Bern when all of a sudden a thought occurred to me: 'If a person falls freely, he will not feel his own weight.' I was startled. This simple thought made a deep impression on me. It impelled me toward a theory of gravitation."

Thus Einstein tells us how he began to form his **general theory of relativity.** The fundamental postulate of this theory about gravitation (the gravitating of objects toward each other) is called the **principle of equivalence,** which says that gravitation and acceleration are equivalent. If a physicist were locked up in a small box as in Fig. 14-19, he would not be able to tell whether the box was at rest on Earth (and subject only to Earth's gravitational force), as in Fig. 14-19a, or accelerating through interstellar space at 9.8 m/s² (and subject only to the force producing that acceleration), as in Fig. 14-19b. In both situations he would feel the same and would read the same value for his weight on a scale. Moreover, if he watched an object fall past him, the object would have the same acceleration relative to him in both situations.

Curvature of Space

We have thus far explained gravitation as due to a force between masses. Einstein showed that, instead, gravitation is due to a curvature (or shape) of space that is caused by the masses. (As is discussed later in this book, space and time are entangled, so the curvature of which Einstein spoke is really a curvature of *spacetime,* the combined four dimensions of our universe.)

Picturing how space (such as vacuum) can have curvature is difficult. An analogy might help: Suppose that from orbit we watch a race in which two boats begin on the equator with a separation of 20 km and head due south (Fig. 14-20a). To the sailors, the boats travel along flat, parallel paths. However, with time the boats draw together until, nearer the south pole, they touch. The sailors in the boats can interpret this drawing together in terms of a force acting on the boats. However, we can see that the boats draw together simply because of the curvature of Earth's surface. We can see this because we are viewing the race from "outside" that surface.

Figure 14-20b shows a similar race: Two horizontally separated apples are dropped from the same height above Earth. Although the apples may appear to travel along parallel paths, they actually move toward each other because they both fall toward Earth's center. We can interpret the motion of the apples in terms of the gravitational force on the apples from Earth. We can also interpret the motion in terms of a curvature of the space near Earth, due to the presence of Earth's mass. This time we cannot see the curvature because we cannot get "outside" the curved space, as we got "outside" the curved Earth in the boat example. However, we can depict the curvature with a drawing like Fig. 14-20c; there the apples would move along a surface that curves toward Earth because of Earth's mass.

(a)

(b)

Fig. 14-19 (a) A physicist in a box resting on Earth sees a cantaloupe falling with acceleration a = 9.8 m/s². (b) If he and the box accelerate in deep space at 9.8 m/s², the cantaloupe has the same acceleration relative to him. It is not possible, by doing experiments within the box, for the physicist to tell which situation he is in. For example, the platform scale on which he stands reads the same weight in both situations.

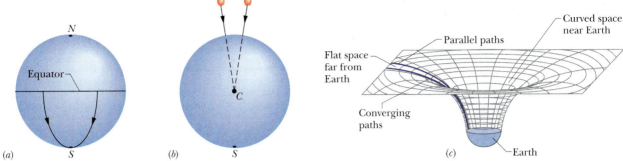

Fig. 14-20 (*a*) Two objects moving along lines of longitude toward the south pole converge because of the curvature of Earth's surface. (*b*) Two objects falling freely near Earth move along lines that converge toward the center of Earth because of the curvature of space near Earth. (*c*) Far from Earth (and other masses), space is flat and parallel paths remain parallel. Close to Earth, the parallel paths begin to converge because space is curved by Earth's mass.

When light passes near Earth, its path bends slightly because of the curvature of space there, an effect called *gravitational lensing*. When it passes a more massive structure, like a galaxy or a black hole having large mass, its path can be bent more. If such a massive structure is between us and a quasar (an extremely bright, extremely distant source of light), the light from the quasar can bend around the massive structure and toward us (Fig. 14-21*a*). Then, because the light seems to be coming to us from a number of slightly different directions in the sky, we see the same quasar in all those different directions. In some situations, the quasars we see blend together to form a giant luminous arc, which is called an *Einstein ring* (Fig. 14-21*b*).

Should we attribute gravitation to the curvature of spacetime due to the presence of masses or to a force between masses? Or should we attribute it to the actions of a type of fundamental particle called a *graviton,* as conjectured in some modern physics theories? We do not know.

Fig. 14-21 (*a*) Light from a distant quasar follows curved paths around a galaxy or a large black hole because the mass of the galaxy or black hole has curved the adjacent space. If the light is detected, it appears to have originated along the backward extensions of the final paths (dashed lines). (*b*) The Einstein ring known as MG1131+0456 on the computer screen of a telescope. The source of the light (actually, radio waves, which are a form of invisible light) is far behind the large, unseen galaxy that produces the ring; a portion of the source appears as the two bright spots seen along the ring.

REVIEW & SUMMARY

The Law of Gravitation   Any particle in the universe attracts any other particle with a **gravitational force** whose magnitude is

$$F = G \frac{m_1 m_2}{r^2} \qquad \text{(Newton's law of gravitation),} \qquad (14\text{-}1)$$

where m_1 and m_2 are the masses of the particles, r is their separation, and $G \, (= 6.67 \times 10^{-11} \; \text{N} \cdot \text{m}^2/\text{kg}^2)$ is the *gravitational constant*.

Gravitational Behavior of Uniform Spherical Shells
Equation 14-1 holds only for particles. The gravitational force between extended bodies must generally be found by adding (integrating) the individual forces on individual particles within the bodies. However, if either of the bodies is a uniform spherical shell or a spherically symmetric solid, the net gravitational force it exerts on an *external* object may be computed as if all the mass of the shell or body were located at its center.

Superposition   Gravitational forces obey the **principle of superposition;** that is, if n particles interact, the net force $\vec{F}_{1,\text{net}}$ on a particle labeled as particle 1 is the sum of the forces on it from all the other particles taken one at a time:

$$\vec{F}_{1,\text{net}} = \sum_{i=2}^{n} \vec{F}_{1i}, \qquad (14\text{-}4)$$

in which the sum is a vector sum of the forces $\vec{F}_{1i}$ on particle 1 from particles 2, 3, $\cdots$, n. The gravitational force $\vec{F}_1$ on a particle from an extended body is found by dividing the body into units of differential mass dm, each of which produces a differential force $d\vec{F}$ on the particle, and then integrating to find the sum of those forces:

$$\vec{F}_1 = \int d\vec{F}. \qquad (14\text{-}5)$$

Gravitational Acceleration   The *gravitational acceleration* a_g of a particle (of mass m) is due solely to the gravitational force acting on it. When the particle is at distance r from the center of a uniform, spherical body of mass M, the magnitude F of the gravitational force on the particle is given by Eq. 14-1. Thus, by Newton's second law,

$$F = m a_g, \qquad (14\text{-}9)$$

which gives

$$a_g = \frac{GM}{r^2}. \qquad (14\text{-}10)$$

Free-Fall Acceleration and Weight   The actual free-fall acceleration $\vec{g}$ of a particle near Earth differs slightly from the gravitational acceleration $\vec{a}_g$, and the particle's weight (equal to mg) differs from the magnitude of the gravitational force acting on the

particle as computed with Eq. 14-1, because Earth is not uniform or spherical and because Earth rotates.

Gravitation Within a Spherical Shell   A uniform shell of matter exerts no net gravitational force on a particle located inside it. This means that if a particle is located inside a uniform solid sphere at distance r from its center, the gravitational force exerted on the particle is due only to the mass M_{ins} that lies inside a sphere of radius r. This mass is given by

$$M_{\text{ins}} = \rho \, \frac{4 \pi r^3}{3}, \qquad (14\text{-}17)$$

where ρ is the density of the sphere.

Gravitational Potential Energy   The gravitational potential energy $U(r)$ of a system of two particles, with masses M and m and separated by a distance r, is the negative of the work that would be done by the gravitational force of either particle acting on the other if the separation between the particles were changed from infinite (very large) to r. This energy is

$$U = -\frac{GMm}{r} \qquad \text{(gravitational potential energy).} \qquad (14\text{-}20)$$

Potential Energy of a System   If a system contains more than two particles, its total gravitational potential energy U is the sum of terms representing the potential energies of all the pairs. As an example, for three particles, of masses m_1, m_2, and m_3,

$$U = -\left(\frac{G m_1 m_2}{r_{12}} + \frac{G m_1 m_3}{r_{13}} + \frac{G m_2 m_3}{r_{23}} \right). \qquad (14\text{-}21)$$

Escape Speed   An object will escape the gravitational pull of an astronomical body of mass M and radius R (to reach an infinite distance) if the object's speed near the body's surface is at least equal to the **escape speed,** given by

$$v = \sqrt{\frac{2GM}{R}}. \qquad (14\text{-}27)$$

Kepler's Laws   Gravitational attraction holds the solar system together and makes possible orbiting Earth satellites, both natural and artificial. Such motions are governed by Kepler's three laws of planetary motion, all of which are direct consequences of Newton's laws of motion and gravitation:

1. ***The law of orbits.*** All planets move in elliptical orbits with the Sun at one focus.

2. ***The law of areas.*** A line joining any planet to the Sun sweeps out equal areas in equal times as the planet orbits the Sun. (This statement is equivalent to conservation of angular momentum.)

3. ***The law of periods.*** The square of the period T of any planet about the Sun is proportional to the cube of the semimajor axis

a of the orbit. For circular orbits with radius *r*, the semimajor axis *a* is replaced by *r* and the law is written as

$$T^2 = \left(\frac{4\pi^2}{GM}\right)r^3 \qquad \text{(law of periods)}, \qquad (14\text{-}33)$$

where *M* is the mass of the attracting body—the Sun in the case of the solar system. This equation is generally valid for elliptical planetary orbits, when the semimajor axis *a* is inserted in place of the circular radius *r*.

Energy in Planetary Motion When a planet or satellite with mass *m* moves in a circular orbit with radius *r*, its potential energy *U* and kinetic energy *K* are given by

$$U = -\frac{GMm}{r} \quad \text{and} \quad K = \frac{GMm}{2r}. \qquad (14\text{-}20, 14\text{-}42)$$

The mechanical energy *E* = *K* + *U* is then

$$E = -\frac{GMm}{2r}. \qquad (14\text{-}44)$$

For an elliptical orbit of semimajor axis *a*,

$$E = -\frac{GMm}{2a}. \qquad (14\text{-}46)$$

Einstein's View of Gravitation Einstein pointed out that gravitation and acceleration are equivalent. This **principle of equivalence** led him to a theory of gravitation (the **general theory of relativity**) that explains gravitational effects in terms of a curvature of space.

QUESTIONS

1. In Fig. 14-22, two particles, of masses *m* and 2*m*, are fixed in place on an axis. (a) Where on the axis can a third particle of mass 3*m* be placed (other than at infinity) so that the net gravitational force on it from the first two particles is zero: to the left of the first two particles, to their right, between them but closer to the more massive particle, or between them but closer to the less massive particle? (b) Does the answer change if the third particle has, instead, a mass of 16*m*? (c) Is there a point off the axis at which the net force on the third particle would be zero?

Fig. 14-22 Question 1.

2. In Fig. 14-23, a central particle is surrounded by two circular rings of particles, at radii *r* and *R*, with *R* > *r*. All the particles have mass *m*. What are the magnitude and direction of the net gravitational force on the central particle due to the particles in the rings?

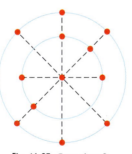

Fig. 14-23 Question 2.

3. In Fig. 14-24, a central particle of mass *M* is surrounded by a square array of other particles, separated by either distance *d* or distance *d*/2 along the perimeter of the square. What are the magnitude and direction of the net gravitational force on the central particle due to the other particles?

Fig. 14-24 Question 3.

4. Figure 14-25 shows four arrangements of a particle of mass *m* and one or more uniform rods of mass *M* and length *L*, each a distance *d* from the particle. Rank the arrangements according to

Fig. 14-25 Question 4.

the magnitude of the net gravitational force on the particle from the rods, greatest first.

5. Figure 14-26 gives the gravitational acceleration a_g for four planets as a function of the radial distance *r* from the center of the

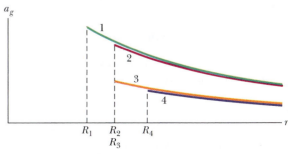

Fig. 14-26 Question 5.

planet, starting at the surface of the planet (at radius R_1, R_2, R_3, or R_4). Plots 1 and 2 coincide for $r \geq R_2$; plots 3 and 4 coincide for $r \geq R_4$. Rank the four planets according to (a) their mass and (b) their density, greatest first.

6. Figure 14-27 shows three uniform spherical planets that are identical in size and mass. The periods of rotation T for the planets are given, and six lettered points are indicated—three points are on the equators of the planets and three points are on the north poles. Rank the points according to the value of the free-fall acceleration g at them, greatest first.

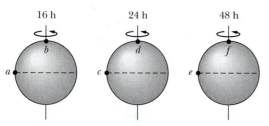

Fig. 14-27 Question 6.

7. From an inertial frame in space, we watch two identical uniform spheres fall toward one another owing to their mutual gravitational attraction. Approximate their initial speed as zero and take the initial gravitational potential energy of the two-sphere system as U_i. When the separation between the two spheres is half the initial separation, what is the kinetic energy of each sphere?

8. Rank the four systems of equal-mass particles in Checkpoint 2 according to the absolute value of the gravitational potential energy of the system, greatest first.

9. Figure 14-28 shows six paths by which a rocket orbiting a moon might move from point a to point b. Rank the paths according to (a) the corresponding change in the gravitational potential energy of the rocket–moon system and (b) the net work done on the rocket by the gravitational force from the moon, greatest first.

Fig. 14-28 Question 9.

10. In Fig. 14-29, a particle of mass m (not shown) is to be moved from an infinite distance to one of the three possible locations a, b, and c. Two other particles, of masses m and $2m$, are fixed in place. Rank the three possible locations according to the work done by the net gravitational force on the moving particle due to the fixed particles, greatest first.

Fig. 14-29 Question 10.

11. In Fig. 14-30, a particle of mass m is initially at point A, at distance d from the center of one uniform sphere and distance $4d$ from the center of another uniform sphere, both of mass $M \gg m$. State whether, if you moved the particle to point D, the following would be positive, negative, or zero: (a) the change in the gravitational potential energy of the particle, (b) the work done by the net gravitational force on the particle, (c) the work done by your force. (d) What are the answers if, instead, the move were from point B to point C?

Fig. 14-30 Question 11.

12. Figure 14-31 gives the masses and separations for three pairs of stars that each form a binary star system. (a) Locate the point about which the stars of each pair orbit. (b) Rank the pairs according to the magnitude of the centripetal acceleration of the stars, greatest first.

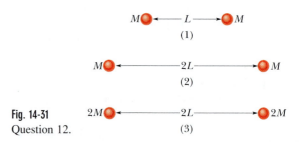

Fig. 14-31 Question 12.

EXERCISES & PROBLEMS

SEC. 14-2   Newton's Law of Gravitation

1E. What must the separation be between a 5.2 kg particle and a 2.4 kg particle for their gravitational attraction to have a magnitude of 2.3×10^{-12} N? ssm

2E. Some believe that the positions of the planets at the time of birth influence the newborn. Others deride this belief and claim that the gravitational force exerted on a baby by the obstetrician is greater than that exerted by the planets. To check this claim, calculate and compare the magnitude of the gravitational force exerted on a 3 kg baby (a) by a 70 kg obstetrician who is 1 m away and roughly approximated as a point mass, (b) by the massive planet Jupiter ($m = 2 \times 10^{27}$ kg) at its closest approach to Earth ($= 6 \times 10^{11}$ m), and (c) by Jupiter at its greatest distance from Earth ($= 9 \times 10^{11}$ m). (d) Is the claim correct?

3E. One of the *Echo* satellites consisted of an inflated spherical aluminum balloon 30 m in diameter and of mass 20 kg. Suppose a meteor having a mass of 7.0 kg passes within 3.0 m of the surface of the satellite. What is the magnitude of the gravitational force on the meteor from the satellite at the closest approach? ssm

4E. The Sun and Earth each exert a gravitational force on the Moon. What is the ratio F_{Sun}/F_{Earth} of these two forces? (The average Sun–Moon distance is equal to the Sun–Earth distance.)

5E. A mass M is split into two parts, m and $M - m$, which are then separated by a certain distance. What ratio m/M maximizes the magnitude of the gravitational force between the parts? ilw

SEC. 14-3 Gravitation and the Principle of Superposition

6E. A spaceship is on a straight-line path between Earth and its moon. At what distance from Earth is the net gravitational force on the spaceship zero?

7E. How far from Earth must a space probe be along a line toward the Sun so that the Sun's gravitational pull on the probe balances Earth's pull? ssm

8P. Three 5.0 kg spheres are located in the xy plane as shown in Fig. 14-32. What is the magnitude of the net gravitational force on the sphere at the origin due to the other two spheres?

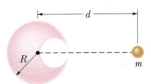

Fig. 14-32 Problem 8.

9P. In Fig. 14-33a, four spheres form the corners of a square whose side is 2.0 cm long. What are the magnitude and direction of the net gravitational force from them on a central sphere with mass $m_5 = 250$ kg?

10P. In Fig. 14-33b, two spheres of mass m and a third sphere of mass M form an equilateral triangle, and a fourth sphere of mass m_4 is at the center of the triangle. The net gravitational force on that central sphere from the three other spheres is zero. (a) What is M in terms of m? (b) If we double the value of m_4, what then is the magnitude of the net gravitational force on the central sphere?

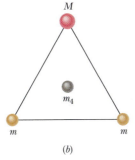

Fig. 14-33 Problems 9 and 10.

11P. The masses and coordinates of three spheres are as follows: 20 kg, $x = 0.50$ m, $y = 1.0$ m; 40 kg, $x = -1.0$ m, $y = -1.0$ m; 60 kg, $x = 0$ m, $y = -0.50$ m. What is the magnitude of the gravitational force on a 20 kg sphere located at the origin due to the other spheres? ilw

12P. Four uniform spheres, with masses $m_A = 400$ kg, $m_B = 350$ kg, $m_C = 2000$ kg, and $m_D = 500$ kg, have (x, y) coordinates of (0, 50 cm), (0, 0), (−80 cm, 0), and (40 cm, 0), respectively. What is the net gravitational force on sphere B due to the other spheres?

13P. Figure 14-34 shows a spherical hollow inside a lead sphere of radius R; the surface of the hollow passes through the center of the sphere and "touches" the right side of the sphere. The mass of the sphere before hollowing was M. With what gravitational force does the hollowed-out lead sphere attract a small sphere of mass m that lies at a distance d from the center of the lead sphere, on the straight line connecting the centers of the spheres and of the hollow? ssm

Fig. 14-34 Problem 13.

SEC. 14-4 Gravitation Near Earth's Surface

14E. You weigh 530 N at sidewalk level outside the World Trade Center in New York City. Suppose that you ride from this level to the top of one of its 410 m towers. Ignoring Earth's rotation, how much less would you weigh there (because you are slightly farther from the center of Earth)?

15E. At what altitude above Earth's surface would the gravitational acceleration be 4.9 m/s^2? ssm

16E. (a) What will an object weigh on the Moon's surface if it weighs 100 N on Earth's surface? (b) How many Earth radii must this same object be from the center of Earth if it is to weigh the same as it does on the Moon?

17P. The fastest possible rate of rotation of a planet is that for which the gravitational force on material at the equator just barely provides the centripetal force needed for the rotation. (Why?) (a) Show that the corresponding shortest period of rotation is

$$T = \sqrt{\frac{3\pi}{G\rho}},$$

where ρ is the uniform density of the spherical planet. (b) Calculate the rotation period assuming a density of 3.0 g/cm^3, typical of many planets, satellites, and asteroids. No astronomical object has ever been found to be spinning with a period shorter than that determined by this analysis. ssm

18P. One model for a certain planet has a core of radius R and mass M surrounded by an outer shell of inner radius R, outer radius $2R$, and mass $4M$. If $M = 4.1 \times 10^{24}$ kg and $R = 6.0 \times 10^6$ m, what is the gravitational acceleration of a particle at points (a) R and (b) $3R$ from the center of the planet?

19P. A body is suspended from a spring scale in a ship sailing along the equator with speed v. (a) Show that the scale reading will be very close to $W_0(1 \pm 2\omega v/g)$, where ω is the angular speed of Earth and W_0 is the scale reading when the ship is at rest. (b) Explain the $\pm$ sign. ssm www

20P. The radius R_h and mass M_h of a black hole are related by $R_h = 2GM_h/c^2$, where c is the speed of light. Assume that the gravitational acceleration a_g of an object at a distance $r_o =$

1.001R_h from the center of a black hole is given by Eq. 14-10 (it is, for large black holes). (a) Find an expression for a_g at r_o in terms of M_h. (b) Does a_g at r_o increase or decrease with an increase of M_h? (c) What is a_g at r_o for a very large black hole whose mass is 1.55×10^{12} times the solar mass of 1.99×10^{30} kg? (d) If the astronaut of Sample Problem 14-3 is at r_o with her feet toward this black hole, what is the difference in gravitational acceleration between her head and her feet? (e) Is the tendency to stretch the astronaut severe?

21P. Certain neutron stars (extremely dense stars) are believed to be rotating at about 1 rev/s. If such a star has a radius of 20 km, what must be its minimum mass so that material on its surface remains in place during the rapid rotation? `ilw`

SEC. 14-5 Gravitation Inside Earth

22E. Two concentric shells of uniform density having masses M_1 and M_2 are situated as shown in Fig. 14-35. Find the magnitude of the net gravitational force on a particle of mass m, due to the shells, when the particle is located at (a) point A, at distance $r = a$ from the center, (b) point B at $r = b$, and (c) point C at $r = c$. The distance r is measured from the center of the shells.

Fig. 14-35 Exercise 22.

23P. A solid sphere of uniform density has a mass of 1.0×10^4 kg and a radius of 1.0 m. What is the magnitude of the gravitational force due to the sphere on a particle of mass m located at a distance of (a) 1.5 m and (b) 0.50 m from the center of the sphere? (c) Write a general expression for the magnitude of the gravitational force on the particle at a distance $r \leq 1.0$ m from the center of the sphere.

24P. A uniform solid sphere of radius R produces a gravitational acceleration of a_g on its surface. At what two distances from the center of the sphere is the gravitational acceleration $a_g/3$? (*Hint:* Consider distances both inside and outside the sphere.)

25P. Figure 14-36 shows, not to scale, a cross section through the interior of Earth. Rather than being uniform throughout, Earth is divided into three zones: an outer *crust*, a *mantle*, and an inner *core*. The dimensions of these zones and the masses contained within them are shown on the figure. Earth has a total mass of 5.98×10^{24} kg and a radius of 6370 km. Ignore rotation and assume that Earth is spherical. (a) Calculate a_g at the surface. (b) Suppose that a bore hole (the *Mohole*) is driven to the crust–mantle interface at a depth of 25 km; what would be the value of a_g at the bottom of the hole? (c) Suppose that Earth were a uniform sphere with the same total mass and size. What would be the value of a_g at a depth of 25 km? (Precise measurements of a_g are sensitive probes of the interior structure of Earth, although results can be clouded by local density variations.) `ssm`

SEC. 14-6 Gravitational Potential Energy

26E. (a) What is the gravitational potential energy of the two-particle system in Exercise 1? If you triple the separation between the particles, how much work is done (b) by the gravitational force between the particles and (c) by you?

27E. (a) In Problem 12, remove sphere A and calculate the gravitational potential energy of the remaining three-particle system. (b) If A is then put back in place, is the potential energy of the four-particle system more or less than that of the system in (a)? (c) In (a), is the work done by you to remove A positive or negative? (d) In (b), is the work done by you to replace A positive or negative?

28E. In Problem 5, what ratio m/M gives the least gravitational potential energy for the system?

29E. The mean diameters of Mars and Earth are 6.9×10^3 km and 1.3×10^4 km, respectively. The mass of Mars is 0.11 times Earth's mass. (a) What is the ratio of the mean density of Mars to that of Earth? (b) What is the value of the gravitational acceleration on Mars? (c) What is the escape speed on Mars? `ssm`

30E. Calculate the amount of energy required to escape from (a) Earth's moon and (b) Jupiter relative to that required to escape from Earth.

31P. The three spheres in Fig. 14-37, with masses $m_A = 800$ g, $m_B = 100$ g, and $m_C = 200$ g, have their centers on a common line, with $L = 12$ cm and $d = 4.0$ cm. You move sphere B along the line until its center-to-center separation from C is $d = 4.0$ cm. How much work is done on sphere B (a) by you and (b) by the net gravitational force on B due to spheres A and C? `ssm`

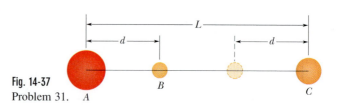

Fig. 14-37
Problem 31. A

32P. Zero, a hypothetical planet, has a mass of 5.0×10^{23} kg, a radius of 3.0×10^6 m, and no atmosphere. A 10 kg space probe is to be launched vertically from its surface. (a) If the probe is launched with an initial energy of 5.0×10^7 J, what will be its kinetic energy when it is 4.0×10^6 m from the center of Zero? (b) If the probe is to achieve a maximum distance of 8.0×10^6 m from the center of Zero, with what initial kinetic energy must it be launched from the surface of Zero?

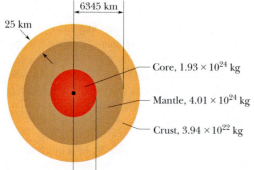

Fig. 14-36
Problem 25.

33P. A rocket is accelerated to speed $v = 2\sqrt{gR_e}$ near Earth's surface (where Earth's radius is R_e), and it then coasts upward. (a) Show that it will escape from Earth. (b) Show that very far from Earth its speed will be $v = \sqrt{2gR_e}$.   ssm

34P. Planet Roton, with a mass of 7.0×10^{24} kg and a radius of 1600 km, gravitationally attracts a meteorite that is initially at rest relative to the planet, at a great enough distance to take as infinite. The meteorite falls toward the planet. Assuming the planet is airless, find the speed of the meteorite when it reaches the planet's surface.

35P. (a) What is the escape speed on a spherical asteroid whose radius is 500 km and whose gravitational acceleration at the surface is 3.0 m/s²? (b) How far from the surface will a particle go if it leaves the asteroid's surface with a radial speed of 1000 m/s? (c) With what speed will an object hit the asteroid if it is dropped from 1000 km above the surface?   ssm   www

36P. A 150.0 kg rocket moving radially outward from Earth has a speed of 3.70 km/s when its engine shuts off 200 km above Earth's surface. (a) Assuming negligible air drag, find the rocket's kinetic energy when the rocket is 1000 km above Earth's surface. (b) What maximum height above the surface is reached by the rocket?

37P. Two neutron stars are separated by a distance of 10^{10} m. They each have a mass of 10^{30} kg and a radius of 10^5 m. They are initially at rest with respect to each other. As measured from that rest frame, how fast are they moving when (a) their separation has decreased to one-half its initial value and (b) they are about to collide?   ssm   www

38P. In deep space, sphere A of mass 20 kg is located at the origin of an x axis and sphere B of mass 10 kg is located on the axis at $x = 0.80$ m. Sphere B is released from rest while sphere A is held at the origin. (a) What is the gravitational potential energy of the two-sphere system as B is released? (b) What is the kinetic energy of B when it has moved 0.20 m toward A?

39P. A projectile is fired vertically from Earth's surface with an initial speed of 10 km/s. Neglecting air drag, how far above the surface of Earth will it go?   ilw

SEC. 14-7   Planets and Satellites: Kepler's Laws

40E. The mean distance of Mars from the Sun is 1.52 times that of Earth from the Sun. From Kepler's law of periods, calculate the number of years required for Mars to make one revolution about the Sun; compare your answer with the value given in Appendix C.

41E. The Martian satellite Phobos travels in an approximately circular orbit of radius 9.4×10^6 m with a period of 7 h 39 min. Calculate the mass of Mars from this information.   ssm

42E. Determine the mass of Earth from the period T (27.3 days) and the radius r (3.82×10^5 km) of the Moon's orbit about Earth. Assume the Moon orbits the center of Earth rather than the center of mass of the Earth–Moon system.

43E. Our Sun, with mass 2.0×10^{30} kg, revolves about the center of the Milky Way galaxy, which is 2.2×10^{20} m away, once every 2.5×10^8 years. Assuming that each of the stars in the galaxy has a mass equal to that of our Sun, that the stars are distributed uniformly in a sphere about the galactic center, and that our Sun is essentially at the edge of that sphere, estimate roughly the number of stars in the galaxy.   ssm

44E. A satellite is placed in a circular orbit about Earth with a radius equal to one-half the radius of the Moon's orbit. What is its period of revolution in lunar months? (A lunar month is the period of revolution of the Moon.)

45E. (a) What linear speed must an Earth satellite have to be in a circular orbit at an altitude of 160 km? (b) What is the period of revolution?   ssm

46E. The Sun's center is at one focus of Earth's orbit. How far from this focus is the other focus, (a) in meters and (b) in terms of the solar radius, 6.96×10^8 m? The eccentricity of Earth's orbit is 0.0167, and the semimajor axis is 1.50×10^{11} m.

47E. A satellite, moving in an elliptical orbit, is 360 km above Earth's surface at its farthest point and 180 km above at its closest point. Calculate (a) the semimajor axis and (b) the eccentricity of the orbit. (*Hint:* See Sample Problem 14-6.)   ssm

48E. A satellite hovers over a certain spot on the equator of (rotating) Earth. What is the altitude of its orbit (called a *geosynchronous orbit*)?

49E. A comet that was seen in April 574 by Chinese astronomers on a day known by them as the Woo Woo day was spotted again in May 1994. Assume the time between observations is the period of the Woo Woo day comet and take its eccentricity as 0.11. What are (a) the semimajor axis of the comet's orbit and (b) its greatest distance from the Sun in terms of the mean orbital radius R_P of Pluto?

50E. In 1993 the spacecraft *Galileo* sent home an image (Fig. 14-38) of asteroid 243 Ida and a tiny orbiting moon (now known as Dactyl), the first confirmed example of an asteroid–moon system. In the image, the moon, which is 1.5 km wide, is 100 km from the center of the asteroid, which is 55 km long. The shape of the moon's orbit is not well known; assume it is circular with a period of 27 h. (a) What is the mass of the asteroid? (b) The volume of the asteroid, measured from the *Galileo* images, is 14 100 km³. What is the density of the asteroid?

Fig. 14-38 Exercise 50. An image from the spacecraft *Galileo* shows a tiny moon orbiting asteroid 243 Ida.

51P. In 1610, Galileo used his telescope to discover four prominent moons around Jupiter. Their mean orbital radii a and periods T are as follows:

Name	a (10^8 m)	T (days)
Io	4.22	1.77
Europa	6.71	3.55
Ganymede	10.7	7.16
Callisto	18.8	16.7

(a) Plot log a (y axis) against log T (x axis) and show that you get a straight line. (b) Measure the slope of the line and compare it with the value that you expect from Kepler's third law. (c) Find the mass of Jupiter from the intercept of this line with the y axis.

52P. A 20 kg satellite has a circular orbit with a period of 2.4 h and a radius of 8.0×10^6 m around a planet of unknown mass. If the magnitude of the gravitational acceleration on the surface of the planet is 8.0 m/s², what is the radius of the planet?

53P. In a certain binary-star system, each star has the same mass as our Sun, and they revolve about their center of mass. The distance between them is the same as the distance between Earth and the Sun. What is their period of revolution in years? ilw

54P. A certain triple-star system consists of two stars, each of mass m, revolving about a central star of mass M in the same circular orbit of radius r (Fig. 14-39). The two stars are always at opposite ends of a diameter of the circular orbit. Derive an expression for the period of revolution of the stars.

Fig. 14-39 Problem 54.

55P*. Three identical stars of mass M are located at the vertices of an equilateral triangle with side L. At what speed must they move if they all revolve under the influence of one another's gravitational force in a circular orbit circumscribing the triangle while still preserving the equilateral triangle? ssm   www

SEC. 14-8   Satellites: Orbits and Energy

56E. Consider two satellites, A and B, both of mass m, moving in the same circular orbit of radius r around Earth, of mass M_E, but in opposite senses of rotation and therefore on a collision course (see Fig. 14-40). (a) In terms of G, M_E, m, and r, find the total mechanical energy $E_A + E_B$ of the two-satellite-plus-Earth system before collision. (b) If the collision is completely inelastic so that the wreckage remains as one piece of tangled material (mass = $2m$), find the total mechanical energy immediately after collision. (c) Describe the subsequent motion of the wreckage.

Fig. 14-40 Exercise 56.

57E. An asteroid, whose mass is 2.0×10^{-4} times the mass of Earth, revolves in a circular orbit around the Sun at a distance that is twice Earth's distance from the Sun. (a) Calculate the period of revolution of the asteroid in years. (b) What is the ratio of the kinetic energy of the asteroid to that of Earth? ssm

58P. Two Earth satellites, A and B, each of mass m, are to be launched into circular orbits about Earth's center. Satellite A is to orbit at an altitude of 6370 km. Satellite B is to orbit at an altitude of 19 110 km. The radius of Earth R_E is 6370 km. (a) What is the ratio of the potential energy of satellite B to that of satellite A, in orbit? (b) What is the ratio of the kinetic energy of satellite B to that of satellite A, in orbit? (c) Which satellite has the greater total energy if each has a mass of 14.6 kg? By how much?

59P. Show that if an object is in an elliptical orbit with semimajor axis a about a planet of mass M, then its distance r from the planet and speed v are related by

$$v^2 = GM \left(\frac{2}{r} - \frac{1}{a} \right).$$

(*Hint:* Use the law of conservation of mechanical energy and Eq. 14-46.) ssm

60P. Use the result of Problem 59 and data contained in Sample Problem 14-6 to calculate (a) the speed v_p of comet Halley at perihelion and (b) its speed v_a at aphelion. (c) Using the law of conservation of angular momentum relative to the Sun, find the ratio of the comet's perihelion distance R_p to its aphelion distance R_a in terms of v_p and v_a.

61P. (a) Does it take more energy to get a satellite up to 1500 km above Earth than to put it in circular orbit once it is there? (Take Earth's radius to be 6370 km.) (b) What about 3185 km? (c) What about 4500 km?

62P. One way to attack a satellite in Earth orbit is to launch a swarm of pellets in the same orbit as the satellite but in the opposite direction. Suppose a satellite in a circular orbit 500 km above Earth's surface collides with a pellet having mass 4.0 g. (a) What is the kinetic energy of the pellet in the reference frame of the satellite just before the collision? (b) What is the ratio of this kinetic energy to the kinetic energy of a 4.0 g bullet from a modern army rifle with a muzzle speed of 950 m/s?

63P. What are (a) the speed and (b) the period of a 220 kg satellite in an approximately circular orbit 640 km above the surface of Earth? Suppose the satellite loses mechanical energy at the average rate of 1.4×10^5 J per orbital revolution. Adopting the reasonable approximation that the satellite's orbit becomes a "circle of slowly diminishing radius," determine the satellite's (c) altitude, (d) speed, and (e) period at the end of its 1500th revolution. (f) What is the magnitude of the average retarding force on the satellite? Is angular momentum around Earth's center conserved for (g) the satellite and (h) the satellite–Earth system? ssm

SEC. 14-9   Einstein and Gravitation

64E. In Fig. 14-19b, the scale on which the 60 kg physicist stands reads 220 N. How long will the cantaloupe take to reach the floor if the physicist drops it from rest (relative to himself), 2.1 m from the floor?

NEW PROBLEMS

N1. Figure 14N-1 is a graph of the kinetic energy K of an asteroid versus its distance r from Earth's center, as the asteroid falls directly in toward that center. (a) What is the (approximate) mass of the asteroid? (b) What is its speed at $r = 1.945 \times 10^7$ m?

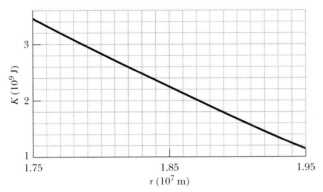

Fig. 14N-1 Problem N1.

N2. Figure 14N-2 gives the potential energy function $U(r)$ of a projectile, plotted outward from a planet's surface at radius R_s. If the projectile is launched radially outward from the surface with a mechanical energy of -2.0×10^9 J, what are (a) its kinetic energy at radius $r = 1.25R_s$ and (b) its *turning point* (see Section 8-5) in terms of R_s?

Fig. 14N-2 Problems N2 and N6.

N3. Four identical 1.5 kg particles are placed at the corners of a square with sides equal to 20 cm. What is the magnitude of the net gravitational force on any one of the particles due to the others?

N4. Figure 14N-3 shows four particles, each of 20.0 g, that form a square with an edge length of $d = 0.600$ m. If the edge length is reduced to 0.200 m, by how much is the gravitational potential energy of the four-particle system changed, and is it increased or decreased?

Fig. 14N-3 Problem N4.

N5. Two Earth satellites, A and B, each of mass m, are to be launched into circular orbits about Earth's center. Satellite A is to orbit at an altitude that equals Earth's radius R_E. Satellite B is to orbit at an altitude of $3R_E$. (a) What is the ratio of the potential energy of satellite B to that of satellite A, in orbit? (b) What is the ratio of the kinetic energy of satellite B to that of satellite A, in orbit? (c) Which satellite has the greater total energy if each has a mass of 14.6 kg? By how much?

N6. Figure 14N-2 gives the potential energy function $U(r)$ of a projectile, plotted outward from a planet's surface at radius R_s. What least kinetic energy is required of a projectile launched at the surface if the projectile is to "escape" the planet?

N7. *One dimension.* In Fig. 14N-4, two point particles are fixed on an x axis, separated by distance d. Particle A has mass m_A and particle B has mass $3.00m_A$. A third particle C, of mass $75.0m_A$, is to be placed near particles A and B. In terms of distance d, where should it be placed so that the net gravitational force on particle A from particles B and C is zero?

Fig. 14N-4 Problem N7.

N8. *Two dimensions.* In Fig. 14N-5, three point particles are fixed in place on an xy plane. Particle A has mass m_A, particle B has mass $2.00m_A$, and particle C has mass $3.00m_A$. A fourth particle D, with mass $4.00m_A$, is to be placed near the other three particles. In terms of distance d, at what xy coordinates should particle D be placed so that the net gravitational force on particle A from particles B, C, and D is zero?

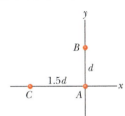

Fig. 14N-5 Problem N8.

N9*. *Three dimensions (requiring a bit of physics courage).* Three point particles are fixed in place in an xyz coordinate system. Particle A, at the origin, has mass m_A. Particle B, at xyz coordinates $(2.00d, 1.00d, 2.00d)$, has mass $2.00m_A$, and particle C, at coordinates $(-1.00d, 2.00d, -3.00d)$, has mass $3.00m_A$. A fourth particle D, with mass $4.00m_A$, is to be placed near the other particles. In terms of distance d, at what xyz coordinates should particle D be placed so that the net gravitational force on particle A from particles B, C, and D is zero?

N10. Three point particles are fixed in position in an xy plane. Two of them, particle A of mass 6.00 g and particle B of mass 12.0 g, are shown in Fig. 14N-6, with a separation of $d_{AB} = 0.500$ m at angle $\theta = 30°$. Particle C, with mass 8.00 g, is not shown. The net gravitational force acting on particle A due to particles B and C is 2.77×10^{-14} N at an angle of $-163.8°$ from the positive direction of the x axis. What are the xy coordinates of particle C?

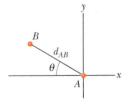

Fig. 14N-6 Problem N10.

N11. What is the percentage change in the acceleration of Earth toward the Sun when the alignment of Earth, Sun, and Moon changes from an eclipse of the Sun (with the Moon between Earth and Sun) to an eclipse of the Moon (Earth between Moon and Sun)?

N12. An object of mass m is initially held in place at radial distance $r = 3R_E$ from the center of Earth, where R_E is the radius of Earth. Let M_E be the mass of Earth. A force is applied to the object to move it to a radial distance $r = 4R_E$, where it again is held in place. Calculate the work done by the applied force during the move by integrating the force magnitude.

N13. A particle of comet dust with mass m is a distance R from Earth's center and a distance r from the Moon's center. If Earth's mass is M_E and the Moon's mass is M_m, what is the total gravitational potential energy of the particle–Earth system and the particle–Moon system?

N14. In Fig. 14N-7a, particle A is fixed in place at $x = -0.20$ m on the x axis and particle B, with a mass of 1.0 kg, is fixed in place at the origin. Particle C (not shown) can be moved along the x axis, between particle B and $x = \infty$. Figure 14N-7b shows the x component $F_{net,x}$ of the net gravitational force on particle B due to particles A and C, as a function of position x of particle C. The plot actually extends to the right, approaching an asymptote of -4.17×10^{-10} N as $x \to \infty$. What are the masses of (a) particle A and (b) particle C?

(a)   (b)

Fig. 14N-7 Problem N14.

N15. Figure 14N-8 shows two identical spheres, each with mass 2.00 kg and radius $R = 0.0200$ m, that initially touch, somewhere in deep space. Suppose the spheres are blown apart such that they initially separate at the relative speed 1.05×10^{-4} m/s. They then slow due to the gravitational force between them.

Fig. 14N-8 Problem N15.

Center-of-mass frame: Assume that we are in an inertial reference frame that is stationary with respect to the center of mass of the two-sphere system. Use the principle of conservation of mechanical energy ($K_f + U_f = K_i + U_i$) to find the following when the center-to-center separation is $10R$: (a) the kinetic energy of each sphere and (b) the speed of sphere B relative to sphere A.

Sphere frame: Next assume that we are in a reference frame attached to sphere A (we ride on the body). Now we see sphere B move away from us. From this reference frame, again use $K_f + U_f = K_i + U_i$ to find the following when the center-to-center separation is $10R$: (c) the kinetic energy of sphere B and (d) the speed of sphere B relative to sphere A. (e) Why are the answers to (b) and (d) different? Which answer is correct?

N16. The radius R_h of a black hole is the radius of a mathematical sphere, called the event horizon, that is centered on the black hole. Information from events inside the event horizon cannot reach the outside world. According to Einstein's general theory of relativity, $R_h = 2GM/c^2$, where M is the mass of the black hole and c is the speed of light.

Suppose that you wish to study black holes near them, at a radial distance of $50R_h$. However, you do not want the difference in gravitational acceleration between your feet and your head to exceed 10 m/s² when you are feet down (or head down) toward the black hole. (a) As a multiple of our sun's mass, what is the limit to the mass of the black hole you can tolerate at the given radial distance? (You need to estimate your height.) (b) Is the limit an upper limit (you can tolerate smaller masses) or a lower limit (you can tolerate larger masses)?

N17. A satellite is in elliptical orbit with a period of 8.00×10^4 s about a planet of mass 7.00×10^{24} kg. At aphelion, at radius 4.5×10^7 m, its angular speed is 7.158×10^{-5} rad/s. What is its angular speed at perihelion?

N18. Figure 14N-9a shows a particle A that can be moved along a y axis from an infinite distance to the origin. That origin lies at the midpoint between particles B and C, which have identical masses, and the y axis is a perpendicular bisector between them. Distance D is 0.3057 m. Figure 14N-9b shows the potential energy U of the three-particle system as a function of the position of particle A along the y axis. The curve actually extends rightward and approaches an asymptote of -2.7×10^{-11} J as $y \to \infty$. What are the masses of (a) particles B and C and (b) particle A?

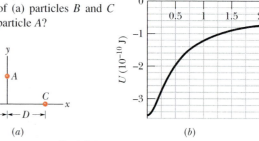

(a)   (b)

Fig. 14N-9 Problem N18.

N19. Assume that a planet is a sphere of radius R with a uniform density and (somehow) has a narrow radial tunnel through its center (Fig. 14-8). Also assume that we can position an apple anywhere along the tunnel or outside the sphere. Let F_R be the magnitude of the gravitational force on the apple when it is located at the planet's surface. How far from the surface is there a point where the magnitude of the gravitational force on the apple is $\frac{1}{2}F_R$ if we move the apple (a) away from the planet and (b) into the tunnel?

N20. A projectile is shot directly away from Earth from its surface at radius R_E. Neglect the rotation of Earth. In terms of R_E, what radial distance does the projectile reach if (a) its initial speed is 0.500 of the escape speed from Earth and (b) its initial kinetic energy is 0.500 of the kinetic energy required to escape Earth? (c) What is the least initial mechanical energy required at launch if the projectile is to escape Earth?

N21*. Several planets (Jupiter, Saturn, Uranus) possess nearly circular surrounding rings, perhaps composed of material that failed to form a satellite. In addition, many galaxies contain ring-like structures. Consider a homogeneous ring of mass M and radius R. (a) What gravitational attraction does it exert on a particle of mass m located a distance x from the center of the ring along its axis? See Fig. 14N-10. (b) Suppose the particle falls from rest as a result of the attraction of the ring of matter. Find an expression for the speed with which it passes through the center of the ring.

Fig. 14N-10 Problem N21.

N22. A satellite of mass 125 kg is in a circular orbit of radius 7.00×10^6 m about a planet, with a period of 8050 s. What is the mechanical energy of the satellite?

15 Fluids

The force exerted by water on the body of a descending diver increases noticeably, even for a relatively shallow descent to the bottom of a swimming pool. However, in 1975, using scuba gear with a special gas mixture for breathing, William Rhodes emerged from a chamber that had been lowered 300 m into the Gulf of Mexico, and he then swam to a record depth of 350 m. Strangely, a novice scuba diver practicing in a swimming pool might be in more danger from the force exerted by the water than was Rhodes. Occasionally, novice scuba divers die because they have neglected that danger.

What is this potentially lethal risk?

The answer is in this chapter.

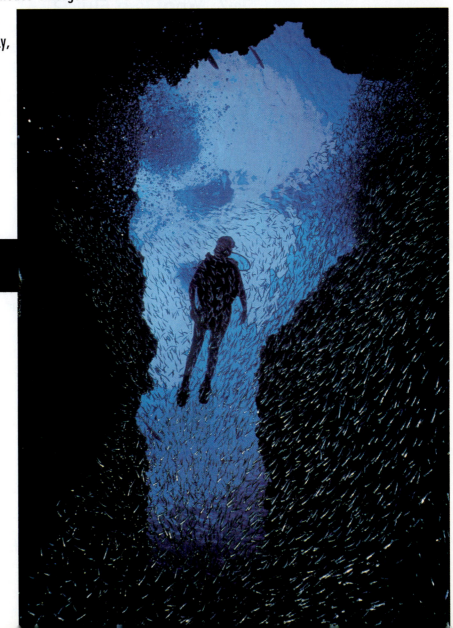

15-1   Fluids and the World Around Us

Fluids—which include both liquids and gases—play a central role in our daily lives. We breathe and drink them, and a rather vital fluid circulates in the human cardiovascular system. There are the fluid ocean and the fluid atmosphere.

In a car, there are fluids in the tires, the gas tank, the radiator, the combustion chambers of the engine, the exhaust manifold, the battery, the air-conditioning system, the windshield wiper reservoir, the lubrication system, and the hydraulic system. (*Hydraulic* means operated via a liquid.) The next time you see a large piece of earthmoving machinery, count the hydraulic cylinders that permit the machine to do its work. Large jet planes have scores of them.

We use the kinetic energy of a moving fluid in windmills, and the potential energy of another fluid in hydroelectric power plants. Given time, fluids carve the landscape. We often travel great distances just to watch fluids move. Perhaps it is time to see what physics can tell us about fluids.

15-2   What Is a Fluid?

A **fluid,** in contrast to a solid, is a substance that can flow. Fluids conform to the boundaries of any container in which we put them. They do so because a fluid cannot sustain a force that is tangential to its surface. (In the more formal language of Section 13-6, a fluid is a substance that flows because it cannot withstand a shearing stress. It can, however, exert a force in the direction perpendicular to its surface.) Some materials, such as pitch, take a long time to conform to the boundaries of a container, but they do so eventually; thus, we classify them as fluids.

You may wonder why we lump liquids and gases together and call them fluids. After all (you may say), liquid water is as different from steam as it is from ice. Actually, it is not. Ice, like other crystalline solids, has its constituent atoms organized in a fairly rigid three-dimensional array called a crystalline lattice. In neither steam nor liquid water, however, is there any such orderly long-range arrangement.

15-3   Density and Pressure

When we discuss rigid bodies, we are concerned with particular lumps of matter, such as wooden blocks, baseballs, or metal rods. Physical quantities that we find useful, and in whose terms we express Newton's laws, are *mass* and *force*. We might speak, for example, of a 3.6 kg block acted on by a 25 N force.

With fluids, we are more interested in the extended substance, and in properties that can vary from point to point in that substance. It is more useful to speak of **density** and **pressure** than of mass and force.

Density

To find the density ρ of a fluid at any point, we isolate a small volume element ΔV around that point and measure the mass Δm of the fluid contained within that element. The **density** is then

$$\rho = \frac{\Delta m}{\Delta V}. \tag{15-1}$$

In theory, the density at any point in a fluid is the limit of this ratio as the volume element ΔV at that point is made smaller and smaller. In practice, we assume that a fluid sample is large compared to atomic dimensions and thus is "smooth" (with uniform density), rather than "lumpy" with atoms. This assumption allows us to

TABLE 15-1   Some Densities

Material or Object	Density (kg/m³)
Interstellar space	10^{-20}
Best laboratory vacuum	10^{-17}
Air: 20°C and 1 atm pressure	1.21
20°C and 50 atm	60.5
Styrofoam	1×10^2
Ice	0.917×10^3
Water: 20°C and 1 atm	0.998×10^3
20°C and 50 atm	1.000×10^3
Seawater: 20°C and 1 atm	1.024×10^3
Whole blood	1.060×10^3
Iron	7.9×10^3
Mercury (the metal)	13.6×10^3
Earth: average	5.5×10^3
core	9.5×10^3
crust	2.8×10^3
Sun: average	1.4×10^3
core	1.6×10^5
White dwarf star (core)	10^{10}
Uranium nucleus	3×10^{17}
Neutron star (core)	10^{18}
Black hole (1 solar mass)	10^{19}

(a)

(b)

Fig. 15-1 (a) A fluid-filled vessel containing a small pressure sensor, shown in (b). The pressure is measured by the relative position of the movable piston in the sensor.

write Eq. 15-1 as

$$\rho = \frac{m}{V} \qquad \text{(uniform density),} \qquad (15\text{-}2)$$

where m and V are the mass and volume of the sample.

Density is a scalar property; its SI unit is the kilogram per cubic meter. Table 15-1 shows the densities of some substances and the average densities of some objects. Note that the density of a gas (see Air in the table) varies considerably with pressure, but the density of a liquid (see Water) does not; that is, gases are readily *compressible* but liquids are not.

Pressure

Let a small pressure-sensing device be suspended inside a fluid-filled vessel, as in Fig. 15-1a. The sensor (Fig. 15-1b) consists of a piston of area ΔA riding in a close-fitting cylinder and resting against a spring. A readout arrangement allows us to record the amount by which the (calibrated) spring is compressed by the surrounding fluid, thus indicating the magnitude ΔF of the force that acts normal to the piston. We define the **pressure** on the piston from the fluid as

$$p = \frac{\Delta F}{\Delta A}. \qquad (15\text{-}3)$$

In theory, the pressure at any point in the fluid is the limit of this ratio as the area ΔA of the piston, centered on that point, is made smaller and smaller. However, if the force is uniform over a flat area A, we can write Eq. 15-3 as

$$p = \frac{F}{A} \qquad \text{(pressure of uniform force on flat area),} \qquad (15\text{-}4)$$

TABLE 15-2   Some Pressures

	Pressure (Pa)
Center of the Sun	2×10^{16}
Center of Earth	4×10^{11}
Highest sustained laboratory pressure	1.5×10^{10}
Deepest ocean trench (bottom)	1.1×10^{8}
Spike heels on a dance floor	1×10^{6}
Automobile tire[a]	2×10^{5}
Atmosphere at sea level	1.0×10^{5}
Normal blood pressure[a,b]	1.6×10^{4}
Best laboratory vacuum	10^{-12}

[a]Pressure in excess of atmospheric pressure.

[b]The systolic pressure, corresponding to 120 torr on the physician's pressure gauge.

where F is the magnitude of the normal force on area A. (When we say a force is uniform over an area, we mean that it is evenly distributed over every point of the area.)

We find by experiment that at a given point in a fluid at rest, the pressure p defined by Eq. 15-3 has the same value no matter how the pressure sensor is oriented. Pressure is a scalar, having no directional properties. It is true that the force acting on the piston of our pressure sensor is a vector, but Eq. 15-3 involves only the *magnitude* of that force, a scalar quantity.

The SI unit of pressure is the newton per square meter, which is given a special name, the **pascal** (Pa). In metric countries, tire pressure gauges are calibrated in kilopascals. The pascal is related to some other common (non-SI) pressure units as follows:

$$1 \text{ atm} = 1.01 \times 10^5 \text{ Pa} = 760 \text{ torr} = 14.7 \text{ lb/in.}^2.$$

The *atmosphere* (atm) is, as the name suggests, the approximate average pressure of the atmosphere at sea level. The *torr* (named for Evangelista Torricelli, who invented the mercury barometer in 1674) was formerly called the *millimeter of mercury* (mm Hg). The pound per square inch is often abbreviated psi. Table 15-2 shows some pressures.

Sample Problem 15-1

A living room has floor dimensions of 3.5 m and 4.2 m and a height of 2.4 m.

(a) What does the air in the room weigh when the air pressure is 1.0 atm?

SOLUTION: The Key Ideas here are these: (1) The air's weight is equal to mg, where m is its mass. (2) Mass m is related to the air density ρ and the air's volume V by Eq. 15-2 ($\rho = m/V$). Putting these two ideas together and taking the density of air at 1.0 atm from Table 15-1, we find

$$mg = (\rho V)g$$
$$= (1.21 \text{ kg/m}^3)(3.5 \text{ m} \times 4.2 \text{ m} \times 2.4 \text{ m})(9.8 \text{ m/s}^2)$$
$$= 418 \text{ N} \approx 420 \text{ N}. \quad \text{(Answer)}$$

This is the weight of about 110 cans of Pepsi.

(b) What is the magnitude of the atmosphere's force on the floor of the room?

SOLUTION: The Key Idea here is that the atmosphere pushes down on the floor with a force of magnitude F that is uniform over the floor. Thus, it produces a pressure that is related to F and the flat area A of the floor by Eq. 15-4 ($p = F/A$), which gives us

$$F = pA = (1.0 \text{ atm})\left(\frac{1.01 \times 10^5 \text{ N/m}^2}{1.0 \text{ atm}}\right)(3.5 \text{ m})(4.2 \text{ m})$$

$$= 1.5 \times 10^6 \text{ N}. \quad \text{(Answer)}$$

This enormous force is equal to the weight of the column of air that covers the floor and extends all the way to the top of the atmosphere.

15-4   Fluids at Rest

Figure 15-2a shows a tank of water—or other liquid—open to the atmosphere. As every diver knows, the pressure *increases* with depth below the air–water interface. The diver's depth gauge, in fact, is a pressure sensor much like that of Fig. 15-1b. As every mountaineer knows, the pressure *decreases* with altitude as one ascends into the atmosphere. The pressures encountered by the diver and the mountaineer are usually called *hydrostatic pressures,* because they are due to fluids that are static (at rest). Here we want to find an expression for hydrostatic pressure as a function of depth or altitude.

Let us look first at the increase in pressure with depth below the water's surface. We set up a vertical y axis in the tank, with its origin at the air–water interface and the positive direction upward. We next consider a water sample contained in an imaginary right circular cylinder of horizontal base (or face) area A, such that y_1 and

(b)

Fig. 15-2 (a) A tank of water in which a sample of water is contained in an imaginary cylinder of horizontal base area A. Force $\vec{F}_1$ acts at the top surface of the cylinder; force $\vec{F}_2$ acts at the bottom surface of the cylinder; the gravitational force on the water in the cylinder is represented by $m\vec{g}$. (b) A free-body diagram of the water sample.

y_2 (both of which are *negative* numbers) are the depths below the surface of the upper and lower cylinder faces, respectively.

Figure 15-2b shows a free-body diagram for the water in the cylinder. The water is in *static equilibrium;* that is, it is stationary and the forces on it balance. Three forces act on it vertically: Force $\vec{F}_1$ acts at the top surface of the cylinder and is due to the water above the cylinder. Similarly, force $\vec{F}_2$ acts at the bottom surface of the cylinder and is due to the water below the cylinder. The gravitational force on the water in the cylinder is represented by $m\vec{g}$, where m is the mass of the water in the cylinder. The balance of these forces is written as

$$F_2 = F_1 + mg. \tag{15-5}$$

We want to transform Eq. 15-5 into an equation involving pressures. From Eq. 15-4, we know that

$$F_1 = p_1 A \quad \text{and} \quad F_2 = p_2 A. \tag{15-6}$$

The mass m of the water in the cylinder is, from Eq. 15-2, $m = \rho V$, where the cylinder's volume V is the product of its face area A and its height $y_1 - y_2$. Thus, m is equal to $\rho A(y_1 - y_2)$. Substituting this and Eq. 15-6 into Eq. 15-5, we find

$$p_2 A = p_1 A + \rho A g(y_1 - y_2)$$

or

$$p_2 = p_1 + \rho g(y_1 - y_2). \tag{15-7}$$

This equation can be used to find pressure both in a liquid (as a function of depth) and in the atmosphere (as a function of altitude or height). For the former, suppose we seek the pressure p at a depth h below the liquid surface. Then we choose level 1 to be the surface, level 2 to be a distance h below it (as in Fig. 15-3), and p_0 to represent the atmospheric pressure on the surface. We then substitute

$$y_1 = 0, \quad p_1 = p_0 \quad \text{and} \quad y_2 = -h, \quad p_2 = p$$

into Eq. 15-7, which becomes

$$p = p_0 + \rho g h \qquad \text{(pressure at depth } h\text{).} \tag{15-8}$$

Note that the pressure at a given depth in the liquid depends on that depth but not on any horizontal dimension.

▶ The pressure at a point in a fluid in static equilibrium depends on the depth of that point but not on any horizontal dimension of the fluid or its container.

Thus, Eq. 15-8 holds no matter what the shape of the container. If the bottom surface of the container is at depth h, then Eq. 15-8 gives the pressure p there.

In Eq. 15-8, p is said to be the total pressure or **absolute pressure** at level 2. To see why, note in Fig. 15-3 that the pressure p at level 2 consists of two contributions: (1) p_0, the pressure due to the atmosphere, which bears down on the liquid, and (2) $\rho g h$, the pressure due to the liquid above level 2, which bears down on level 2. In general, the difference between an absolute pressure and an atmospheric pressure is called the **gauge pressure.** (The name comes from the use of a gauge to measure this difference in pressures.) For the situation of Fig. 15-3, the gauge pressure is $\rho g h$.

Equation 15-7 also holds above the liquid surface: It gives the atmospheric pressure at a given distance above level 1 in terms of the atmospheric pressure p_1 at level 1 (*assuming* that the atmospheric density is uniform over that distance). For example, to find the atmospheric pressure at a distance d above level 1 in

Fig. 15-3 The pressure p increases with depth h below the liquid surface according to Eq. 15-8.

Fig. 15-3, we substitute

$$y_1 = 0, \quad p_1 = p_0 \quad \text{and} \quad y_2 = d, \quad p_2 = p.$$

Then with $\rho = \rho_{\text{air}}$, we obtain

$$p = p_0 - \rho_{\text{air}} g d.$$

✔**CHECKPOINT 1:**
The figure shows four containers of olive oil. Rank them according to the pressure at depth h, greatest first.

(a)   (b)   (c)   (d)

Sample Problem 15-2

A novice scuba diver practicing in a swimming pool takes enough air from his tank to fully expand his lungs before abandoning the tank at depth L and swimming to the surface. He ignores instructions and fails to exhale during his ascent. When he reaches the surface, the difference between the external pressure on him and the air pressure in his lungs is 9.3 kPa. From what depth does he start? What potentially lethal danger does he face?

SOLUTION: The **Key Idea** here is that when the diver fills his lungs at depth L, the external pressure on him (and thus the air pressure within his lungs) is greater than normal and given by Eq. 15-8 as

$$p = p_0 + \rho g L,$$

where p_0 is atmospheric pressure and ρ is the water's density (998 kg/m^3, from Table 15-1). As he ascends, the external pressure on him decreases, until it is atmospheric pressure p_0 at the surface. His blood pressure also decreases, until it is normal. However, because he does not exhale, the air pressure in his lungs remains

at the value it had at depth L. At the surface, the pressure difference between the higher pressure in his lungs and the lower pressure on his chest is

$$\Delta p = p - p_0 = \rho g L,$$

from which we find

$$L = \frac{\Delta p}{\rho g} = \frac{9300 \text{ Pa}}{(998 \text{ kg/m}^3)(9.8 \text{ m/s}^2)}$$
$$= 0.95 \text{ m.} \qquad \text{(Answer)}$$

This is not deep! Yet, the pressure difference of 9.3 kPa (about 9% of atmospheric pressure) is sufficient to rupture the diver's lungs and force air from them into the depressurized blood, which then carries the air to the heart, killing the diver. If the diver follows instructions and gradually exhales as he ascends, he allows the pressure in his lungs to equalize with the external pressure, and then there is no danger.

Sample Problem 15-3

The U-tube in Fig. 15-4 contains two liquids in static equilibrium: Water of density ρ_w (= 998 kg/m^3) is in the right arm, and oil of unknown density ρ_x is in the left. Measurement gives $l = 135$ mm and $d = 12.3$ mm. What is the density of the oil?

SOLUTION: One **Key Idea** here is that the pressure p_{int} at the oil–water interface in the left arm depends on the density ρ_x and height of the oil above the interface. A second **Key Idea** is that the water in the right arm *at the same level* must be at the same pressure p_{int}. The reason is that, because the water is in static equilibrium, pressures at points in the water at the same level must be the same even if the points are separated horizontally.

In the right arm, the interface is a distance l below the free surface of the *water* and we have, from Eq. 15-8,

$$p_{\text{int}} = p_0 + \rho_w g l \qquad \text{(right arm)}.$$

In the left arm, the interface is a distance $l + d$ below the free surface of the *oil* and we have, again from Eq. 15-8,

$$p_{\text{int}} = p_0 + \rho_x g(l + d) \qquad \text{(left arm)}.$$

Equating these two expressions and solving for the unknown density yield

$$\rho_x = \rho_w \frac{l}{l + d} = (998 \text{ kg/m}^3) \frac{135 \text{ mm}}{135 \text{ mm} + 12.3 \text{ mm}}$$
$$= 915 \text{ kg/m}^3. \qquad \text{(Answer)}$$

Note that the answer does not depend on the atmospheric pressure p_0 or the free-fall acceleration g.

Fig. 15-4 Sample Problem 15-3. The oil in the left arm stands higher than the water in the right arm because the oil is less dense than the water. Both fluid columns produce the same pressure p_{int} at the level of the interface.

15-5 Measuring Pressure

The Mercury Barometer

Figure 15-5*a* shows a very basic *mercury barometer,* a device used to measure the pressure of the atmosphere. The long glass tube is filled with mercury and inverted with its open end in a dish of mercury, as the figure shows. The space above the mercury column contains only mercury vapor, whose pressure is so small at ordinary temperatures that it can be neglected.

We can use Eq. 15-7 to find the atmospheric pressure p_0 in terms of the height h of the mercury column. We choose level 1 of Fig. 15-2 to be that of the air–mercury interface and level 2 to be that of the top of the mercury column, as labeled in Fig. 15-5*a*. We then substitute

$$y_1 = 0, \quad p_1 = p_0 \quad \text{and} \quad y_2 = h, \quad p_2 = 0$$

into Eq. 15-7, finding that

$$p_0 = \rho g h, \tag{15-9}$$

where ρ is the density of the mercury.

For a given pressure, the height h of the mercury column does not depend on the cross-sectional area of the vertical tube. The fanciful mercury barometer of Fig. 15-5*b* gives the same reading as that of Fig. 15-5*a*; all that counts is the vertical distance h between the mercury levels.

Equation 15-9 shows that, for a given pressure, the height of the column of mercury depends on the value of g at the location of the barometer and on the density of mercury, which varies with temperature. The column height (in millimeters) is numerically equal to the pressure (in torr) *only* if the barometer is at a place where g has its accepted standard value of 9.80665 m/s² *and* the temperature of the mercury is 0°C. If these conditions do not prevail (and they rarely do), small corrections must be made before the height of the mercury column can be transformed into a pressure.

The Open-Tube Manometer

An *open-tube manometer* (Fig. 15-6) measures the gauge pressure p_g of a gas. It consists of a U-tube containing a liquid, with one end of the tube connected to the

(a)　　　　(b)

Fig. 15-5 (*a*) A mercury barometer. (*b*) Another mercury barometer. The distance h is the same in both cases.

Fig. 15-6 An open-tube manometer, connected to measure the gauge pressure of the gas in the tank on the left. The right arm of the U-tube is open to the atmosphere.

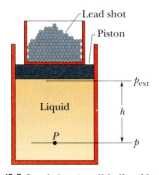

Fig. 15-7 Lead shot (small balls of lead) loaded onto the piston create a pressure p_{ext} at the top of the enclosed (incompressible) liquid. If p_{ext} is increased, by adding more lead shot, the pressure increases by the same amount at all points within the liquid.

vessel whose gauge pressure we wish to measure and the other end open to the atmosphere. We can use Eq. 15-7 to find the gauge pressure in terms of the height h shown in Fig. 15-6. Let us choose levels 1 and 2 as shown in Fig. 15-6. We then substitute

$$y_1 = 0, \quad p_1 = p_0 \quad \text{and} \quad y_2 = -h, \quad p_2 = p$$

into Eq. 15-7, finding that

$$p_g = p - p_0 = \rho g h, \tag{15-10}$$

where ρ is the density of the liquid in the tube. The gauge pressure p_g is directly proportional to h.

The gauge pressure can be positive or negative, depending on whether $p > p_0$ or $p < p_0$. In inflated tires or the human circulatory system, the (absolute) pressure is greater than atmospheric pressure, so the gauge pressure is a positive quantity, sometimes called the *overpressure*. If you suck on a straw to pull fluid up the straw, the (absolute) pressure in your lungs is actually less than atmospheric pressure. The gauge pressure in your lungs is then a negative quantity.

15-6   Pascal's Principle

When you squeeze one end of a tube to get toothpaste out the other end, you are watching **Pascal's principle** in action. This principle is also the basis for the Heimlich maneuver, in which a sharp pressure increase properly applied to the abdomen is transmitted to the throat, forcefully ejecting food lodged there. The principle was first stated clearly in 1652 by Blaise Pascal (for whom the unit of pressure is named):

> A change in the pressure applied to an enclosed incompressible fluid is transmitted undiminished to every portion of the fluid and to the walls of its container.

Demonstrating Pascal's Principle

Consider the case in which the incompressible fluid is a liquid contained in a tall cylinder, as in Fig. 15-7. The cylinder is fitted with a piston on which a container of lead shot rests. The atmosphere, container, and shot put pressure p_{ext} on the piston and thus on the liquid. The pressure p at any point P in the liquid is then

$$p = p_{ext} + \rho g h. \tag{15-11}$$

Let us add a little more lead shot to the container to increase p_{ext} by an amount Δp_{ext}. The quantities ρ, g, and h in Eq. 15-11 are unchanged, so the pressure change at P is

$$\Delta p = \Delta p_{ext}. \tag{15-12}$$

This pressure change is independent of h, so it must hold for all points within the liquid, as Pascal's principle states.

Pascal's Principle and the Hydraulic Lever

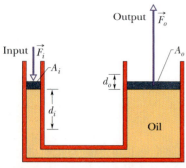

Fig. 15-8 A hydraulic arrangement that can be used to magnify a force $\vec{F}_i$. The work done is, however, not magnified and is the same for both the input and output forces.

Figure 15-8 shows how Pascal's principle can be made the basis of a hydraulic lever. In operation, let an external force of magnitude F_i be directed downward on the left-hand (or input) piston, whose area is A_i. An incompressible liquid in the device then produces an upward force of magnitude F_o on the righthand (or output) piston, whose area is A_o. To keep the system in equilibrium, there must be a downward force of magnitude F_o on the output piston from an external load (not shown). The force $\vec{F}_i$ applied on the left and the downward force $\vec{F}_o$ from the load on the right produce a change Δp in the pressure of the liquid that is given by

$$\Delta p = \frac{F_i}{A_i} = \frac{F_o}{A_o},$$

so

$$F_o = F_i \frac{A_o}{A_i}. \tag{15-13}$$

Equation 15-13 shows that the output force F_o on the load must be greater than the input force F_i if $A_o > A_i$, as is the case in Fig. 15-8.

If we move the input piston downward a distance d_i, the output piston moves upward a distance d_o, such that the same volume V of the incompressible liquid is displaced at both pistons. Then

$$V = A_i d_i = A_o d_o,$$

which we can write as

$$d_o = d_i \frac{A_i}{A_o}. \tag{15-14}$$

This shows that, if $A_o > A_i$ (as in Fig. 15-8), the output piston moves a smaller distance than the input piston moves.

From Eqs. 15-13 and 15-14 we can write the output work as

$$W = F_o d_o = \left(F_i \frac{A_o}{A_i} \right) \left(d_i \frac{A_i}{A_o} \right) = F_i d_i, \tag{15-15}$$

which shows that the work W done *on* the input piston by the applied force is equal to the work W done *by* the output piston in lifting the load placed on it.

The advantage of a hydraulic lever is this:

> ► With a hydraulic lever, a given force applied over a given distance can be transformed to a greater force applied over a smaller distance.

The product of force and distance remains unchanged so that the same work is done. However, there is often tremendous advantage in being able to exert the larger force. Most of us, for example, cannot lift an automobile directly but can with a hydraulic jack, even though we have to pump the handle farther than the automobile rises. In this device, the displacement d_i is accomplished not in a single stroke but over a series of small strokes.

15-7 Archimedes' Principle

Figure 15-9 shows a student in a swimming pool, manipulating a very thin plastic sack (of negligible mass) that is filled with water. She finds that the sack and its contained water are in static equilibrium, tending neither to rise nor to sink. The downward gravitational force $\vec{F}_g$ on the contained water must be balanced by a net upward force from the water surrounding the sack.

This net upward force is a **buoyant force** $\vec{F}_b$. It exists because the pressure in the surrounding water increases with depth below the surface. Thus, the pressure near the bottom of the sack is greater than the pressure near the top. Then the forces on the sack due to this pressure are greater in magnitude near the bottom of the sack than near the top. Some of the forces are represented in Fig. 15-10a, where the space occupied by the sack has been left empty. Note that the force vectors drawn near the bottom of that space (with upward components) have longer lengths than those drawn near the top of the sack (with downward components). If we vectorially add all the forces on the sack from the water, the horizontal components cancel and the

Fig. 15-9 A thin-walled plastic sack of water is in static equilibrium in the pool. The gravitational force on it must be balanced by a net upward force on the sack from the surrounding water.

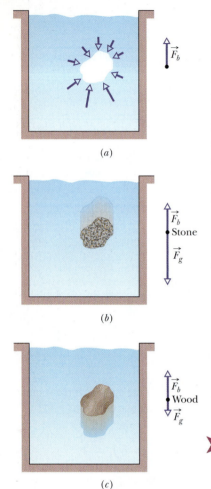

(a)

(b)

(c)

Fig. 15-10 (a) The water surrounding the hole in the water produces a net upward buoyant force on whatever fills the hole. (b) For a stone of the same volume as the hole, the gravitational force exceeds the buoyant force in magnitude. (c) For a lump of wood of the same volume, the gravitational force is less than the buoyant force in magnitude.

vertical components add to yield the upward buoyant force $\vec{F}_b$ on the sack. (Force $\vec{F}_b$ is shown to the right of the pool in Fig. 15-10a.)

Because the sack of water is in static equilibrium, the magnitude of $\vec{F}_b$ is equal to the magnitude $m_f g$ of the gravitational force $\vec{F}_g$ on the sack of water: $F_b = m_f g$. (Subscript f refers to *fluid*, here the water.) In words, the magnitude of the buoyant force is equal to the weight of the water in the sack.

In Fig. 15-10b, we have replaced the sack of water with a stone that exactly fills the hole in Fig. 15-10a. The stone is said to *displace* the water, meaning that it occupies space that would otherwise be occupied by water. We have changed nothing about the shape of the hole, so the forces at the hole's surface must be the same as when the water-filled sack was in place. Thus, the same upward buoyant force that acted on the water-filled sack now acts on the stone; that is, the magnitude F_b of the buoyant force is equal to $m_f g$, the weight of the water displaced by the stone.

Unlike the water-filled sack, the stone is not in static equilibrium. The downward gravitational force $\vec{F}_g$ on the stone is greater in magnitude than the upward buoyant force, as is shown in the free-body diagram to the right of the pool in Fig. 15-10b. The stone thus accelerates downward, sinking to the bottom of the pool.

Let us next exactly fill the hole in Fig. 15-10a with a block of light-weight wood, as in Fig. 15-10c. Again, nothing has changed about the forces at the hole's surface, so the magnitude F_b of the buoyant force is still equal to $m_f g$, the weight of the displaced water. Like the stone, the block is not in static equilibrium. However, this time the gravitational force $\vec{F}_g$ is lesser in magnitude than the buoyant force (as shown to the right of the pool), and so the block accelerates upward, rising to the top surface of the water.

Our results with the sack, stone, and block apply to all fluids and are summarized in **Archimedes' principle:**

> ▶ When a body is fully or partially submerged in a fluid, a buoyant force $\vec{F}_b$ from the surrounding fluid acts on the body. The force is directed upward and has a magnitude equal to the weight $m_f g$ of the fluid that has been displaced by the body.

The buoyant force on a body in a fluid has the magnitude

$$F_b = m_f g \qquad \text{(buoyant force),} \qquad (15\text{-}16)$$

where m_f is the mass of the fluid that is displaced by the body.

In the late evening of August 21, 1986, something (possibly a volcanic tremor) disturbed Cameroon's Lake Nyos, which has a high concentration of dissolved carbon dioxide. The disturbance caused that gas to form bubbles. Being lighter than the surrounding fluid (the water), those bubbles were buoyed to the surface, where they released the carbon dioxide. The gas, being heavier than the surrounding fluid (now the air), rushed down the mountainside like a river, asphyxiating 1700 persons and the scores of animals seen here.

Floating

When we release a block of light-weight wood just above the water in a pool, it moves into the water because the gravitational force on it pulls it downward. As the block displaces more and more water, the magnitude F_b of the upward buoyant force acting on it increases. Eventually, F_b is large enough to equal the magnitude F_g of the downward gravitational force on the block, and the block comes to rest. The block is then in static equilibrium and is said to be *floating* in the water. In general,

> ➤ When a body floats in a fluid, the magnitude F_b of the buoyant force on the body is equal to the magnitude F_g of the gravitational force on the body.

We can write this statement as

$$F_b = F_g \qquad \text{(floating).} \tag{15-17}$$

From Eq. 15-16, we know that $F_b = m_f g$. Thus,

> ➤ When a body floats in a fluid, the magnitude F_g of the gravitational force on the body is equal to the weight $m_f g$ of the fluid that has been displaced by the body.

We can write this statement as

$$F_g = m_f g \qquad \text{(floating).} \tag{15-18}$$

In other words, a floating body displaces its own weight of fluid.

Apparent Weight in a Fluid

If we place a stone on a scale that is calibrated to measure weight, then the reading on the scale is the stone's weight. However, if we do this underwater, the upward buoyant force on the stone from the water decreases the reading. That reading is then an apparent weight. In general, an apparent weight is related to the actual weight of a body and the buoyant force on the body by

$$\begin{pmatrix} \text{apparent} \\ \text{weight} \end{pmatrix} = \begin{pmatrix} \text{actual} \\ \text{weight} \end{pmatrix} - \begin{pmatrix} \text{magnitude of} \\ \text{buoyant force} \end{pmatrix},$$

which we can write as

$$\text{weight}_{\text{app}} = \text{weight} - F_b \qquad \text{(apparent weight).} \tag{15-19}$$

If, in some strange test of strength, you had to lift a heavy stone, you could do it more easily with the stone underwater. Then your applied force would need to exceed only the stone's apparent weight, not its larger actual weight, because the upward buoyant force would help you lift the stone.

The magnitude of the buoyant force on a floating body is equal to the body's weight. Equation 15-19 thus tells us that a floating body has an apparent weight of zero—the body would produce a reading of zero on a scale. (When astronauts prepare to perform a complex task in space, they practice the task floating underwater, where their apparent weight is zero as it is in space.)

✔**CHECKPOINT 2:**   A penguin floats first in a fluid of density ρ_0, then in a fluid of density $0.95\rho_0$, and then in a fluid of density $1.1\rho_0$. (a) Rank the densities according to the magnitude of the buoyant force on the penguin, greatest first. (b) Rank the densities according to the amount of fluid displaced by the penguin, greatest first.

Sample Problem 15-4

What fraction of the volume of an iceberg floating in seawater is visible?

SOLUTION: Let V_i be the total volume of the iceberg. The nonvisible portion is below water and thus is equal to the volume V_f of the fluid (the seawater) displaced by the iceberg. We seek the fraction (call it frac)

$$\text{frac} = \frac{V_i - V_f}{V_i} = 1 - \frac{V_f}{V_i}, \tag{15-20}$$

but we know neither volume. A **Key Idea** here is that, because the iceberg is floating, Eq. 15-18 ($F_g = m_f g$) applies. We can write that equation as

$$m_i g = m_f g,$$

from which we see that $m_i = m_f$. Thus, the mass of the iceberg is equal to the mass of the displaced fluid (seawater). Although we know neither mass, we can relate them to the densities of ice and seawater given in Table 15-1 by using Eq. 15-2 ($\rho = m/V$). Because $m_i = m_f$, we can write

$$\rho_i V_i = \rho_f V_f$$

or

$$\frac{V_f}{V_i} = \frac{\rho_i}{\rho_f}.$$

Substituting this into Eq. 15-20 and then using the known densities, we find

$$\text{frac} = 1 - \frac{\rho_i}{\rho_f} = 1 - \frac{917 \text{ kg/m}^3}{1024 \text{ kg/m}^3}$$

$$= 0.10 \text{ or } 10\%. \qquad\text{(Answer)}$$

Sample Problem 15-5

A spherical, helium-filled balloon has a radius R of 12.0 m. The balloon, support cables, and basket have a mass m of 196 kg. What maximum load M can the balloon support while it floats at an altitude at which the helium density ρ_{He} is 0.160 kg/m³ and the air density ρ_{air} is 1.25 kg/m³? Assume that the volume of air displaced by the load, support cables, and basket is negligible.

SOLUTION: The **Key Idea** here is that the balloon, cables, basket, load, *and* the contained helium form a floating body, with a total mass of $m + M + m_{\text{He}}$, where m_{He} is the mass of the contained helium. Then the magnitude of the total gravitational force on this body must be equal to the weight of the air displaced by the body (the air is the fluid in which this body floats). Let m_{air} be the mass of the air displaced by the body. From Eq. 15-18 ($F_g = m_f g$), we have

$$(m + M + m_{\text{He}})g = m_{\text{air}}g$$

or

$$M = m_{\text{air}} - m_{\text{He}} - m. \tag{15-21}$$

We do not know m_{He} and m_{air} but we do know the corresponding densities, so we can use Eq. 15-2 ($\rho = m/V$) to rewrite Eq. 15-21 in terms of those densities. First we note that, because the load, support cables, and basket displace a negligible amount of air, the volume of the displaced air is equal to the volume V ($= \frac{4}{3}\pi R^3$) of the spherical balloon. Then Eq. 15-21 becomes

$$M = \rho_{\text{air}}V - \rho_{\text{He}}V - m$$
$$= (\tfrac{4}{3}\pi R^3)(\rho_{\text{air}} - \rho_{\text{He}}) - m$$
$$= (\tfrac{4}{3}\pi)(12.0 \text{ m})^3(1.25 \text{ kg/m}^3 - 0.160 \text{ kg/m}^3) - 196 \text{ kg}$$
$$= 7694 \text{ kg} \approx 7690 \text{ kg}. \qquad\text{(Answer)}$$

15-8   Ideal Fluids in Motion

The motion of *real fluids* is very complicated and not yet fully understood. Instead, we shall discuss the motion of an **ideal fluid,** which is simpler to handle mathematically and yet provides useful results. Here are four assumptions that we make about our ideal fluid; they all are concerned with *flow:*

1. **Steady flow** In steady (or *laminar*) flow, the velocity of the moving fluid at any fixed point does not change with time, either in magnitude or in direction. The gentle flow of water near the center of a quiet stream is steady; that in a chain of rapids is not. Figure 15-11 shows a transition from steady flow to *nonsteady* (or *turbulent*) flow for a rising stream of smoke. The speed of the smoke particles increases as they rise and, at a certain critical speed, the flow changes from steady to nonsteady (that is, from laminar to *nonlaminar* flow).

2. **Incompressible flow** We assume, as we have already done for fluids at rest, that our ideal fluid is incompressible; that is, its density has a constant, uniform value.

3. **Nonviscous flow** Roughly speaking, the viscosity of a fluid is a measure of how resistive the fluid is to flow. For example, thick honey is more resistive to flow than water, and so honey is said to be more viscous than water. Viscosity is the

Fig. 15-11 At a certain point, the rising flow of smoke and heated gas changes from steady to turbulent.

Fig. 15-12 The steady flow of a fluid around a cylinder, as revealed by a dye tracer that was injected into the fluid upstream of the cylinder.

Fig. 15-13 Smoke reveals streamlines in airflow past a car in a wind-tunnel test.

Fig. 15-14 A fluid element P traces out a streamline as it moves. The velocity vector of the element is tangent to the streamline at every point.

fluid analog of friction between solids; both are mechanisms by which the kinetic energy of moving objects can be transferred to thermal energy. In the absence of friction, a block could glide at constant speed along a horizontal surface. In the same way, an object moving through a nonviscous fluid would experience no *viscous drag force*—that is, no resistive force due to viscosity; it could move at constant speed through the fluid. The British scientist Lord Rayleigh noted that in an ideal fluid a ship's propeller would not work but, on the other hand, a ship (once set into motion) would not need a propeller!

4. *Irrotational flow* Although it need not concern us further, we also assume that the flow is *irrotational*. To test for this property, let a tiny grain of dust move with the fluid. Although this test body may (or may not) move in a circular path, in irrotational flow the test body will not rotate about an axis through its own center of mass. For a loose analogy, the motion of a Ferris wheel is rotational; that of its passengers is irrotational.

We can make the flow of a fluid visible by adding a *tracer*. This might be a dye injected into many points across a liquid stream (Fig. 15-12) or smoke particles added to a gas flow (Figs. 15-11 and 15-13). Each bit of a tracer follows a *streamline*, which is the path that a tiny element of the fluid would take as the fluid flows. Recall from Chapter 2 that the velocity of a particle is always tangent to the path taken by the particle. Here the particle is the fluid element, and its velocity $\vec{v}$ is always tangent to a streamline (Fig. 15-14). For this reason, two streamlines can never intersect; if they did, then an element arriving at their intersection would have two different velocities simultaneously, an impossibility.

15-9 The Equation of Continuity

You may have noticed that you can increase the speed of the water emerging from a garden hose by partially closing the hose opening with your thumb. Apparently the speed v of the water depends on the cross-sectional area A through which the water flows.

Here we wish to derive an expression that relates v and A for the steady flow of an ideal fluid through a tube with varying cross section, like that in Fig. 15-15. The flow there is toward the right, and the tube segment shown (part of a longer

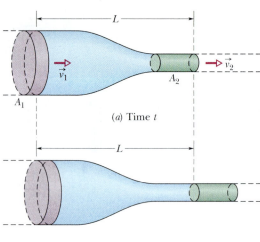

Fig. 15-15 Fluid flows from left to right at a steady rate through a tube segment of length L. The fluid's speed is v_1 at the left side and v_2 at the right side. The tube's cross-sectional area is A_1 at the left side and A_2 at the right side. From time t in (a) to time $t + \Delta t$ in (b), the amount of fluid shown in purple enters at the left side and the equal amount of fluid shown in green emerges at the right side.

(a) Time t

(b) Time $t + \Delta t$

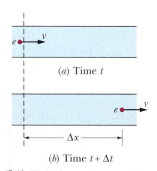

(a) Time t

(b) Time $t + \Delta t$

Fig. 15-16 Fluid flows at a constant speed v through a tube. (a) At time t, fluid element e is about to pass the dashed line. (b) At time $t + \Delta t$, element e is a distance $\Delta x = v\,\Delta t$ from the dashed line.

tube) has length L. The fluid has speeds v_1 at the left end of the segment and v_2 at the right end. The tube has cross-sectional areas A_1 at the left end and A_2 at the right end. Suppose that in a time interval Δt a volume ΔV of fluid enters the tube segment at its left end (that volume is colored purple in Fig. 15-15a). Then, because the fluid is incompressible, an identical volume ΔV must emerge from the right end of the segment (it is colored green in Fig. 15-15b).

We can use this common volume ΔV to relate the speeds and areas. To do so, we first consider Fig. 15-16, which shows a side view of a tube of *uniform* cross-sectional area A. In Fig. 15-16a, a fluid element e is about to pass through the dashed line drawn across the tube width. The element's speed is v, so during a time interval Δt, the element moves along the tube a distance $\Delta x = v\,\Delta t$. The volume ΔV of fluid that has passed through the dashed line in that time interval Δt is

$$\Delta V = A\,\Delta x = Av\,\Delta t. \tag{15-22}$$

Applying Eq. 15-22 to both the left and right ends of the tube segment in Fig. 15-15, we have

$$\Delta V = A_1 v_1\,\Delta t = A_2 v_2\,\Delta t$$

or $$A_1 v_1 = A_2 v_2 \qquad \text{(equation of continuity).} \tag{15-23}$$

This relation between speed and cross-sectional area is called the **equation of continuity** for the flow of an ideal fluid. It tells us that the flow speed increases when we decrease the cross-sectional area through which the fluid flows (as when we partially close off a garden hose with a thumb).

Equation 15-23 applies not only to an actual tube but also to any so-called *tube of flow*, or imaginary tube whose boundary consists of streamlines. Such a tube acts like a real tube because no fluid element can cross a streamline; thus, all the fluid within a tube of flow must remain within its boundary. Figure 15-17 shows a tube of flow in which the cross-sectional area increases from area A_1 to area A_2 along the flow direction. From Eq. 15-23 we know that, with the increase in area, the speed must decrease, as is indicated by the greater spacing between streamlines at the right in Fig. 15-17. Similarly, you can see that in Fig. 15-12 the speed of the flow is greatest just above and just below the cylinder.

We can rewrite Eq. 15-23 as

$$R_V = Av = \text{a constant} \qquad \text{(volume flow rate, equation of continuity),} \tag{15-24}$$

Fig. 15-17 A tube of flow is defined by the streamlines that form the boundary. The volume flow rate must be the same for all cross sections of the tube of flow.

in which R_V is the **volume flow rate** of the fluid (volume per unit time). Its SI unit is the cubic meter per second (m^3/s). If the density ρ of the fluid is uniform, we can multiply Eq. 15-24 by that density to get the **mass flow rate** R_m (mass per unit time):

$$R_m = \rho R_V = \rho A v = \text{a constant} \qquad \text{(mass flow rate).} \qquad (15\text{-}25)$$

The SI unit of mass flow rate is the kilogram per second (kg/s). Equation 15-25 says that the mass that flows into the tube segment of Fig. 15-15 each second must be equal to the mass that flows out of that segment each second.

✔**CHECKPOINT 3:**   The figure shows a pipe and gives the volume flow rate (in cm^3/s) and the direction of flow for all but one section. What are the volume flow rate and the direction of flow for that section?

Sample Problem 15-6

The cross-sectional area A_0 of the aorta (the major blood vessel emerging from the heart) of a normal resting person is 3 cm^2, and the speed v_0 of the blood through it is 30 cm/s. A typical capillary (diameter $\approx$ 6 μm) has a cross-sectional area A of 3×10^{-7} cm^2 and a flow speed v of 0.05 cm/s. How many capillaries does such a person have?

SOLUTION: The Key Idea here is that all the blood that passes through the capillaries must have passed through the aorta. Therefore, the volume flow rate through the aorta must equal the total volume flow rate through the capillaries. Let us assume that the capillaries are identical, with the given cross-sectional area A and flow speed v. Then, from Eq. 15-24 we have

$$A_0 v_0 = nAv,$$

where n is the number of capillaries. Solving for n yields

$$n = \frac{A_0 v_0}{Av} = \frac{(3\ cm^2)(30\ cm/s)}{(3 \times 10^{-7}\ cm^2)(0.05\ cm/s)}$$

$$= 6 \times 10^9 \text{ or 6 billion.} \qquad \text{(Answer)}$$

You can easily show that the combined cross-sectional area of the capillaries is about 600 times the cross-sectional area of the aorta.

Sample Problem 15-7

Figure 15-18 shows how the stream of water emerging from a faucet "necks down" as it falls. The indicated cross-sectional areas are $A_0 = 1.2\ cm^2$ and $A = 0.35\ cm^2$. The two levels are separated by a vertical distance $h = 45$ mm. What is the volume flow rate from the tap?

SOLUTION: The Key Idea here is simply that the volume flow rate through the higher cross section must be the same as that through the lower cross section. Thus, from Eq. 15-24, we have

$$A_0 v_0 = Av, \qquad (15\text{-}26)$$

where v_0 and v are the water speeds at the levels corresponding to A_0 and A. From Eq. 2-16 we can also write, because the water is falling freely with acceleration g,

$$v^2 = v_0^2 + 2gh. \qquad (15\text{-}27)$$

Eliminating v between Eqs. 15-26 and 15-27 and solving for v_0, we obtain

$$v_0 = \sqrt{\frac{2ghA^2}{A_0^2 - A^2}}$$

$$= \sqrt{\frac{(2)(9.8\ m/s^2)(0.045\ m)(0.35\ cm^2)^2}{(1.2\ cm^2)^2 - (0.35\ cm^2)^2}}$$

$$= 0.286\ m/s = 28.6\ cm/s.$$

From Eq. 15-24, the volume flow rate R_V is then

$$R_V = A_0 v_0 = (1.2\ cm^2)(28.6\ cm/s)$$

$$= 34\ cm^3/s. \qquad \text{(Answer)}$$

Fig. 15-18 Sample Problem 15-7. As water falls from a tap, its speed increases. Because the flow rate must be the same at all cross sections, the stream must "neck down."

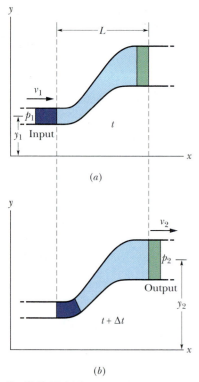

Fig. 15-19 Fluid flows at a steady rate through a length L of a tube, from the input end at the left to the output end at the right. From time t in (a) to time $t + \Delta t$ in (b), the amount of fluid shown in purple enters the input end and the equal amount shown in green emerges from the output end.

15-10 Bernoulli's Equation

Figure 15-19 represents a tube through which an ideal fluid is flowing at a steady rate. In a time interval Δt, suppose that a volume of fluid ΔV, colored purple in Fig. 15-19a, enters the tube at the left (or input) end and an identical volume, colored green in Fig. 15-19b, emerges at the right (or output) end. The emerging volume must be the same as the entering volume because the fluid is incompressible, with an assumed constant density ρ.

Let y_1, v_1, and p_1 be the elevation, speed, and pressure of the fluid entering at the left, and y_2, v_2, and p_2 be the corresponding quantities for the fluid emerging at the right. By applying the principle of conservation of energy to the fluid, we shall show that these quantities are related by

$$p_1 + \tfrac{1}{2}\rho v_1^2 + \rho g y_1 = p_2 + \tfrac{1}{2}\rho v_2^2 + \rho g y_2. \qquad (15\text{-}28)$$

We can also write this equation as

$$p + \tfrac{1}{2}\rho v^2 + \rho g y = \text{a constant} \qquad \text{(Bernoulli's equation).} \qquad (15\text{-}29)$$

Equations 15-28 and 15-29 are equivalent forms of **Bernoulli's equation,** after Daniel Bernoulli, who studied fluid flow in the 1700s.* Like the equation of continuity (Eq. 15-24), Bernoulli's equation is not a new principle but simply the reformulation of a familiar principle in a form more suitable to fluid mechanics. As a check, let us apply Bernoulli's equation to fluids at rest, by putting $v_1 = v_2 = 0$ in Eq. 15-28. The result is

$$p_2 = p_1 + \rho g(y_1 - y_2),$$

which is Eq. 15-7 with a slight change in notation.

A major prediction of Bernoulli's equation emerges if we take y to be a constant ($y = 0$, say) so that the fluid does not change elevation as it flows. Equation 15-28 then becomes

$$p_1 + \tfrac{1}{2}\rho v_1^2 = p_2 + \tfrac{1}{2}\rho v_2^2, \qquad (15\text{-}30)$$

which tells us that:

▶ If the speed of a fluid element increases as it travels along a horizontal streamline, the pressure of the fluid must decrease, and conversely.

Put another way, where the streamlines are relatively close together (that is, where the velocity is relatively great), the pressure is relatively low, and conversely.

The link between a change in speed and a change in pressure makes sense if you consider a fluid element. When the element nears a narrow region, the higher pressure behind it accelerates it so that it then has a greater speed in the narrow region. When it nears a wide region, the higher pressure ahead of it decelerates it so that it then has a lesser speed in the wide region.

Bernoulli's equation is strictly valid only to the extent that the fluid is ideal. If viscous forces are present, thermal energy will be involved. We take no account of this in the derivation that follows.

Proof of Bernoulli's Equation

Let us take as our system the entire volume of the (ideal) fluid shown in Fig. 15-19. We shall apply the principle of conservation of energy to this system as it moves

---

*For irrotational flow (which we assume), the constant in Eq. 15-29 has the same value for all points within the tube of flow; the points do not have to lie along the same streamline. Similarly, the points 1 and 2 in Eq. 15-28 can lie anywhere within the tube of flow.

from its initial state (Fig. 15-19a) to its final state (Fig. 15-19b). The fluid lying between the two vertical planes separated by a distance L in Fig. 15-19 does not change its properties during this process; we need be concerned only with changes that take place at the input and output ends.

We apply energy conservation in the form of the work–kinetic energy theorem,

$$W = \Delta K, \tag{15-31}$$

which tells us that the change in the kinetic energy of our system must equal the net work done on the system. The change in kinetic energy results from the change in speed between the ends of the tube and is

$$\Delta K = \tfrac{1}{2}\Delta m \, v_2^2 - \tfrac{1}{2}\Delta m \, v_1^2$$
$$= \tfrac{1}{2}\rho \, \Delta V \, (v_2^2 - v_1^2), \tag{15-32}$$

in which $\Delta m \, (= \rho \, \Delta V)$ is the mass of the fluid that enters at the input end and leaves at the output end during a small time interval Δt.

The work done on the system arises from two sources. The work W_g done by the gravitational force $(\Delta m \, \vec{g})$ on the fluid of mass Δm during the vertical lift of the mass from the input level to the output level is

$$W_g = -\Delta m \, g(y_2 - y_1)$$
$$= -\rho g \, \Delta V \, (y_2 - y_1). \tag{15-33}$$

This work is negative because the upward displacement and the downward gravitational force have opposite directions.

Work must also be done *on* the system (at the input end) to push the entering fluid into the tube and *by* the system (at the output end) to push forward the fluid that is located ahead of the emerging fluid. In general, the work done by a force of magnitude F, acting on a fluid sample contained in a tube of area A to move the fluid through a distance Δx, is

$$F \, \Delta x = (pA)(\Delta x) = p(A \, \Delta x) = p \, \Delta V.$$

The work done on the system is then $p_1 \, \Delta V$, and the work done by the system is $-p_2 \, \Delta V$. Their sum W_p is

$$W_p = -p_2 \, \Delta V + p_1 \, \Delta V$$
$$= -(p_2 - p_1) \, \Delta V. \tag{15-34}$$

The work–kinetic energy theorem of Eq. 15-31 now becomes

$$W = W_g + W_p = \Delta K.$$

Substituting from Eqs. 15-32, 15-33, and 15-34 yields

$$-\rho g \, \Delta V \, (y_2 - y_1) - \Delta V \, (p_2 - p_1) = \tfrac{1}{2}\rho \, \Delta V \, (v_2^2 - v_1^2).$$

This, after a slight rearrangement, matches Eq. 15-28, which we set out to prove.

✔CHECKPOINT 4:   Water flows smoothly through the pipe shown in the figure, descending in the process. Rank the four numbered sections of pipe according to (a) the volume flow rate R_V through them, (b) the flow speed v through them, and (c) the water pressure p within them, greatest first.

Sample Problem 15-8

Ethanol of density $\rho = 791$ kg/m^3 flows smoothly through a horizontal pipe that tapers (as in Fig. 15-15) in cross-sectional area from $A_1 = 1.20 \times 10^{-3}$ m^2 to $A_2 = A_1/2$. The pressure difference between the wide and narrow sections of pipe is 4120 Pa. What is the volume flow rate R_V of the ethanol?

SOLUTION: One Key Idea here is that, because the fluid flowing through the wide section of pipe must entirely pass through the narrow section, the volume flow rate R_V must be the same in the two sections. Thus, from Eq. 15-24,

$$R_V = v_1 A_1 = v_2 A_2. \qquad (15\text{-}35)$$

However, with two unknown speeds, we cannot evaluate this equation for R_V.

A second Key Idea is that, because the flow is smooth, we can apply Bernoulli's equation. From Eq. 15-28, we can write

$$p_1 + \tfrac{1}{2}\rho v_1^2 + \rho g y = p_2 + \tfrac{1}{2}\rho v_2^2 + \rho g y, \qquad (15\text{-}36)$$

where subscripts 1 and 2 refer to the wide and narrow sections of pipe, respectively, and y is their common elevation. This equation hardly seems to help because it does not contain the desired volume flow R_V and it contains the unknown speeds v_1 and v_2.

However, there is a neat way to make it work for us: First, we can use Eq. 15-35 and the fact that $A_2 = A_1/2$ to write

$$v_1 = \frac{R_V}{A_1} \quad \text{and} \quad v_2 = \frac{R_V}{A_2} = \frac{2R_V}{A_1}. \qquad (15\text{-}37)$$

Then we can substitute these expressions into Eq. 15-36 to eliminate the unknown speeds and introduce the desired volume flow rate. Doing this and solving for R_V yield

$$R_V = A_1 \sqrt{\frac{2(p_1 - p_2)}{3\rho}}. \qquad (15\text{-}38)$$

We still have a decision to make: We know that the pressure difference between the two sections is 4120 Pa, but does that mean that $p_1 - p_2$ is 4120 Pa or -4120 Pa? We could guess the former is true, or otherwise the square root in Eq. 15-38 would give us an imaginary number. Instead of guessing, however, let's try some reasoning. From Eq. 15-35 we see that speed v_2 in the narrow section (small A_2) must be greater than speed v_1 in the wider section (larger A_1). Recall that if the speed of a fluid increases as it travels along a horizontal path (as here), the pressure of the fluid must decrease. Thus, p_1 is greater than p_2, and $p_1 - p_2 = 4120$ Pa. Inserting this and known data into Eq. 15-38 gives

$$R_V = 1.20 \times 10^{-3} \text{ m}^2 \sqrt{\frac{(2)(4120 \text{ Pa})}{(3)(791 \text{ kg/m}^3)}}$$

$$= 2.24 \times 10^{-3} \text{ m}^3/\text{s}. \qquad \text{(Answer)}$$

Sample Problem 15-9

In the old West, a desperado fires a bullet into an open water tank (Fig. 15-20), creating a hole a distance h below the water surface. What is the speed v of the water emerging from the hole?

SOLUTION: A Key Idea is that this situation is essentially that of water moving (downward) with speed v_0 through a wide pipe (the tank) of cross-sectional area A and then moving (horizontally) with speed v through a narrow pipe (the hole) of cross-sectional area a. Another Key Idea is that, because the water flowing through the wide pipe must entirely pass through the narrow pipe, the volume

Fig. 15-20 Sample Problem 15-9. Water pours through a hole in a water tank, at a distance h below the water surface. The pressure at the water surface and at the hole is atmospheric pressure p_0.

flow rate R_V must be the same in the two "pipes." Then from Eq. 15-24,

$$R_V = av = A v_0$$

and thus

$$v_0 = \frac{a}{A} v.$$

Because $a \ll A$, we see that $v_0 \ll v$.

A third Key Idea is that we can also relate v to v_0 (and to h) through Bernoulli's equation (Eq. 15-28). We take the level of the hole as our reference level for measuring elevations (and thus gravitational potential energy). Noting that the pressure at the top of the tank and at the bullet hole is the atmospheric pressure p_0 (because both places are exposed to the atmosphere), we write Eq. 15-28 as

$$p_0 + \tfrac{1}{2}\rho v_0^2 + \rho g h = p_0 + \tfrac{1}{2}\rho v^2 + \rho g(0). \qquad (15\text{-}39)$$

(Here the top of the tank is represented by the left side of the equation, and the hole by the right side. The zero on the right indicates that the hole is at our reference level.) Before we solve Eq. 15-39 for v, we can use our result that $v_0 \ll v$ to simplify it: We assume that v_0^2, and thus the term $\tfrac{1}{2}\rho v_0^2$ in Eq. 15-39, is negligible compared to the other terms, and we drop it. Solving the remaining equation for v then yields

$$v = \sqrt{2gh}. \qquad \text{(Answer)}$$

This is the same speed that an object would have when falling a height h from rest.

REVIEW & SUMMARY

Density   The **density** ρ of any material is defined as its mass per unit volume:

$$\rho = \frac{\Delta m}{\Delta V}. \qquad (15\text{-}1)$$

Usually, where a material sample is large compared with atomic dimensions, we can write Eq. 15-1 as

$$\rho = \frac{m}{V}. \qquad (15\text{-}2)$$

Fluid Pressure   A **fluid** is a substance that can flow; it conforms to the boundaries of its container because it cannot withstand shearing stress. It can, however, exert a force perpendicular to its surface. That force is described in terms of **pressure** p:

$$p = \frac{\Delta F}{\Delta A}, \qquad (15\text{-}3)$$

in which ΔF is the force acting on a surface element of area ΔA. If the force is uniform over a flat area, Eq. 15-3 can be written as

$$p = \frac{F}{A}. \qquad (15\text{-}4)$$

The force resulting from fluid pressure at a particular point in a fluid has the same magnitude in all directions. *Gauge pressure* is the difference between the actual pressure (or *absolute pressure*) at a point and the atmospheric pressure.

Pressure Variation with Height and Depth   Pressure in a fluid at rest varies with vertical position y. For y measured positive upward,

$$p_2 = p_1 + \rho g(y_1 - y_2). \qquad (15\text{-}7)$$

The pressure in a fluid is the same for all points at the same level. If h is the *depth* of a fluid sample below some reference level at which the pressure is p_0, Eq. 15-7 becomes

$$p = p_0 + \rho gh, \qquad (15\text{-}8)$$

where p is the pressure in the sample.

Pascal's Principle   *Pascal's principle*, which can be derived from Eq. 15-7, states that a change in the pressure applied to an enclosed fluid is transmitted undiminished to every portion of the fluid and to the walls of the containing vessel.

Archimedes' Principle   When a body is fully or partially submerged in a fluid, a buoyant force $\vec{F}_b$ from the surrounding fluid acts on the body. The force is directed upward and has a magnitude given by

$$F_b = m_f g, \qquad (15\text{-}16)$$

where m_f is the mass of the fluid that has been displaced by the body.

When a body floats in a fluid, the magnitude F_b of the (upward) buoyant force on the body is equal to the magnitude F_g of the (downward) gravitational force on the body. The apparent weight of a body on which a buoyant force acts is related to its actual weight by

$$\text{weight}_{app} = \text{weight} - F_b. \qquad (15\text{-}19)$$

Flow of Ideal Fluids   An *ideal fluid* is incompressible and lacks viscosity, and its flow is steady and irrotational. A **streamline** is the path followed by an individual fluid particle. A *tube of flow* is a bundle of streamlines. The flow within any tube of flow obeys the **equation of continuity:**

$$R_V = Av = \text{a constant}, \qquad (15\text{-}24)$$

in which R_V is the **volume flow rate**, A is the cross-sectional area of the tube of flow at any point, and v is the speed of the fluid at that point, assumed to be constant across A. The **mass flow rate** R_m is

$$R_m = \rho R_V = \rho Av = \text{a constant}. \qquad (15\text{-}25)$$

Bernoulli's Equation   Applying the principle of conservation of mechanical energy to the flow of an ideal fluid leads to **Bernoulli's equation:**

$$p + \tfrac{1}{2}\rho v^2 + \rho gy = \text{a constant} \qquad (15\text{-}29)$$

along any tube of flow.

QUESTIONS

1. Figure 15-21 shows a tank filled with water. Five horizontal floors and ceilings are indicated; all have the same area and are located at distances L, $2L$, or $3L$ below the top of the tank. Rank the floors and ceilings according to the force on them due to the water, greatest first.

2. *The Teapot Effect:* When water is poured slowly from a teapot

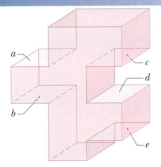

Fig. 15-21   Question 1.

spout, it can double back under the spout for a considerable distance before detaching and falling. (The water layer is held against the underside of the spout by atmospheric pressure.) In Fig. 15-22, within the water layer in the spout, point a is at the top of the layer

Fig. 15-22   Question 2.

and point *b* is at the bottom of the layer; within the water layer below the spout, point *c* is at the top of the layer and point *d* is at the bottom of the layer. Rank those four points according to the gauge pressure in the water there, most positive first, most negative last.

3. Figure 15-23 shows four situations in which a red liquid and a gray liquid are in a **U**-tube. In one situation the liquids cannot be in static equilibrium. (a) Which situation is that? (b) For the other three situations, assume static equilibrium. For each, is the density of the red liquid greater than, less than, or equal to the density of the gray liquid?

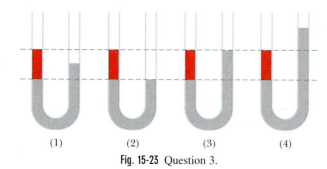

(1)   (2)   (3)   (4)

Fig. 15-23 Question 3.

4. Three hydraulic levers such as that in Fig. 15-8 are used to lift identical loads (at the output side) by identical distances. The levels are identical on the input side but differ in the area of the output piston: Lever 1 has a piston of area *A*, lever 2 has a piston of area 2*A*, and lever 3 has a piston of area 3*A*. For the lift, rank the levers according to (a) the required work at the input side, (b) the required magnitude of the force (assumed to be constant) at the input side, and (c) the displacement of the piston at the input side, greatest first.

5. We fully submerge an irregular 3 kg lump of material in a certain fluid. The fluid that would have been in the space now occupied by the lump has a mass of 2 kg. (a) When we release the lump, does it move upward, move downward, or remain in place? (b) If we next fully submerge the lump in a less dense fluid and again release it, what does it do?

6. Figure 15-24 shows four solid objects floating in corn syrup. Rank the objects according to their density, greatest first.

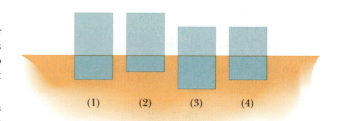

(1)   (2)   (3)   (4)

Fig. 15-24 Question 6.

7. Figure 15-25 shows three identical open-top containers filled to the brim with water; toy ducks float in two of them. Rank the containers and contents according to their weight, greatest first.

(a)   (b)   (c)

Fig. 15-25 Question 7.

8. A block floats in a pail of water in a stationary elevator. Does the block float higher, lower, or at the same level when the elevator cab (a) moves upward at constant speed, (b) moves downward at constant speed, (c) accelerates upward, and (d) accelerates downward with an acceleration magnitude less than *g*?

9. A boat with an anchor on board floats in a swimming pool that is somewhat wider than the boat. Does the water level in the pool move upward, move downward, or remain the same if the anchor is (a) dropped into the water or (b) thrown onto the surrounding ground? (c) Does the water level in the pool move upward, move downward, or remain the same if, instead, a cork is dropped from the boat into the water, where it floats?

10. Figure 15-26 shows three straight pipes through which water flows. The figure gives the speed of the water in each pipe and the cross-sectional area of each pipe. Rank the pipes according to the volume of water that passes through the cross-sectional area per minute, greatest first.

(a)   (b)   (c)

Fig. 15-26 Question 10.

EXERCISES & PROBLEMS

SEC. 15-3 Density and Pressure

1E. Find the pressure increase in the fluid in a syringe when a nurse applies a force of 42 N to the syringe's circular piston, which has a radius of 1.1 cm. **ssm**

2E. Three liquids that will not mix are poured into a cylindrical container. The volumes and densities of the liquids are 0.50 L, 2.6 g/cm^3; 0.25 L, 1.0 g/cm^3; and 0.40 L, 0.80 g/cm^3. What is the force on the bottom of the container due to these liquids? One liter = 1 L = 1000 cm^3. (Ignore the contribution due to the atmosphere.)

3E. An office window has dimensions 3.4 m by 2.1 m. As a result of the passage of a storm, the outside air pressure drops to 0.96 atm, but inside the pressure is held at 1.0 atm. What net force pushes out on the window? **ssm**

4E. You inflate the front tires on your car to 28 psi. Later, you measure your blood pressure, obtaining a reading of 120/80, the readings being in mm Hg. In metric countries (which is to say, most of the world), these pressures are customarily reported in kilopascals (kPa). In kilopascals, what are (a) your tire pressure and (b) your blood pressure?

5E. A fish maintains its depth in fresh water by adjusting the air content of porous bone or air sacs to make its average density the same as that of the water. Suppose that with its air sacs collapsed, a fish has a density of 1.08 g/cm^3. To what fraction of its expanded body volume must the fish inflate the air sacs to reduce its density to that of water? **ilw**

6P. An airtight container having a lid with negligible mass and an area of 77 cm^2 is partially evacuated. If a 480 N force is required to pull the lid off the container and the atmospheric pressure is 1.0×10^5 Pa, what is the air pressure in the container before it is opened? **ssm**

7P. In 1654 Otto von Guericke, inventor of the air pump, gave a demonstration before the noblemen of the Holy Roman Empire in which two teams of eight horses could not pull apart two evacuated brass hemispheres. (a) Assuming that the hemispheres have thin walls, so that R in Fig. 15-27 may be considered both the inside and outside radius, show that the force $\vec{F}$ required to pull apart the hemispheres has magnitude $F = \pi R^2 \, \Delta p$, where Δp is the difference between the pressures outside and inside the sphere. (b) Taking R as 30 cm, the inside pressure as 0.10 atm, and the outside pressure as 1.00 atm, find the force magnitude the teams of horses would have had to exert to pull apart the hemispheres. (c) Explain why one team of horses could have proved the point just as well if the hemispheres were attached to a sturdy wall. **ssm** **www**

Fig. 15-27 Problem 7.

SEC. 15-4   Fluids at Rest

8E. Calculate the hydrostatic difference in blood pressure between the brain and the foot in a person of height 1.83 m. The density of blood is 1.06×10^3 kg/m^3.

9E. The sewage outlet of a house constructed on a slope is 8.2 m below street level. If the sewer is 2.1 m below street level, find the minimum pressure difference that must be created by the sewage pump to transfer waste of average density 900 kg/m^3 from outlet to sewer. **ssm**

10E. Figure 15-28 displays the *phase diagram* of carbon, showing the ranges of temperature and pressure in which carbon will crystallize either as diamond or graphite. What is the minimum depth at which diamonds can form if the temperature at that

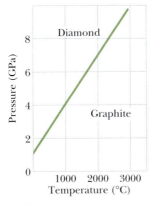

Fig. 15-28 Exercise 10.

depth is 1000°C and the rocks there have density 3.1 g/cm^3? Assume that, as in a fluid, the pressure at any level is due to the gravitational force on the material lying above that level.

11E. A swimming pool has the dimensions 24 m × 9.0 m × 2.5 m. When it is filled with water, what is the force (resulting from the water alone) on (a) the bottom, (b) each short side, and (c) each long side? (d) If you are concerned with the possibility that the concrete walls and floor will collapse, is it appropriate to take the atmospheric pressure into account? Why? **ilw**

12E. (a) Assuming the density of seawater is 1.03 g/cm^3, find the total weight of water on top of a nuclear submarine at a depth of 200 m if its (horizontal cross-sectional) hull area is 3000 m^2. (b) In atmospheres, what water pressure would a diver experience at this depth? Do you think that occupants of a damaged submarine at this depth could escape without special equipment?

13E. Crew members attempt to escape from a damaged submarine 100 m below the surface. What force must be applied to a pop-out hatch, which is 1.2 m by 0.60 m, to push it out at that depth? Assume that the density of the ocean water is 1025 kg/m^3. **ssm**

14E. A cylindrical barrel has a narrow tube fixed to the top, as shown (with dimensions) in Fig. 15-29. The vessel is filled with water to the top of the tube. Calculate the ratio of the hydrostatic force on the bottom of the barrel to the gravitational force on the water contained inside the barrel. Why is that ratio not equal to one? (You need not consider the atmospheric pressure.)

15P. Two identical cylindrical vessels with their bases at the same level each contain a liquid of density ρ. The area of each base is A, but in one vessel the liquid height is h_1, and in the other it is h_2. Find the work done by the gravitational force in equalizing the levels when the two vessels are connected. **ssm**

Fig. 15-29 Exercise 14.

16P. In analyzing certain geological features, it is often appropriate to assume that the pressure at some horizontal *level of compensation,* deep inside Earth, is the same over a large region and is equal to the pressure due to the gravitational force on the overlying material. Thus, the pressure on the level of compensation is given by the fluid pressure formula. This model requires, for one thing, that mountains have *roots* of continental rock extending into the denser mantle (Fig. 15-30). Consider a mountain

Fig. 15-30 Problem 16.

6.0 km high. The continental rocks have a density of 2.9 g/cm^3, and beneath the continent the mantle has a density of 3.3 g/cm^3. Calculate the depth D of the root. (*Hint:* Set the pressure at points a and b equal; the depth y of the level of compensation will cancel out.)

17P. Figure 15-31 shows the juncture of ocean and continent. Find the depth h of the ocean using the level-of-compensation technique presented in Problem 16. **ssm   www**

Fig. 15-31 Problem 17.

18P. The **L**-shaped tank shown in Fig. 15-32 is filled with water and is open at the top. If $d = 5.0$ m, what are (a) the force on face A and (b) the force on face B due to the water?

19P. Water stands at a depth D behind the vertical upstream face of a dam, as shown in Fig. 15-33. Let W be the width of the dam. Find (a) the net horizontal force on the dam from the gauge pressure of the water and (b) the net torque due to that force (and thus gauge pressure) about a line through O parallel to the width of the dam. (c) Find the moment arm of the net horizontal force about the line through O. **ssm**

Fig. 15-32 Problem 18.

Fig. 15-33 Problem 19.

SEC. 15-5   Measuring Pressure

20E. To suck lemonade of density 1000 kg/m^3 up a straw to a maximum height of 4.0 cm, what minimum gauge pressure (in atmospheres) must you produce in your lungs?

21E. What would be the height of the atmosphere if the air density (a) were uniform and (b) decreased linearly to zero with height? Assume that at sea level the air pressure is 1.0 atm and the air density is 1.3 kg/m^3. **ssm**

SEC. 15-6   Pascal's Principle

22E. A piston of small cross-sectional area a is used in a hydraulic press to exert a small force $\vec{f}$ on the enclosed liquid. A connecting pipe leads to a larger piston of cross-sectional area A (Fig. 15-34). (a) What force magnitude F will the larger piston sustain without moving? (b) If the small piston has a diameter of 3.80 cm and the large piston one of 53.0 cm, what force magnitude on the small piston will balance a 20.0 kN force on the large piston? **ssm**

Fig. 15-34 Exercises 22 and 23.

23E. In the hydraulic press of Exercise 22, through what distance must the large piston be moved to raise the small piston a distance of 0.85 m? **ssm**

SEC. 15-7   Archimedes' Principle

24E. A boat floating in fresh water displaces water weighing 35.6 kN. (a) What is the weight of the water that this boat would displace if it were floating in salt water with a density of 1.10×10^3 kg/m^3? (b) Would the volume of the displaced water change? If so, by how much?

25E. An iron anchor of density 7870 kg/m^3 appears 200 N lighter in water than in air. (a) What is the volume of the anchor? (b) How much does it weigh in air? **ssm**

26E. In Fig. 15-35, a cubical object of dimensions $L = 0.600$ m on a side and with a mass of 450 kg is suspended by a rope in an open tank of liquid of density 1030 kg/m^3. (a) Find the magnitude of the total downward force on the top of the object from the liquid and the atmosphere, assuming that atmospheric pressure is 1.00 atm. (b) Find the magnitude of the total upward force on the bottom of the object. (c) Find the tension in the rope. (d) Calculate the magnitude of the buoyant force on the object using Archimedes' principle. What relation exists among all these quantities?

Fig. 15-35 Exercise 26.

27E. A block of wood floats in fresh water with two-thirds of its volume submerged. In oil the block floats with 0.90 of its volume submerged. Find the density of (a) the wood and (b) the oil. **ssm**

28E. A blimp is cruising slowly at low altitude, filled as usual with helium gas. Its maximum useful payload, including crew and cargo, is 1280 kg. The volume of the helium-filled interior space is 5000 m^3. The density of helium gas is 0.16 kg/m^3, and the density of hydrogen is 0.081 kg/m^3. How much more payload could the blimp carry if you replaced the helium with hydrogen? (Why not do it?)

29P. A hollow sphere of inner radius 8.0 cm and outer radius 9.0 cm floats half-submerged in a liquid of density 800 kg/m^3. (a) What is the mass of the sphere? (b) Calculate the density of the material of which the sphere is made. **ssm   www**

30P. About one-third of the body of a person floating in the Dead Sea will be above the water line. Assuming that the human body density is 0.98 g/cm^3, find the density of the water in the Dead Sea. (Why is it so much greater than 1.0 g/cm^3?)

31P. A hollow spherical iron shell floats almost completely submerged in water. The outer diameter is 60.0 cm, and the density of iron is 7.87 g/cm^3. Find the inner diameter. **ilw**

32P. A block of wood has a mass of 3.67 kg and a density of 600 kg/m^3. It is to be loaded with lead so that it will float in water with 0.90 of its volume submerged. What mass of lead is needed (a) if the lead is attached to the top of the wood and (b) if the lead is attached to the bottom of the wood? The density of lead is 1.13×10^4 kg/m^3.

33P. An iron casting containing a number of cavities weighs 6000 N in air and 4000 N in water. What is the total volume of all the cavities in the casting? The density of iron (that is, a sample with no cavities) is 7.87 g/cm³. ssm

34P. Assume the density of brass weights to be 8.0 g/cm³ and that of air to be 0.0012 g/cm³. What percent error arises from neglecting the buoyancy of air in weighing an object of mass m and density ρ on a beam balance, as in Fig. 5-6?

35P. (a) What is the minimum area of the top surface of a slab of ice 0.30 m thick floating on fresh water that will hold up an automobile of mass 1100 kg? (b) Does it matter where the car is placed on the block of ice? ssm

36P. Three children, each of weight 356 N, make a log raft by lashing together logs of diameter 0.30 m and length 1.80 m. How many logs will be needed to keep them afloat in fresh water? Take the density of the logs to be 800 kg/m³.

37P. A metal rod of length 80 cm and mass 1.6 kg has a uniform cross-sectional area of 6.0 cm². Due to a nonuniform density, the center of mass of the rod is 20 cm from one end of the rod. The rod is suspended in a horizontal position in water by ropes attached to both ends (Fig. 15-36). (a) What is the tension in the rope closer to the center of mass? (b) What is the tension in the rope farther from the center of mass? (*Hint:* The buoyancy force on the rod effectively acts at the rod's center.) ssm

Fig. 15-36 Problem 37.

38P. A car has a total mass of 1800 kg. The volume of air space in the passenger compartment is 5.00 m³. The volume of the motor and front wheels is 0.750 m³, and the volume of the rear wheels, gas tank, and trunk is 0.800 m³; water cannot enter these areas. The car is parked on a hill; the handbrake cable snaps and the car rolls down the hill into a lake (Fig. 15-37). (a) At first, no water enters the passenger compartment. How much of the car, in cubic meters, is below the water surface with the car floating as shown? (b) As water slowly enters, the car sinks. How many cubic meters of water are in the car as it disappears below the water surface? (The car, with a heavy load in the trunk, remains horizontal.)

5.00 m³
0.750 m³   0.800 m³
Fig. 15-37
Problem 38.

SEC. 15-9 The Equation of Continuity

39E. A garden hose with an internal diameter of 1.9 cm is connected to a (stationary) lawn sprinkler that consists merely of an enclosure with 24 holes, each 0.13 cm in diameter. If the water in the hose has a speed of 0.91 m/s, at what speed does it leave the sprinkler holes? ssm

40E. Figure 15-38 shows the merging of two streams to form a river. One stream has a width of 8.2 m, depth of 3.4 m, and current speed of 2.3 m/s. The other stream is 6.8 m wide and 3.2 m deep, and flows at 2.6 m/s. The width of the river is 10.5 m, and the current speed is 2.9 m/s. What is its depth?

Fig. 15-38 Exercise 40.

41P. Water is pumped steadily out of a flooded basement at a speed of 5.0 m/s through a uniform hose of radius 1.0 cm. The hose passes out through a window 3.0 m above the waterline. What is the power of the pump? ssm

42E. The water flowing through a 1.9 cm (inside diameter) pipe flows out through three 1.3 cm pipes. (a) If the flow rates in the three smaller pipes are 26, 19, and 11 L/min, what is the flow rate in the 1.9 cm pipe? (b) What is the ratio of the speed of water in the 1.9 cm pipe to that in the pipe carrying 26 L/min?

SEC. 15-10 Bernoulli's Equation

43E. Water is moving with a speed of 5.0 m/s through a pipe with a cross-sectional area of 4.0 cm². The water gradually descends 10 m as the pipe increases in area to 8.0 cm². (a) What is the speed at the lower level? (b) If the pressure at the upper level is 1.5×10^5 Pa, what is the pressure at the lower level? ssm

44E. Models of torpedoes are sometimes tested in a horizontal pipe of flowing water, much as a wind tunnel is used to test model airplanes. Consider a circular pipe of internal diameter 25.0 cm and a torpedo model, aligned along the axis of the pipe, with a diameter of 5.00 cm. The model is to be tested with water flowing past it at 2.50 m/s. (a) With what speed must the water flow in the part of the pipe that is unconstricted by the model? (b) What will the pressure difference be between the constricted and unconstricted parts of the pipe?

45E. A water pipe having a 2.5 cm inside diameter carries water into the basement of a house at a speed of 0.90 m/s and a pressure of 170 kPa. If the pipe tapers to 1.2 cm and rises to the second floor 7.6 m above the input point, what are (a) the speed and (b) the water pressure at the second floor? ilw

46E. A water intake at a pump storage reservoir (Fig. 15-39) has a cross-sectional area of 0.74 m². The water flows in at a speed of 0.40 m/s. At the generator building 180 m below the intake point, the cross-sectional area is smaller than at the intake and the water flows out at 9.5 m/s. What is the difference in pressure, in megapascals, between inlet and outlet?

Fig. 15-39 Exercise 46.

47E. A tank of large area is filled with water to a depth $D = 0.30$ m. A hole of cross-sectional area $A = 6.5$ cm² in the bottom of the tank allows water to drain out. (a) What is the rate at which water flows out, in cubic meters per second? (b) At what distance below the bottom of the tank is the cross-sectional area of the stream equal to one-half the area of the hole?   ssm

48E. Air flows over the top of an airplane wing of area A with speed v_t and past the underside of the wing (also of area A) with speed v_u. Show that in this simplified situation Bernoulli's equation predicts that the magnitude L of the upward lift force on the wing will be

$$L = \tfrac{1}{2}\rho A(v_t^2 - v_u^2),$$

where ρ is the density of the air.

49E. If the speed of flow past the lower surface of an airplane wing is 110 m/s, what speed of flow over the upper surface will give a pressure difference of 900 Pa between upper and lower surfaces? Take the density of air to be 1.30×10^{-3} g/cm³, and see Exercise 48.   ssm

50E. Suppose that two tanks, 1 and 2, each with a large opening at the top, contain different liquids. A small hole is made in the side of each tank at the same depth h below the liquid surface, but the hole in tank 1 has half the cross-sectional area of the hole in tank 2. (a) What is the ratio ρ_1/ρ_2 of the densities of the liquids if the mass flow rate is the same for the two holes? (b) What is the ratio of the volume flow rates from the two tanks? (c) To what height above the hole in the second tank should liquid be added or drained to equalize the volume flow rates?

51P. In Fig. 15-40, water flows through a horizontal pipe, and then out into the atmosphere at a speed of 15 m/s. The diameters of the left and right sections of the pipe are 5.0 cm and 3.0 cm, respectively. (a) What volume of water flows into the atmosphere during a 10 min period? In the left section of the pipe, what are (b) the speed v_2, and (c) the gauge pressure?

Fig. 15-40 Problem 51.

52P. An opening of area 0.25 cm² in an otherwise closed beverage keg is 50 cm below the level of the liquid (of density 1.0 g/cm³) in the keg. What is the speed of the liquid flowing through the opening if the gauge pressure in the air space above the liquid is (a) zero and (b) 0.40 atm?

53P. The fresh water behind a reservoir dam is 15 m deep. A horizontal pipe 4.0 cm in diameter passes through the dam 6.0 m below the water surface, as shown in Fig. 15-41. A plug secures the pipe opening. (a) Find the magnitude of the frictional force between plug and pipe wall. (b) The plug is removed. What volume of water flows out of the pipe in 3.0 h?   ilw

Fig. 15-41 Problem 53.

54P. A tank is filled with water to a height H. A hole is punched in one of the walls at a depth h below the water surface (Fig. 15-42). (a) Show that the distance x from the base of the tank to the point at which the resulting stream strikes the floor is given by $x = 2\sqrt{h(H - h)}$. (b) Could a hole be punched at another depth to produce a second stream that would have the same range? If so, at what depth? (c) At what depth should the hole be placed to make the emerging stream strike the ground at the maximum distance from the base of the tank?

Fig. 15-42 Problem 54.

55P. A *venturi meter* is used to measure the flow speed of a fluid in a pipe. The meter is connected between two sections of the pipe (Fig. 15-43); the cross-sectional area A of the entrance and exit of the meter matches the pipe's cross-sectional area. Between the entrance and exit, the fluid flows from the pipe with speed V and then through a narrow "throat" of cross-sectional area a with speed v. A manometer connects the wider portion of the meter to the narrower portion. The change in the fluid's speed is accompanied by a change Δp in the fluid's pressure, which causes a height difference h of the liquid in the two arms of the manometer. (Here Δp means pressure in the throat minus pressure in the pipe.) (a) By applying Bernoulli's equation and the equation of continuity to points 1 and 2 in Fig. 15-43, show that

$$V = \sqrt{\frac{2a^2\,\Delta p}{\rho(a^2 - A^2)}},$$

where ρ is the density of the fluid. (b) Suppose that the fluid is fresh water, that the cross-sectional areas are 64 cm² in the pipe

Fig. 15-43 Problems 55 and 56.

and 32 cm^2 in the throat, and that the pressure is 55 kPa in the pipe and 41 kPa in the throat. What is the rate of water flow in cubic meters per second? ssm www

56P. Consider the venturi tube of Problem 55 and Fig. 15-43 without the manometer. Let A equal $5a$. Suppose that the pressure p_1 at A is 2.0 atm. Compute the values of (a) V at A and (b) v at a that would make the pressure p_2 at a equal to zero. (c) Compute the corresponding volume flow rate if the diameter at A is 5.0 cm. The phenomenon that occurs at a when p_2 falls to nearly zero is known as cavitation. The water vaporizes into small bubbles.

57P. A pitot tube (Fig. 15-44) is used to determine the airspeed of an airplane. It consists of an outer tube with a number of small holes B (four are shown) that allow air into the tube; that tube is connected to one arm of a U-tube. The other arm of the U-tube is connected to hole A at the front end of the device, which points in the direction the plane is headed. At A the air becomes stagnant so that $v_A = 0$. At B, however, the speed of the air presumably equals the airspeed v of the aircraft. (a) Use Bernoulli's equation to show that

$$v = \sqrt{\frac{2\rho gh}{\rho_{air}}},$$

where ρ is the density of the liquid in the U-tube and h is the difference in the fluid levels in that tube. (b) Suppose that the tube contains alcohol and indicates a level difference h of 26.0 cm. What is the plane's speed relative to the air? The density of the air is 1.03 kg/m^3 and that of alcohol is 810 kg/m^3.

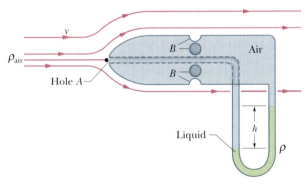

Fig. 15-44 Problems 57 and 58.

58P. A pitot tube (see Problem 57) on a high-altitude aircraft measures a differential pressure of 180 Pa. What is the airspeed if the density of the air is 0.031 kg/m^3?

Additional Problems

59. The dinosaur *Diplodocus* was enormous, with long neck and tail and a mass that was great enough to test its leg strength. According to conjecture, *Diplodocus* waded in water, perhaps up to its head, so that buoyancy could offset its weight and lighten the load on its legs. To check the conjecture, take the density of *Diplodocus* to be 0.90 that of water, and assume that its mass was the published estimate of 1.85×10^4 kg. (a) What then would be its actual weight? Find its apparent weight when it had the follow-

ing fractions of its volume submerged: (b) 0.50, (c) 0.80, and (d) 0.90. When almost fully submerged, with only its head above water, its lungs would have been about 8.0 m below the water surface. (e) At that depth, what would be the difference between the (external) water pressure and the pressure of the air in its lungs? For the dinosaur to breathe in, its lung muscles would have had to expand its lungs against this pressure difference. It probably could not do so against a pressure difference of more than 8 kPa. (f) Did *Diplodocus* wade as conjectured?

60. When you cough, you expel air at high speed through the trachea and upper bronchi so that the air will remove excess mucus lining the pathway. You produce the high speed by this procedure: You breathe in a large amount of air, trap it by closing the glottis (the narrow opening in the larynx), increase the air pressure by contracting the lungs, partially collapse the trachea and upper bronchi to narrow the pathway, and then expel the air through the pathway by suddenly reopening the glottis. Assume that during the expulsion the volume flow rate is 7.0×10^{-3} m^3/s. In terms of the speed of sound $v_s = 343$ m/s, what is the air speed through the trachea if the trachea diameter (a) remains its normal diameter of 14 mm and (b) contracts to a diameter of 5.2 mm?

61. Suppose that your body has a uniform density of 0.95 times that of water. (a) If you float in a swimming pool, what fraction of your body's volume is above the water surface?

Quicksand is a fluid that is produced when water is forced up into sand, moving the sand grains away from one another so they are no longer locked together by friction. Pools of quicksand can occur when water drains underground from hills into valleys with sand pockets. (b) If you step into a deep pool of quicksand with a density 1.6 times that of water, what fraction of your body's volume is then above the quicksand surface? (c) In particular, are you submerged enough to be unable to breathe? The viscosity of quicksand dramatically increases during any quick movements (the fluid is said to be *thixotropic*). Thus, if you struggle to escape from quicksand, the quicksand holds you tighter. (d) How might you escape without someone helping?

62. In a sink with a flat bottom, turn on a sink faucet so that a smoothly flowing (laminar) stream strikes the bottom. The water spreads from the impact point in a shallow layer but then, at a certain radius r_J from the impact point, it suddenly increases in depth. This depth change, called a *hydraulic jump*, forms a prominent circle around the impact point. Inside the circle, the speed v_1 of the spreading water is constant and is equal to its speed in the falling stream just before impact.

In a certain experiment, the radius of the falling stream is 1.3 mm just before impact, the volume flow rate R_V is 7.9 cm^3/s, the jump radius r_J is 2.0 cm, and the depth just after the jump is 2.0 mm. (a) What is speed v_1? (b) For $r < r_J$, express the water depth d as a function of the radial distance r from the impact point. (c) Does the depth of the water increase or decrease with r? (d) What is the depth just before the water reaches the hydraulic jump? (e) What is the speed v_2 of the water just after the jump? What are the kinetic energy densities (f) just before and (g) just after the jump? (h) What is the change in pressure on the sink due to the jump? (i) Does Bernoulli's equation apply to a streamline through the jump?

NEW PROBLEMS

N1. Fresh water flows horizontally from pipe section 1 of cross-sectional area A_1 into pipe section 2 of cross-sectional area A_2. Figure 15N-1 gives a plot of the pressure difference $p_2 - p_1$ versus the inverse area squared A_1^{-2} that would be expected for a volume flow rate of a certain value if the water flow were laminar under all circumstances. What are the values of (a) A_2 and (b) the volume flow rate for the conditions of the figure?

Fig. 15N-1 Problem N1.

N2. In Fig. 15N-2, water flows smoothly from the left-hand pipe section (radius $2.00R$), through the middle section (radius R), and into the right-hand section (radius $3.00R$). The speed of the water in the middle section is 0.500 m/s. What is the net work done on 0.400 m^3 of the water as it moves from the left-hand section to the right-hand section?

$2.00R \quad R \quad 3.00R$

Fig. 15N-2 Problem N2.

N3. Figure 15N-3 shows an imaginary upright cylinder of water located inside a bucket filled with water. The cylinder has height h of 3.00 cm and a (horizontal) face area of 2.00 cm^2 at the top and bottom. The bucket (and thus also the cylinder of water) is moving upward with an acceleration $\vec{a}$ of magnitude $g/2$. Let $\vec{F}_1$ be the force on the top face of the cylinder and $\vec{F}_2$ be the force on the bottom face. What is the magnitude of $\vec{F}_2 - \vec{F}_1$?

Fig. 15N-3 Problem N3.

N4. Figure 15N-4 shows two sections of an old pipe system that runs through a hill. On each side of the hill, the pipe radius is 2.00 cm. However, the radius of the pipe inside the hill is no longer known. To determine it, hydraulic engineers first establish that water flows through the left-hand and right-hand sections at 2.50 m/s. Then they release a dye in the water at point A and find that it takes 88.8 s to reach point B. What is the radius (or average radius) of the pipe within the hill?

A ? B

30 m   30 m

110 m

Fig. 15N-4 Problem N4.

N5. If a bubble in sparkling water accelerates upward at the rate of 0.225 m/s^2 and has the radius 0.500 mm, what is its mass? Assume that the drag force on the bubble is negligible.

N6. A 5.00 kg object is released from rest while fully submerged in a liquid. The liquid displaced by the presence of the object has a mass of 3.00 kg. How far and in what direction does the object move in 0.200 s, assuming that it moves freely and that the drag force on the object from the liquid is negligible?

N7. A rectangular block with a height of 8.00 cm floats in liquid 1 with a height of 6.00 cm above the liquid surface. What height of the block will be above the liquid surface when the block floats in liquid 2, which has 0.500 times the density of liquid 1?

N8. A small solid ball is released from rest while fully submerged in a fluid and then its kinetic energy is measured when it has moved by 4.0 cm in the fluid. Figure 15N-5 gives the results after many fluids are used: The kinetic energy K is plotted versus the fluid density ρ_{liq}. What are (a) the density and (b) the volume of the ball?

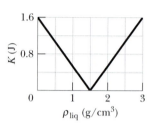

Fig. 15N-5 Problem N8.

N9. The maximum depth d_{max} that a diver can snorkel is set by the density of the water and the fact that human lungs can function against a maximum pressure difference (between inside and outside the chest cavity) of 0.050 atm. What is the difference in d_{max} for fresh water and the water of the Dead Sea (the saltiest natural water in the world, with a density of $1.5 \times 10^3 \text{ kg/m}^3$)?

N10. Suppose that you release a small ball from rest at a depth of 0.600 m below the surface in a pool of water. If the density of the ball is 0.300 that of water and if the drag force on the ball from the water is negligible, how high above the water surface will the ball shoot as it emerges from the water? (Neglect any transfer of energy to the splashing and waves produced by the emerging ball.)

N11. A large aquarium of height 5.00 m is filled with fresh water to a depth of 2.00 m. One wall of the aquarium consists of thick plastic 8.00 m wide. By how much does the total force on that wall increase if the aquarium is next filled to a depth of 4.00 m?

N12. In Fig. 15N-6, a plastic tube with cross-sectional area 5.00 cm^2 is in the shape of an L. The tube is filled with water until the short arm (of length $d = 0.800$ m) is full. Then the short arm is sealed and more water is gradually poured into the longer arm. If the seal will pop off when the force on it exceeds 9.80 N, what (total) height of water in the longer arm will put the seal on the verge of popping off?

Fig. 15N-6 Problems N12 and N14.

N13. Figure 15N-7 shows an iron ball suspended by thread of negligible mass from an upright cylinder that floats partially submerged in water. The cylinder has a height of 6.00 cm, a face area

Fig. 15N-7 Problem N13.

of 12.0 cm² on the top and bottom, and a density of 0.30 g/cm³, and 2.00 cm of its height is above the water surface. What is the radius of the iron ball?

N14. Figure 15N-6 shows a modified U-tube: the right arm is shorter than the left arm. The open end of the right arm is height $d = 10.0$ cm above the laboratory bench. The radius throughout the tube is 1.50 cm. Water is gradually poured into the open end of the left arm until it begins to flow out the open end of the right arm. Then a liquid of density 0.80 g/cm³ is gradually added to the left arm until its height in that arm is 8.0 cm (it does not mix with the water). How much water flows out of the right arm?

N15. A vertical cylinder holds liquid 1 (with density 1.20 g/cm³ and height 8.00 cm) and liquid 2 (with density 2.00 g/cm³ and height 4.00 cm). The liquids are initially separated, with liquid 1 above liquid 2, but then they are well stirred. Assuming that liquids 1 and 2 are uniformly mixed and that the total volume does not change, what is the density of the liquid mixture?

N16. In Fig. 15N-8, a spring of spring constant 3.00×10^4 N/m is placed between the output piston of a hydraulic lever and a rigid beam, and an empty container with negligible mass is placed on the input piston. The input piston has area A_i and the output piston has area $18.0A_i$. Initially the spring is at its rest length. How many kilograms of sand must be (slowly) poured into the container to compress the spring by 5.00 cm?

Fig. 15N-8 Problem N16.

N17. The tension in a string holding a solid block below the surface of a liquid (of density greater than the solid) is T_0 when the containing vessel (Fig. 15N-9) is at rest. Show that when the vessel has an upward vertical acceleration of magnitude a, the tension T is equal to $T_0(1 + a/g)$.

Fig. 15N-9 Problem N17.

N18. You have been asked to review plans for a swimming pool in a new hotel. The water is to be supplied to the hotel by a horizontal main pipe of radius $R_1 = 6.00$ cm, with water under pressure of 2.00 atm. A vertical pipe of radius $R_2 = 1.00$ cm is to carry the water to a height of 9.40 m, where the water is to pour out freely into a square pool of width 10.0 m and (proposed) water depth of 2.00 m. (a) How much time will be required to fill the pool? (b) If more than a few days is considered unacceptable and less than a few hours is considered dangerous, is the filling time acceptable and safe?

N19. A river 20 m wide and 4.0 m deep drains a 3000 km² land area in which the average precipitation is 48 cm/y. One-fourth of this rainfall returns to the atmosphere by evaporation, but the remainder ultimately drains into the river. What is the average speed of the river current?

N20. In Fig. 15N-10a, a rectangular block is gradually pushed face-down into a liquid. The block has height d; on the bottom and top the face area is $A = 5.67$ cm². Figure 15N-10b gives the apparent weight W_{app} of the block as a function of the depth h of its lower face. What is the density of the liquid?

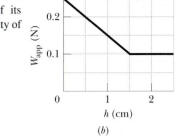

(a)   (b)

Fig. 15N-10 Problem N20.

N21. Modern racecars come with a variety of airfoils to help hold them on the track, especially in fast turns where the cars tend to slide out of the turn. Another technique involves channeling air through an opening in the front of the car, down under the car's body, and then out behind the car. The air effectively flows through a pipe that is narrow in one section (here, the space below the car). Suppose the front opening has an area of 0.75 m² and the space between the track and the bottom of the car has a vertical area of 0.15 m². If the car is moving at 240 km/h and the pressure above the car is 1.0 atm, approximately what is the pressure difference between the top and bottom of the car, pushing down on the car?

N22. Caught in an avalanche, a skier is fully submerged in the flowing snow, which has a density of 96 kg/m³. Assume that the average density of the skier, clothing, and skiing equipment is 1020 kg/m³. What percentage of the gravitational force on the skier is offset by the buoyant force from the snow?

N23. Partially fill a tall drinking cup with water to a depth h. Cut a square of sturdy paper so that it is somewhat wider than the opening to the cup. Place the paper over the opening (Fig. 15N-11a). Spread the fingers of your left hand over the paper, pressing it against the cup's rim as widely apart as possible. Grab the cup with your right hand and then as rapidly as you can, invert the cup with your left hand still pressing the paper against the rim. Chances are good that you can then remove your left hand without the water pouring from the cup (Fig. 15N-11b). The paper bulges downward but stays against the rim. If $h = 11.0$ cm, what is the gauge pressure in the air that is now trapped in the cup above the water?

(a)   (b)

Fig. 15N-11 Problem N23.

N24. A flotation device is in the shape of a right cylinder, with a height of 0.500 m and a face area of 4.00 m² on top and bottom, and its density is 0.400 times that of fresh water. It is initially held fully submerged in fresh water, with its top face at the water surface. Then it is allowed to ascend gradually until it begins to float. How much work does the buoyant force do on the device during the ascent?

N25. A glass ball of radius 2.00 cm sits at the bottom of a container of milk that has a density of 1.03 g/cm³. The normal force on the ball from the container's lower surface has magnitude 9.48×10^{-2} N. What is the mass of the ball?

16 Oscillations

On September 19, 1985, seismic waves from an earthquake that originated along the west coast of Mexico caused terrible and widespread damage in Mexico City, about 400 km from the origin.

Why did the seismic waves cause such extensive damage in Mexico City but relatively little on the way there?

The answer is in this chapter.

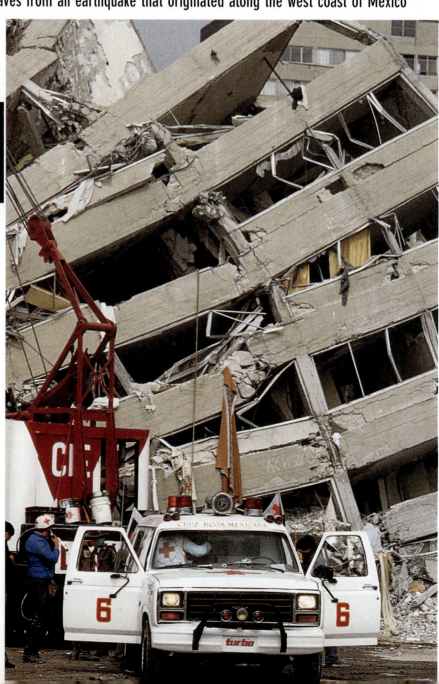

16-1 Oscillations

We are surrounded by oscillations—motions that repeat themselves. There are swinging chandeliers, boats bobbing at anchor, and the surging pistons in the engines of cars. There are oscillating guitar strings, drums, bells, diaphragms in telephones and speaker systems, and quartz crystals in wristwatches. Less evident are the oscillations of the air molecules that transmit the sensation of sound, the oscillations of the atoms in a solid that convey the sensation of temperature, and the oscillations of the electrons in the antennas of radio and TV transmitters that convey information.

Oscillations in the real world are usually *damped;* that is, the motion dies out gradually, transferring mechanical energy to thermal energy by the action of frictional forces. Although we cannot totally eliminate such loss of mechanical energy, we can replenish the energy from some source. As an example, you know that by swinging your legs or torso you can "pump" a swing to maintain or increase the oscillations. In doing this, you transfer biochemical energy to mechanical energy of the oscillating system.

16-2 Simple Harmonic Motion

Figure 16-1a shows a sequence of "snapshots" of a simple oscillating system, a particle moving repeatedly back and forth about the origin of an x axis. In this section we simply describe the motion. Later, we shall discuss how to attain such motion.

One important property of oscillatory motion is its **frequency,** or number of oscillations that are completed each second. The symbol for frequency is f, and its SI unit is the **hertz** (abbreviated Hz), where

$$1 \text{ hertz} = 1 \text{ Hz} = 1 \text{ oscillation per second} = 1 \text{ s}^{-1}. \quad (16\text{-}1)$$

Related to the frequency is the **period** T of the motion, which is the time for one complete oscillation (or **cycle**); that is,

$$T = \frac{1}{f}. \quad (16\text{-}2)$$

Any motion that repeats itself at regular intervals is called **periodic motion** or **harmonic motion.** We are interested here in motion that repeats itself in a particular way—namely, like that in Fig. 16-1a. For such motion the displacement x of the

Fig. 16-1 (a) A sequence of "snapshots" (taken at equal time intervals) showing the position of a particle as it oscillates back and forth about the origin along an x axis, between the limits $+x_m$ and $-x_m$. The vector arrows are scaled to indicate the speed of the particle. The speed is maximum when the particle is at the origin and zero when it is at $\pm x_m$. If the time t is chosen to be zero when the particle is at $+x_m$, then the particle returns to $+x_m$ at $t = T$, where T is the period of the motion. The motion is then repeated. (b) A graph of x as a function of time for the motion of (a).

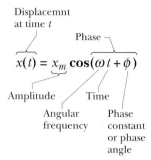

$$x(t) = x_m \cos(\omega t + \phi)$$

Displacement at time t — Phase — Amplitude — Time — Angular frequency — Phase constant or phase angle

Fig. 16-2 A handy reference to the quantities in Eq. 16-3 for simple harmonic motion.

particle from the origin is given as a function of time by

$$x(t) = x_m \cos(\omega t + \phi) \quad \text{(displacement)}, \qquad (16\text{-}3)$$

in which x_m, ω, and ϕ are constants. This motion is called **simple harmonic motion** (SHM), a term that means the periodic motion is a sinusoidal function of time. Equation 16-3, in which the sinusoidal function is a cosine function, is graphed in Fig. 16-1b. (You can get that graph by rotating Fig. 16-1a counterclockwise by 90° and then connecting the successive locations of the particle with a curve.) The quantities that determine the shape of the graph are displayed in Fig. 16-2 with their names. We now shall define those quantities.

The quantity x_m, called the **amplitude** of the motion, is a positive constant whose value depends on how the motion was started. The subscript m stands for *maximum* because the amplitude is the magnitude of the maximum displacement of the particle in either direction. The cosine function in Eq. 16-3 varies between the limits ± 1, so the displacement $x(t)$ varies between the limits $\pm x_m$.

The time-varying quantity $(\omega t + \phi)$ in Eq. 16-3 is called the **phase** of the motion, and the constant ϕ is called the **phase constant** (or **phase angle**). The value of ϕ depends on the displacement and velocity of the particle at time $t = 0$. For the $x(t)$ plots of Fig. 16-3a, the phase constant ϕ is zero.

To interpret the constant ω, called the **angular frequency** of the motion, we first note that the displacement $x(t)$ must return to its initial value after one period T of the motion; that is, $x(t)$ must equal $x(t + T)$ for all t. To simplify this analysis, let us put $\phi = 0$ in Eq. 16-3. From that equation we then can write

$$x_m \cos \omega t = x_m \cos \omega(t + T). \qquad (16\text{-}4)$$

The cosine function first repeats itself when its argument (the phase) has increased by 2π rad, so Eq. 16-4 gives us

$$\omega(t + T) = \omega t + 2\pi$$

or

$$\omega T = 2\pi.$$

Thus, from Eq. 16-2 the angular frequency is

$$\omega = \frac{2\pi}{T} = 2\pi f. \qquad (16\text{-}5)$$

(a)

(b)

Fig. 16-3 In each case, the blue curve is obtained from Eq. 16-3 with $\phi = 0$. (a) The red curve differs from the blue curve *only* in that its amplitude x'_m is greater (the red-curve extremes of displacement are higher and lower). (b) The red curve differs from the blue curve *only* in that its period is $T' = T/2$ (the red curve is compressed horizontally). (c) The red curve differs from the blue curve *only* in that $\phi = -\pi/4$ rad rather than zero (the negative value of ϕ shifts the red curve to the right).

(c)

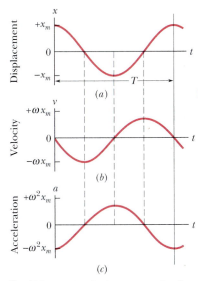

Fig. 16-4 (*a*) The displacement $x(t)$ of a particle oscillating in SHM with phase angle ϕ equal to zero. The period T marks one complete oscillation. (*b*) The velocity $v(t)$ of the particle. (*c*) The acceleration $a(t)$ of the particle.

The SI unit of angular frequency is the radian per second. (To be consistent, then, ϕ must be in radians.) Figure 16-3 compares $x(t)$ for two simple harmonic motions that differ either in amplitude, in period (and thus in frequency and angular frequency), or in phase constant.

✔ **CHECKPOINT 1:** A particle undergoing simple harmonic oscillation of period T (like that in Fig. 16-1) is at $-x_m$ at time $t = 0$. Is it at $-x_m$, at $+x_m$, at 0, between $-x_m$ and 0, or between 0 and $+x_m$ when (a) $t = 2.00T$, (b) $t = 3.50T$, and (c) $t = 5.25T$?

The Velocity of SHM

By differentiating Eq. 16-3, we can find an expression for the velocity of a particle moving with simple harmonic motion; that is,

$$v(t) = \frac{dx(t)}{dt} = \frac{d}{dt}[x_m \cos(\omega t + \phi)]$$

or

$$v(t) = -\omega x_m \sin(\omega t + \phi) \qquad \text{(velocity).} \qquad (16\text{-}6)$$

Figure 16-4*a* is a plot of Eq. 16-3 with $\phi = 0$. Figure 16-4*b* shows Eq. 16-6, also with $\phi = 0$. Analogous to the amplitude x_m in Eq. 16-3, the positive quantity ωx_m in Eq. 16-6 is called the **velocity amplitude** v_m. As you can see in Fig. 16-4*b*, the velocity of the oscillating particle varies between the limits $\pm v_m = \pm \omega x_m$. Note also in that figure that the curve of $v(t)$ is *shifted* (to the left) from the curve of $x(t)$ by one-quarter period; when the magnitude of the displacement is greatest (that is, $x(t) = x_m$), the magnitude of the velocity is least (that is, $v(t) = 0$). When the magnitude of the displacement is least (that is, zero), the magnitude of the velocity is greatest (that is, $v_m = \omega x_m$).

The Acceleration of SHM

Knowing the velocity $v(t)$ for simple harmonic motion, we can find an expression for the acceleration of the oscillating particle by differentiating once more. Thus, we have, from Eq. 16-6,

$$a(t) = \frac{dv(t)}{dt} = \frac{d}{dt}[-\omega x_m \sin(\omega t + \phi)]$$

or

$$a(t) = -\omega^2 x_m \cos(\omega t + \phi) \qquad \text{(acceleration).} \qquad (16\text{-}7)$$

Figure 16-4*c* is a plot of Eq. 16-7 for the case $\phi = 0$. The positive quantity $\omega^2 x_m$ in Eq. 16-7 is called the **acceleration amplitude** a_m; that is, the acceleration of the particle varies between the limits $\pm a_m = \pm \omega^2 x_m$, as Fig. 16-4*c* shows. Note also that the curve of $a(t)$ is shifted (to the left) by $\frac{1}{4}T$ relative to the curve of $v(t)$.

We can combine Eqs. 16-3 and 16-7 to yield

$$a(t) = -\omega^2 x(t), \qquad (16\text{-}8)$$

which is the hallmark of simple harmonic motion:

> In SHM, the acceleration is proportional to the displacement but opposite in sign, and the two quantities are related by the square of the angular frequency.

Thus, as Fig. 16-4 shows, when the displacement has its greatest positive value, the acceleration has its greatest negative value, and conversely. When the displacement is zero, the acceleration is also zero.

Tactic 1: *Phase Angles*

Note the effect of the phase angle ϕ on a plot of $x(t)$. When $\phi = 0$, $x(t)$ has a graph like that in Fig. 16-4a, a typical cosine curve. Increasing ϕ shifts the curve leftward along the t axis. (You might remember this with the symbol ↰ϕ, where the up arrow indicates an increase in ϕ and the left arrow indicates the resulting shift in the curve.) Decreasing ϕ shifts the curve rightward, as in Fig. 16-3c for $\phi = -\pi/4$.

Two plots of SHM with different phase angles are said to have a *phase difference*; each is said to be *phase-shifted* from the other,

or *out of phase* with the other. The curves in Fig. 16-3c, for example, have a phase difference of $\pi/4$ rad.

Because SHM repeats after each period T and the cosine function repeats after each 2π rad, one period T represents a phase difference of 2π rad. In Fig. 16-4, $x(t)$ is phase-shifted to the right from $v(t)$ by one-quarter period, or $-\pi/2$ rad; it is shifted to the right from $a(t)$ by one-half period, or $-\pi$ rad. A phase shift of 2π rad causes a curve of SHM to coincide with itself; that is, it looks unchanged.

16-3 The Force Law for Simple Harmonic Motion

Once we know how the acceleration of a particle varies with time, we can use Newton's second law to learn what force must act on the particle to give it that acceleration. If we combine Newton's second law and Eq. 16-8, we find, for simple harmonic motion,

$$F = ma = -(m\omega^2)x. \tag{16-9}$$

This result—a restoring force that is proportional to the displacement but opposite in sign—is familiar. It is Hooke's law,

$$F = -kx, \tag{16-10}$$

for a spring, the spring constant here being

$$k = m\omega^2. \tag{16-11}$$

We can in fact take Eq. 16-10 as an alternative definition of simple harmonic motion. It says:

▶ Simple harmonic motion is the motion executed by a particle of mass m subject to a force that is proportional to the displacement of the particle but opposite in sign.

The block–spring system of Fig. 16-5 forms a **linear simple harmonic oscillator** (linear oscillator, for short), where "linear" indicates that F is proportional to x rather than to some other power of x. The angular frequency ω of the simple harmonic motion of the block is related to the spring constant k and the mass m of the block by Eq. 16-11, which yields

$$\omega = \sqrt{\frac{k}{m}} \quad \text{(angular frequency).} \tag{16-12}$$

By combining Eqs. 16-5 and 16-12, we can write, for the **period** of the linear oscillator of Fig. 16-5,

$$T = 2\pi\sqrt{\frac{m}{k}} \quad \text{(period).} \tag{16-13}$$

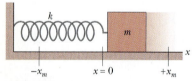

Fig. 16-5 A linear simple harmonic oscillator. The surface is frictionless. Like the particle of Fig. 16-2, the block moves in simple harmonic motion once it has been pulled to the side and released. Its displacement is then given by Eq. 16-3.

Equations 16-12 and 16-13 tell us that a large angular frequency (and thus a small period) goes with a stiff spring (large k) and a light block (small m).

Every oscillating system, be it the linear oscillator of Fig. 16-5, a diving board, or a violin string, has some element of "springiness" and some element of "inertia" or mass, and thus resembles a linear oscillator. In the linear oscillator of Fig. 16-5, these elements are located in separate parts of the system: The springiness is entirely in the spring, which we assume to be massless, and the inertia is entirely in the block, which we assume to be rigid. In a violin string, however, the two elements are both within the string itself, as you will see in Chapter 17.

✔ **CHECKPOINT 2:**   Which of the following relationships between the force F on a particle and the particle's position x implies simple harmonic oscillation: (a) $F = -5x$, (b) $F = -400x^2$, (c) $F = 10x$, (d) $F = 3x^2$?

Sample Problem 16-1

A block whose mass m is 680 g is fastened to a spring whose spring constant k is 65 N/m. The block is pulled a distance $x = 11$ cm from its equilibrium position at $x = 0$ on a frictionless surface and released from rest at $t = 0$.

(a) What are the angular frequency, the frequency, and the period of the resulting motion?

SOLUTION: The **Key Idea** here is that the block–spring system forms a linear simple harmonic oscillator, with the block undergoing SHM. Then the angular frequency is given by Eq. 16-12:

$$\omega = \sqrt{\frac{k}{m}} = \sqrt{\frac{65 \text{ N/m}}{0.68 \text{ kg}}} = 9.78 \text{ rad/s}$$

$$\approx 9.8 \text{ rad/s}. \qquad \text{(Answer)}$$

The frequency follows from Eq. 16-5, which yields

$$f = \frac{\omega}{2\pi} = \frac{9.78 \text{ rad/s}}{2\pi \text{ rad}} = 1.56 \text{ Hz} \approx 1.6 \text{ Hz}. \quad \text{(Answer)}$$

The period follows from Eq. 16-2, which yields

$$T = \frac{1}{f} = \frac{1}{1.56 \text{ Hz}} = 0.64 \text{ s} = 640 \text{ ms}. \qquad \text{(Answer)}$$

(b) What is the amplitude of the oscillation?

SOLUTION: The **Key Idea** here is that, with no friction involved, the mechanical energy of the spring–block system is conserved. The block is released from rest 11 cm from its equilibrium position, with zero kinetic energy and the elastic potential energy of the system at a maximum. Thus, the block will have zero kinetic energy whenever it is again 11 cm from its equilibrium position, which means it will never be farther than 11 cm from that position. Its maximum displacement is 11 cm:

$$x_m = 11 \text{ cm}. \qquad \text{(Answer)}$$

(c) What is the maximum speed v_m of the oscillating block, and where is the block when it occurs?

SOLUTION: The **Key Idea** here is that the maximum speed v_m is the velocity amplitude ωx_m in Eq. 16-6; that is,

$$v_m = \omega x_m = (9.78 \text{ rad/s})(0.11 \text{ m})$$

$$= 1.1 \text{ m/s}. \qquad \text{(Answer)}$$

This maximum speed occurs when the oscillating block is rushing through the origin; compare Figs. 16-4a and 16-4b, where you can see that the speed is a maximum whenever $x = 0$.

(d) What is the magnitude a_m of the maximum acceleration of the block?

SOLUTION: The **Key Idea** this time is that the magnitude a_m of the maximum acceleration is the acceleration amplitude $\omega^2 x_m$ in Eq. 16-7; that is,

$$a_m = \omega^2 x_m = (9.78 \text{ rad/s})^2(0.11 \text{ m})$$

$$= 11 \text{ m/s}^2. \qquad \text{(Answer)}$$

This maximum acceleration occurs when the block is at the ends of its path. At those points, the force acting on the block has its maximum magnitude; compare Figs. 16-4a and 16-4c, where you can see that the magnitudes of the displacement and acceleration are maximum at the same times.

(e) What is the phase constant ϕ for the motion?

SOLUTION: Here the **Key Idea** is that Eq. 16-3 gives the displacement of the block as a function of time. We know that at time $t = 0$, the block is located at $x = x_m$. Substituting these *initial conditions*, as they are called, into Eq. 16-3 and canceling x_m give us

$$1 = \cos \phi. \qquad (16\text{-}14)$$

Taking the inverse cosine then yields

$$\phi = 0 \text{ rad}. \qquad \text{(Answer)}$$

(Any angle that is an integer multiple of 2π rad also satisfies Eq. 16-14; we chose the smallest angle.)

(f) What is the displacement function $x(t)$ for the spring–block system?

SOLUTION: The **Key Idea** here is that $x(t)$ is given in general form by Eq. 16-3. Substituting known quantities into that equation gives us

$$x(t) = x_m \cos(\omega t + \phi)$$

$$= (0.11 \text{ m}) \cos[(9.8 \text{ rad/s})t + 0]$$

$$= 0.11 \cos(9.8t), \qquad \text{(Answer)}$$

where x is in meters and t is in seconds.

Sample Problem 16-2

At $t = 0$, the displacement $x(0)$ of the block in a linear oscillator like that of Fig. 16-5 is -8.50 cm. (Read $x(0)$ as "x at time zero.") The block's velocity $v(0)$ then is -0.920 m/s, and its acceleration $a(0)$ is $+47.0$ m/s^2.

(a) What is the angular frequency ω of this system?

SOLUTION: A **Key Idea** here is that, with the block in SHM, Eqs. 16-3, 16-6, and 16-7 give its displacement, velocity, and acceleration, respectively, and each contains ω. Let's substitute $t = 0$ into each to see whether we can solve any one of them for ω. We find

$$x(0) = x_m \cos \phi, \tag{16-15}$$

$$v(0) = -\omega x_m \sin \phi, \tag{16-16}$$

and

$$a(0) = -\omega^2 x_m \cos \phi. \tag{16-17}$$

In Eq. 16-15, ω has disappeared. In Eqs. 16-16 and 16-17, we know values for the left sides, but we do not know x_m and ϕ. However, if we divide Eq. 16-17 by Eq. 16-15, we neatly eliminate both x_m and ϕ and can then solve for ω as

$$\omega = \sqrt{-\frac{a(0)}{x(0)}} = \sqrt{-\frac{47.0 \text{ m/s}^2}{-0.0850 \text{ m}}}$$

$$= 23.5 \text{ rad/s}. \qquad \text{(Answer)}$$

(b) What are the phase constant ϕ and amplitude x_m?

SOLUTION: The same **Key Idea** as in part (a) also applies here, as do Eqs. 16-15 through 16-17. Now, however, we know ω and want ϕ and x_m. If we divide Eq. 16-16 by Eq. 16-15, we find

$$\frac{v(0)}{x(0)} = \frac{-\omega x_m \sin \phi}{x_m \cos \phi} = -\omega \tan \phi.$$

Solving for $\tan \phi$, we find

$$\tan \phi = -\frac{v(0)}{\omega x(0)} = -\frac{-0.920 \text{ m/s}}{(23.5 \text{ rad/s})(-0.0850 \text{ m})}$$

$$= -0.461.$$

This equation has two solutions:

$$\phi = -25° \quad \text{and} \quad \phi = 180° + (-25°) = 155°.$$

(Normally only the first solution here is displayed by a calculator.) A **Key Idea** in choosing the proper solution is to test them both by using them to compute values for the amplitude x_m. From Eq. 16-15, we find that if $\phi = -25°$, then

$$x_m = \frac{x(0)}{\cos \phi} = \frac{-0.0850 \text{ m}}{\cos(-25°)} = -0.094 \text{ m}.$$

We find similarly that if $\phi = 155°$, then $x_m = 0.094$ m. Because the amplitude of SHM must be a positive constant, the correct phase constant and amplitude here are

$$\phi = 155° \quad \text{and} \quad x_m = 0.094 \text{ m} = 9.4 \text{ cm}. \quad \text{(Answer)}$$

PROBLEM-SOLVING TACTICS

Tactic 2: *Identifying SHM*

In linear SHM the acceleration a and displacement x of the system are related by an equation of the form

$$a = -(\text{a positive constant})x,$$

which says that the acceleration is proportional to the displacement from the equilibrium position but is in the opposite direction. Once you find such an expression for an oscillating system, you can immediately compare it to Eq. 16-8, identify the positive constant as being equal to ω^2, and so quickly get an expression for the angular frequency of the motion. With Eq. 16-5 you then can find the period T and the frequency f.

In some problems you might derive an expression for the force F as a function of displacement x. If the motion is linear SHM, the force and displacement are related by

$$F = -(\text{a positive constant})x,$$

which says that the force is proportional to the displacement but is in the opposite direction. Once you have found such an expression, you can immediately compare it to Eq. 16-10 and identify the positive constant as being k. If you know the mass that is involved, you can then use Eqs. 16-12, 16-13, and 16-5 to find the angular frequency ω, period T, and frequency f.

16-4 Energy in Simple Harmonic Motion

In Chapter 8 we saw that the energy of a linear oscillator transfers back and forth between kinetic energy and potential energy, while the sum of the two—the mechanical energy E of the oscillator—remains constant. We now consider this situation quantitatively.

The potential energy of a linear oscillator like that of Fig. 16-5 is associated entirely with the spring. Its value depends on how much the spring is stretched or compressed—that is, on $x(t)$. We can use Eqs. 8-11 and 16-3 to find

$$U(t) = \tfrac{1}{2}kx^2 = \tfrac{1}{2}kx_m^2 \cos^2(\omega t + \phi). \tag{16-18}$$

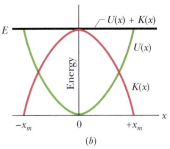

Fig. 16-6 (a) Potential energy $U(t)$, kinetic energy $K(t)$, and mechanical energy E as functions of time t for a linear harmonic oscillator. Note that all energies are positive and that the potential energy and the kinetic energy peak twice during every period. (b) Potential energy $U(x)$, kinetic energy $K(x)$, and mechanical energy E as functions of position x for a linear harmonic oscillator with amplitude x. For $x = 0$ the energy is all kinetic, and for $x = \pm x_m$ it is all potential.

Note carefully that a function written in the form $\cos^2 A$ (as here) means $(\cos A)^2$ and is *not* the same as one written $\cos A^2$, which means $\cos (A^2)$.

The kinetic energy of the system of Fig. 16-5 is associated entirely with the block. Its value depends on how fast the block is moving—that is, on $v(t)$. We can use Eq. 16-6 to find

$$K(t) = \tfrac{1}{2}mv^2 = \tfrac{1}{2}m\omega^2 x_m^2 \sin^2(\omega t + \phi). \qquad (16\text{-}19)$$

If we use Eq. 16-12 to substitute k/m for ω^2, we can write Eq. 16-19 as

$$K(t) = \tfrac{1}{2}mv^2 = \tfrac{1}{2}kx_m^2 \sin^2(\omega t + \phi). \qquad (16\text{-}20)$$

The mechanical energy follows from Eqs. 16-18 and 16-20 and is

$$\begin{aligned}
E &= U + K \\
&= \tfrac{1}{2}kx_m^2 \cos^2(\omega t + \phi) + \tfrac{1}{2}kx_m^2 \sin^2(\omega t + \phi) \\
&= \tfrac{1}{2}kx_m^2[\cos^2(\omega t + \phi) + \sin^2(\omega t + \phi)].
\end{aligned}$$

For any angle α,

$$\cos^2 \alpha + \sin^2 \alpha = 1.$$

Thus, the quantity in the square brackets above is unity and we have

$$E = U + K = \tfrac{1}{2}kx_m^2. \qquad (16\text{-}21)$$

The mechanical energy of a linear oscillator is indeed constant and independent of time. The potential energy and kinetic energy of a linear oscillator are shown as functions of time t in Fig. 16-6a, and as functions of displacement x in Fig. 16-6b.

You might now understand why an oscillating system normally contains an element of springiness and an element of inertia: The former stores its potential energy and the latter stores its kinetic energy.

✔**CHECKPOINT 3:**   In Fig. 16-5, the block has a kinetic energy of 3 J and the spring has an elastic potential energy of 2 J when the block is at $x = +2.0$ cm. (a) What is the kinetic energy when the block is at $x = 0$? What are the elastic potential energies when the block is at (b) $x = -2.0$ cm and (c) $x = -x_m$?

Sample Problem 16-3

(a) What is the mechanical energy E of the linear oscillator of Sample Problem 16-1? (Initially, the block's position is $x = 11$ cm and its speed is $v = 0$. Spring constant k is 65 N/m.)

SOLUTION: The **Key Idea** here is that the mechanical energy E (the sum of the kinetic energy $K = \tfrac{1}{2}mv^2$ of the block and the potential energy $U = \tfrac{1}{2}kx^2$ of the spring) is constant throughout the motion of the oscillator. Thus, we can evaluate E at any point during the motion. Because we are given the initial conditions of the oscillator as $x = 11$ cm and $v = 0$, let us evaluate E for those conditions. We find

$$E = K + U = \tfrac{1}{2}mv^2 + \tfrac{1}{2}kx^2 = 0 + \tfrac{1}{2}(65 \text{ N/m})(0.11 \text{ m})^2$$
$$= 0.393 \text{ J} \approx 0.39 \text{ J.} \qquad \text{(Answer)}$$

(b) What are the potential energy U and kinetic energy K of the oscillator when the block is at $x = \tfrac{1}{2}x_m$? What are they when the block is at $x = -\tfrac{1}{2}x_m$?

SOLUTION: The **Key Idea** here is that, because we are given the location of the block, we can easily find the spring's potential energy with $U = \tfrac{1}{2}kx^2$. For $x = \tfrac{1}{2}x_m$, we have

$$U = \tfrac{1}{2}kx^2 = \tfrac{1}{2}k(\tfrac{1}{2}x_m)^2 = (\tfrac{1}{2})(\tfrac{1}{4})kx_m^2.$$

We can substitute for k and x_m, or we can use the **Key Idea** that the total mechanical energy, which we know from part (a), is $\tfrac{1}{2}kx_m^2$. That idea allows us to write, from the above equation,

$$U = \tfrac{1}{4}(\tfrac{1}{2}kx_m^2) = \tfrac{1}{4}E = \tfrac{1}{4}(0.393 \text{ J}) = 0.098 \text{ J.} \quad \text{(Answer)}$$

Now, using the **Key Idea** of (a) (namely, $E = K + U$), we can write

$$K = E - U = 0.393 \text{ J} - 0.098 \text{ J} \approx 0.30 \text{ J.} \quad \text{(Answer)}$$

By repeating these calculations for $x = -\tfrac{1}{2}x_m$, we would find the same answers for that displacement, consistent with the left–right symmetry of Fig. 16-6b.

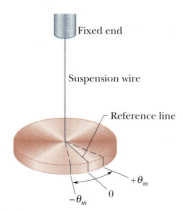

Fixed end

Suspension wire

Reference line

$+\theta_m$

0

$-\theta_m$

Fig. 16-7 An angular simple harmonic oscillator, or torsion pendulum, is an angular version of the linear simple harmonic oscillator of Fig. 16-5. The disk oscillates in a horizontal plane; the reference line oscillates with angular amplitude θ_m. The twist in the suspension wire stores potential energy as a spring does and provides the restoring torque.

16-5 An Angular Simple Harmonic Oscillator

Figure 16-7 shows an angular version of a simple harmonic oscillator; the element of springiness or elasticity is associated with the twisting of a suspension wire rather than the extension and compression of a spring as we previously had. The device is called a **torsion pendulum,** with *torsion* referring to the twisting.

If we rotate the disk in Fig. 16-7 by some angular displacement θ from its rest position (where the reference line is at $\theta = 0$) and release it, it will oscillate about that position in **angular simple harmonic motion.** Rotating the disk through an angle θ in either direction introduces a restoring torque given by

$$\tau = -\kappa\theta. \tag{16-22}$$

Here κ (Greek *kappa*) is a constant, called the **torsion constant,** that depends on the length, diameter, and material of the suspension wire.

Comparison of Eq. 16-22 with Eq. 16-10 leads us to suspect that Eq. 16-22 is the angular form of Hooke's law, and that we can transform Eq. 16-13, which gives the period of linear SHM, into an equation for the period of angular SHM: We replace the spring constant k in Eq. 16-13 with its equivalent, the constant κ of Eq. 16-22, and we replace the mass m in Eq. 16-13 with *its* equivalent, the rotational inertia I of the oscillating disk. These replacements lead to

$$T = 2\pi\sqrt{\frac{I}{\kappa}} \quad \text{(torsion pendulum)}, \tag{16-23}$$

which is the correct equation for the period of an angular simple harmonic oscillator, or torsion pendulum.

Tactic 3: *Identifying Angular SHM*
When a system undergoes angular simple harmonic motion, its angular acceleration α and angular displacement θ are related by an equation of the form

$$\alpha = -(\text{a positive constant})\theta.$$

This equation is the angular equivalent of Eq. 16-8 ($a = -\omega^2 x$). It says that the angular acceleration α is proportional to the angular displacement θ from the equilibrium position but tends to rotate the system in the direction opposite the displacement. If you have an expression of this form, you can identify the positive constant as being ω^2, and then you can determine ω, f, and T.

You can also identify angular SHM if you have an expression for the torque τ in terms of the angular displacement, because that expression must be in the form of Eq. 16-22 ($\tau = -\kappa\theta$) or

$$\tau = -(\text{a positive constant})\theta.$$

This equation is the angular equivalent of Eq. 16-10 ($F = -kx$). It says that the torque τ is proportional to the angular displacement θ from the equilibrium position but tends to rotate the system in the opposite direction. If you have an expression of this form, then you can identify the positive constant as being the system's torsion constant κ. If you know the rotational inertia I of the system, you can then determine T with Eq. 16-23.

Sample Problem 16-4

Figure 16-8a shows a thin rod whose length L is 12.4 cm and whose mass m is 135 g, suspended at its midpoint from a long wire. Its period T_a of angular SHM is measured to be 2.53 s. An irregularly shaped object, which we call object X, is then hung from the same wire, as in Fig. 16-8b, and its period T_b is found to be 4.76 s. What is the rotational inertia of object X about its suspension axis?

SOLUTION: The Key Idea here is that the rotational inertia of either the rod or object X is related to the measured period by Eq. 16-23. In Table 11-2e, the rotational inertia of a thin rod about a perpendic-

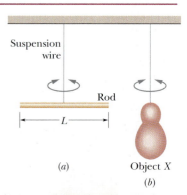

Fig. 16-8 Sample Problem 16-4. Two torsion pendulums, consisting of (a) a wire and a rod and (b) the same wire and an irregularly shaped object.

Suspension wire

Rod

L

(a)

Object X

(b)

ular axis through its midpoint is given as $\frac{1}{12}mL^2$. Thus, we have, for the rod in Fig. 16-8a,

$$I_a = \tfrac{1}{12}mL^2 = (\tfrac{1}{12})(0.135 \text{ kg})(0.124 \text{ m})^2$$
$$= 1.73 \times 10^{-4} \text{ kg} \cdot \text{m}^2.$$

Now let us write Eq. 16-23 twice, once for the rod and once for object X:

$$T_a = 2\pi\sqrt{\frac{I_a}{\kappa}} \quad \text{and} \quad T_b = 2\pi\sqrt{\frac{I_b}{\kappa}}.$$

The constant κ, which is a property of the wire, is the same for both figures; only the periods and the rotational inertias differ.

Let us square each of these equations, divide the second by the first, and solve the resulting equation for I_b. The result is

$$I_b = I_a\frac{T_b^2}{T_a^2} = (1.73 \times 10^{-4} \text{ kg} \cdot \text{m}^2)\frac{(4.76 \text{ s})^2}{(2.53 \text{ s})^2}$$
$$= 6.12 \times 10^{-4} \text{ kg} \cdot \text{m}^2. \qquad \text{(Answer)}$$

16-6 Pendulums

We turn now to a class of simple harmonic oscillators in which the springiness is associated with the gravitational force rather than with the elastic properties of a twisted wire or a compressed or stretched spring.

The Simple Pendulum

If you hang an apple at the end of a long thread fixed at its upper end, and then set the apple swinging back and forth a small distance, you easily see that the apple's motion is periodic. Is it, in fact, simple harmonic motion? If so, what is the period T? To answer, we consider a **simple pendulum,** which consists of a particle of mass m (called the *bob* of the pendulum) suspended from one end of an unstretchable, massless string of length L that is fixed at the other end, as in Fig. 16-9a. The bob is free to swing back and forth in the plane of the page, to the left and right of a vertical line through the pendulum's pivot point.

The forces acting on the bob are the force $\vec{T}$ from the string and the gravitational force $\vec{F}_g$, as shown in Fig. 16-9b where the string makes an angle θ with the vertical. We resolve $\vec{F}_g$ into a radial component $F_g \cos\theta$ and a component $F_g \sin\theta$ that is tangent to the path taken by the bob. This tangential component produces a restoring torque about the pendulum's pivot point, because it always acts opposite the displacement of the bob so as to bring the bob back toward its central location. That location is called the *equilibrium position* ($\theta = 0$), because the pendulum would be at rest there were it not swinging.

From Eq. 11-33 ($\tau = r_\perp F$), we can write this restoring torque as

$$\tau = -L(F_g \sin\theta), \qquad (16\text{-}24)$$

where the minus sign indicates that the torque acts to reduce θ, and L is the moment arm of the force component $F_g \sin\theta$ about the pivot point. Substituting Eq. 16-24 into Eq. 11-36 ($\tau = I\alpha$) and then substituting mg as the magnitude of F_g, we obtain

$$-L(mg \sin\theta) = I\alpha, \qquad (16\text{-}25)$$

where I is the pendulum's rotational inertia about the pivot point and α is its angular acceleration about that point.

We can simplify Eq. 16-25 if we assume the angle θ is small, for then we can approximate $\sin\theta$ with θ (expressed in radian measure). (As an example, if $\theta = 5.00° = 0.0873$ rad, then $\sin\theta = 0.0872$, a difference of only about 0.1%.) With that approximation and some rearranging, we then have

$$\alpha = -\frac{mgL}{I}\theta. \qquad (16\text{-}26)$$

This equation is the angular equivalent of Eq. 16-8, the hallmark of SHM. It tells

Fig. 16-9 (a) A simple pendulum. (b) The forces acting on the bob are the gravitational force $\vec{F}_g$ and the force $\vec{T}$ from the string. The tangential component $F_g \sin\theta$ of the gravitational force is a restoring force that tends to bring the pendulum back to its central position.

us that the angular acceleration α of the pendulum is proportional to the angular displacement θ but opposite in sign. Thus, as the pendulum bob moves to, say, the right as in Fig. 16-9a, its acceleration *to the left* increases until it stops and begins moving to the left. Then, when it is on the left, its acceleration to the right tends to return it to the right, and so on, as it swings back and forth in SHM. More precisely, the motion of a *simple pendulum swinging through only small angles* is approximately SHM. We can state this restriction to small angles another way: The **angular amplitude** θ_m of the motion (the maximum angle of swing) must be small.

Comparing Eqs. 16-26 and Eq. 16-8, we see that the angular frequency of the pendulum is $\omega = \sqrt{mgL/I}$. Next, if we substitute this expression for ω into Eq. 16-5 ($\omega = 2\pi/T$), we see that the period of the pendulum may be written as

$$T = 2\pi\sqrt{\frac{I}{mgL}}. \tag{16-27}$$

All the mass of a simple pendulum is concentrated in the mass m of the particle-like bob, which is at radius L from the pivot point. Thus, we can use Eq. 11-26 ($I = mr^2$) to write $I = mL^2$ for the rotational inertia of the pendulum. Substituting this into Eq. 16-27 and simplifying then yield

$$T = 2\pi\sqrt{\frac{L}{g}} \quad \text{(simple pendulum, small amplitude)} \tag{16-28}$$

as a simpler expression for the period of a simple pendulum swinging through only small angles. (We assume small-angle swinging in the problems of this chapter.)

The Physical Pendulum

A real pendulum, usually called a **physical pendulum,** can have a complicated distribution of mass, much different from that of a simple pendulum. Does a physical pendulum also undergo SHM? If so, what is its period?

Figure 16-10 shows an arbitrary physical pendulum displaced to one side by angle θ. The gravitational force $\vec{F}_g$ acts at its center of mass C, at a distance h from the pivot point O. In spite of their shapes, comparison of Figs. 16-10 and 16-9b reveals only one important difference between an arbitrary physical pendulum and a simple pendulum. For a physical pendulum the restoring component $F_g \sin\theta$ of the gravitational force has a moment arm of distance h about the pivot point, rather than of string length L. In all other respects, an analysis of the physical pendulum would duplicate our analysis of the simple pendulum up through Eq. 16-27. Again, (for small θ_m) we would find that the motion is approximately SHM.

If we replace L with h in Eq. 16-27, we can write the period of a physical pendulum as

$$T = 2\pi\sqrt{\frac{I}{mgh}} \quad \text{(physical pendulum, small amplitude).} \tag{16-29}$$

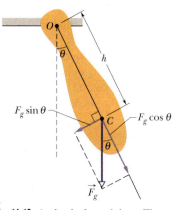

Fig. 16-10 A physical pendulum. The restoring torque is $hF_g \sin\theta$. When $\theta = 0$, center of mass C hangs directly below pivot point O.

As with the simple pendulum, I is the rotational inertia of the pendulum about O. However, now I is not simply mL^2 (it depends on the shape of the physical pendulum), but it is still proportional to m.

A physical pendulum will not swing if it pivots at its center of mass. Formally, this corresponds to putting $h = 0$ in Eq. 16-29. That equation then predicts $T \to \infty$, which implies that such a pendulum will never complete one swing.

Corresponding to any physical pendulum that oscillates about a given pivot point O with period T is a simple pendulum of length L_0 with the same period T. We can find L_0 with Eq. 16-28. The point along the physical pendulum at distance L_0 from point O is called the *center of oscillation* of the physical pendulum for the given suspension point.

Measuring *g*

We can use a physical pendulum to measure the free-fall acceleration g at a particular location on Earth's surface. (Countless thousands of such measurements have been made during geophysical prospecting.)

To analyze a simple case, take the pendulum to be a uniform rod of length L, suspended from one end. For such a pendulum, h in Eq. 16-29, the distance between the pivot point and the center of mass, is $\frac{1}{2}L$. Table 11-2e tells us that the rotational inertia of this pendulum about a perpendicular axis through its center of mass is $\frac{1}{12}mL^2$. From the parallel-axis theorem of Eq. 11-29 ($I = I_{com} + Mh^2$), we then find that the rotational inertia about a perpendicular axis through one end of the rod is

$$I = I_{com} + mh^2 = \tfrac{1}{12}mL^2 + m(\tfrac{1}{2}L)^2 = \tfrac{1}{3}mL^2. \qquad (16\text{-}30)$$

If we put $h = \frac{1}{2}L$ and $I = \frac{1}{3}mL^2$ in Eq. 16-29 and solve for g, we find

$$g = \frac{8\pi^2 L}{3T^2}. \qquad (16\text{-}31)$$

Thus, by measuring L and the period T, we can find the value of g at the pendulum's location. (If precise measurements are to be made, a number of refinements are needed, such as swinging the pendulum in an evacuated chamber.)

Sample Problem 16-5

In Fig. 16-11a, a meter stick swings about a pivot point at one end, at distance h from its center of mass.

(a) What is its period of oscillation T?

SOLUTION: One Key Idea here is that the stick is not a simple pendulum because its mass is not concentrated in a bob at the end opposite the pivot point—so the stick is a physical pendulum. Then its period is given by Eq. 16-29, for which we need the rotational inertia I of the stick about the pivot point. We can treat the stick as a uniform rod of length L and mass m. Then Eq. 16-30 tells us that $I = \frac{1}{3}mL^2$, and the distance h in Eq. 16-29 is $\frac{1}{2}L$. Substituting these quantities into Eq. 16-29, we find

$$T = 2\pi\sqrt{\frac{I}{mgh}} = 2\pi\sqrt{\frac{\tfrac{1}{3}mL^2}{mg(\tfrac{1}{2}L)}} = 2\pi\sqrt{\frac{2L}{3g}} \qquad (16\text{-}32)$$

$$= 2\pi\sqrt{\frac{(2)(1.00\ \text{m})}{(3)(9.8\ \text{m/s}^2)}} = 1.64\ \text{s}. \qquad \text{(Answer)}$$

Note that the result is independent of the pendulum's mass m.

(b) What is the distance L_0 between the pivot point O of the stick and the center of oscillation of the stick?

SOLUTION: The Key Idea here is that we want the length L_0 of the simple pendulum (drawn in Fig. 16-11b) that has the same period as the physical pendulum (the stick) of Fig. 16-11a. Setting

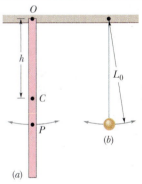

Fig. 16-11 Sample Problem 16-5.
(a) A meter stick suspended from one end as a physical pendulum. (b) A simple pendulum whose length L_0 is chosen so that the periods of the two pendulums are equal. Point P on the pendulum of (a) marks the center of oscillation.

Eqs. 16-28 and 16-32 equal yields

$$T = 2\pi\sqrt{\frac{L_0}{g}} = 2\pi\sqrt{\frac{2L}{3g}}.$$

You can see by inspection that

$$L_0 = \tfrac{2}{3}L = (\tfrac{2}{3})(100\ \text{cm}) = 66.7\ \text{cm}. \qquad \text{(Answer)}$$

In Fig. 16-11a, point P marks this distance from suspension point O. Thus, point P is the stick's center of oscillation for the given suspension point.

✔CHECKPOINT 4:   Three physical pendulums, of masses m_0, $2m_0$, and $3m_0$, have the same shape and size and are suspended at the same point. Rank the masses according to the periods of the pendulums, greatest period first.

Sample Problem 16-6

In Fig. 16-12, a penguin (obviously skilled in aquatic sports) dives from a uniform board that is hinged at the left and attached to a spring at the right. The board has length $L = 2.0$ m and mass $m = 12$ kg; the spring constant k is 1300 N/m. When the penguin dives, it leaves the board and spring oscillating with a small amplitude. Assume that the board is stiff enough not to bend, and find the period T of the oscillations.

SOLUTION: Since a spring is involved, we might guess that the oscillations are SHM, but we shall not assume that. Instead, we use the following Key Idea: If the board is in SHM, then the acceleration and displacement of the oscillating end of the board must be related by an expression in the form of Eq. 16-8 ($a = -\omega^2 x$). If so, we shall be able to find ω and then the desired T from the expression. Let us check by finding the relation between the acceleration and displacement of the board's right end.

Because the board rotates about the hinge as one end oscillates, we are concerned with a torque $\vec{\tau}$ on the board about the hinge. That torque is due to the force $\vec{F}$ on the board from the spring. Because $\vec{F}$ varies with time, $\vec{\tau}$ must also. However, at any given instant we can relate the magnitudes of $\vec{\tau}$ and $\vec{F}$ with Eq. 11-31 ($\tau = rF \sin \phi$). Here we have

$$\tau = LF \sin 90°, \qquad (16\text{-}33)$$

where L is the moment arm of force $\vec{F}$ and 90° is the angle between the moment arm and the force's line of action. Combining Eq. 16-33 with Eq. 11-36 ($\tau = I\alpha$) gives us

$$LF = I\alpha, \qquad (16\text{-}34)$$

where I is the board's rotational inertia about the hinge, and α is its angular acceleration about that point. We may treat the board as a thin rod pivoted about one end. Then, from Eq. 16-30, the board's rotational inertia I is $\frac{1}{3}mL^2$.

Now let us mentally erect a vertical x through the oscillating right end of the board, with the positive direction upward. Then the force on the right end of the board from the spring is $F = -kx$, where x is the vertical displacement of the right end.

Substituting these expressions for I and F into Eq. 16-34 gives us

$$-Lkx = \frac{mL^2\alpha}{3}. \qquad (16\text{-}35)$$

We now have a mixture of linear displacement x (vertically) and rotational acceleration α (about the hinge). We can replace α in Eq. 16-35 with the (linear) acceleration a along the x axis by substituting according to Eq. 11-22 ($a_t = \alpha r$) for tangential acceleration. Here the tangential acceleration is a and the radius of rotation r is L, so $\alpha = a/L$. With that substitution, Eq. 16-35 becomes

$$-Lkx = \frac{mL^2 a}{3L},$$

which yields

$$a = -\frac{3k}{m}x. \qquad (16\text{-}36)$$

Equation 16-36 is, in fact, of the same form as Eq. 16-8 ($a = -\omega^2 x$). Therefore, the board does indeed undergo SHM, and comparison of Eqs. 16-36 and 16-8 shows that

$$\omega^2 = \frac{3k}{m},$$

which gives $\omega = \sqrt{3k/m}$. Using Eq. 16-5 ($\omega = 2\pi/T$) to find T then gives us

$$T = 2\pi\sqrt{\frac{m}{3k}} = 2\pi\sqrt{\frac{12 \text{ kg}}{3(1300 \text{ N/m})}}$$

$$= 0.35 \text{ s.} \qquad \text{(Answer)}$$

Perhaps surprisingly, the period is independent of the board's length L.

Fig. 16-12 Sample Problem 16-6. The dive by the penguin causes the board and spring to oscillate; the board pivots about the hinge at the left.

16-7 Simple Harmonic Motion and Uniform Circular Motion

In 1610, Galileo, using his newly constructed telescope, discovered the four principal moons of Jupiter. Over weeks of observation, each moon seemed to him to be moving back and forth relative to the planet in what today we would call simple harmonic motion; the disk of the planet was the midpoint of the motion. The record of Galileo's observations, written in his own hand, is still available. A. P. French of MIT used Galileo's data to work out the position of the moon Callisto relative to

Fig. 16-13 The angle between Jupiter and its moon Callisto as seen from Earth. The circles are based on Galileo's 1610 measurements. The curve is a best fit, strongly suggesting simple harmonic motion. At Jupiter's mean distance, 10 minutes of arc corresponds to about 2×10^6 km. (Adapted from A. P. French, *Newtonian Mechanics*, W. W. Norton & Company, New York, 1971, p. 288.)

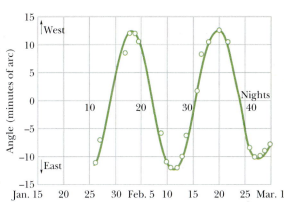

Jupiter. In the results shown in Fig. 16-13, the circles are based on Galileo's observations and the curve is a best fit to the data. The curve strongly suggests Eq. 16-3, the displacement function for SHM. A period of about 16.8 days can be measured from the plot.

Actually, Callisto moves with essentially constant speed in an essentially circular orbit around Jupiter. Its true motion—far from being simple harmonic—is uniform circular motion. What Galileo saw—and what you can see with a good pair of binoculars and a little patience—is the projection of this uniform circular motion on a line in the plane of the motion. We are led by Galileo's remarkable observations to the conclusion that simple harmonic motion is uniform circular motion viewed edge-on. In more formal language:

> Simple harmonic motion is the projection of uniform circular motion on a diameter of the circle in which the latter motion occurs.

Figure 16-14a gives an example. It shows a *reference particle P'* moving in uniform circular motion with (constant) angular speed ω in a *reference circle*. The radius x_m of the circle is the magnitude of the particle's position vector. At any time t, the angular position of the particle is $\omega t + \phi$, where ϕ is its angular position at $t = 0$.

The projection of particle P' onto the x axis is a point P, which we take to be a second particle. The projection of the position vector of particle P' onto the x axis gives the location $x(t)$ of P. Thus, we find

$$x(t) = x_m \cos(\omega t + \phi),$$

which is precisely Eq. 16-3. Our conclusion is correct. If reference particle P' moves in uniform circular motion, its projection particle P moves in simple harmonic motion along a diameter of the circle.

Figure 16-14b shows the velocity $\vec{v}$ of the reference particle. From Eq. 11-18 ($v = \omega r$), the magnitude of the velocity vector is ωx_m; its projection on the x axis is

$$v(t) = -\omega x_m \sin(\omega t + \phi),$$

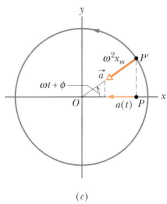

Fig. 16-14 (a) A reference particle P' moving with uniform circular motion in a reference circle of radius x_m. Its projection P on the x axis executes simple harmonic motion. (b) The projection of the velocity $\vec{v}$ of the reference particle is the velocity of SHM. (c) The projection of the radial acceleration $\vec{a}$ of the reference particle is the acceleration of SHM.

which is exactly Eq. 16-6. The minus sign appears because the velocity component of P in Fig. 16-14b is directed to the left, in the negative direction of x.

Figure 16-14c shows the radial acceleration $\vec{a}$ of the reference particle. From Eq. 11-23 ($a_r = \omega^2 r$), the magnitude of the radial acceleration vector is $\omega^2 x_m$; its projection on the x axis is

$$a(t) = -\omega^2 x_m \cos(\omega t + \phi),$$

which is exactly Eq. 16-7. Thus, whether we look at the displacement, the velocity, or the acceleration, the projection of uniform circular motion is indeed simple harmonic motion.

16-8 Damped Simple Harmonic Motion

A pendulum will swing only briefly under water, because the water exerts a drag force on the pendulum that quickly eliminates the motion. A pendulum swinging in air does better, but still the motion dies out eventually, because the air exerts a drag force on the pendulum (and friction acts at its support), transferring energy from the pendulum's motion.

When the motion of an oscillator is reduced by an external force, the oscillator and its motion are said to be **damped.** An idealized example of a damped oscillator is shown in Fig. 16-15, where a block with mass m oscillates vertically on a spring with spring constant k. From the block, a rod extends to a vane (both assumed massless) that is submerged in a liquid. As the vane moves up and down, the liquid exerts an inhibiting drag force on it and thus on the entire oscillating system. With time, the mechanical energy of the block–spring system decreases, as energy is transferred to thermal energy of the liquid and vane.

Let us assume the liquid exerts a **damping force** $\vec{F}_d$ that is proportional in magnitude to the velocity $\vec{v}$ of the vane and block (an assumption that is accurate if the vane moves slowly). Then, for components along the x axis in Fig. 16-15, we have

$$F_d = -bv, \tag{16-37}$$

where b is a **damping constant** that depends on the characteristics of both the vane and the liquid and has the SI unit of kilogram per second. The minus sign indicates that $\vec{F}_d$ opposes the motion.

The force on the block from the spring is $F_s = -kx$. Let us assume that the gravitational force on the block is negligible compared to F_d and F_s. Then we can write Newton's second law for components along the x axis ($F_{net,x} = ma_x$) as

$$-bv - kx = ma. \tag{16-38}$$

Substituting dx/dt for v and d^2x/dt^2 for a and rearranging give us the differential equation

$$m\frac{d^2x}{dt^2} + b\frac{dx}{dt} + kx = 0. \tag{16-39}$$

The solution of this equation is

$$x(t) = x_m e^{-bt/2m} \cos(\omega' t + \phi), \tag{16-40}$$

where x_m is the amplitude and ω' is the angular frequency of the damped oscillator. This angular frequency is given by

$$\omega' = \sqrt{\frac{k}{m} - \frac{b^2}{4m^2}}. \tag{16-41}$$

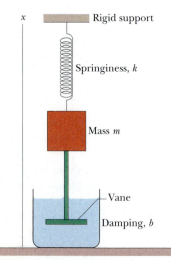

x Rigid support

Springiness, k

Mass m

Vane

Damping, b

Fig. 16-15 An idealized damped simple harmonic oscillator. A vane immersed in a liquid exerts a damping force on the block as the block oscillates parallel to the x axis.

If $b = 0$ (there is no damping), then Eq. 16-41 reduces to Eq. 16-12 ($\omega = \sqrt{k/m}$) for the angular frequency of an undamped oscillator, and Eq. 16-40 reduces to Eq. 16-3 for the displacement of an undamped oscillator. If the damping constant is small but not zero (so that $b \ll \sqrt{km}$), then $\omega' \approx \omega$.

We can regard Eq. 16-40 as a cosine function whose amplitude, which is $x_m e^{-bt/2m}$, gradually decreases with time, as Fig. 16-16 suggests. For an undamped oscillator, the mechanical energy is constant and is given by Eq. 16-21 ($E = \frac{1}{2}kx_m^2$). If the oscillator is damped, the mechanical energy is not constant but decreases with time. If the damping is small, we can find $E(t)$ by replacing x_m in Eq. 16-21 with $x_m e^{-bt/2m}$, the amplitude of the damped oscillations. By doing so, we find that

$$E(t) \approx \tfrac{1}{2}kx_m^2\, e^{-bt/m}, \qquad (16\text{-}42)$$

which tells us that, like the amplitude, the mechanical energy decreases exponentially with time.

✔**CHECKPOINT 5:** Here are three sets of values for the spring constant, damping constant, and mass for the damped oscillator of Fig. 16-15. Rank the sets according to the time required for the mechanical energy to decrease to one-fourth of its initial value, greatest first.

Set 1	$2k_0$	b_0	m_0
Set 2	k_0	$6b_0$	$4m_0$
Set 3	$3k_0$	$3b_0$	m_0

Sample Problem 16-7

For the damped oscillator of Fig. 16-15, $m = 250$ g, $k = 85$ N/m, and $b = 70$ g/s.

(a) What is the period of the motion?

SOLUTION: The **Key Idea** here is that because $b \ll \sqrt{km} = 4.6$ kg/s, the period is approximately that of the undamped oscillator. From Eq. 16-13, we then have

$$T = 2\pi\sqrt{\frac{m}{k}} = 2\pi\sqrt{\frac{0.25\ \text{kg}}{85\ \text{N/m}}} = 0.34\ \text{s}. \quad \text{(Answer)}$$

(b) How long does it take for the amplitude of the damped oscillations to drop to half its initial value?

SOLUTION: Now the **Key Idea** is that the amplitude at time t is displayed in Eq. 16-40 as $x_m e^{-bt/2m}$. It has the value x_m at $t = 0$. Thus, we must find the value of t for which

$$x_m e^{-bt/2m} = \tfrac{1}{2}x_m.$$

Canceling x_m and taking the natural logarithm of the equation that remains, we have $\ln \frac{1}{2}$ on the right side and

$$\ln(e^{-bt/2m}) = -bt/2m$$

on the left side. Thus,

$$t = \frac{-2m \ln \frac{1}{2}}{b} = \frac{-(2)(0.25\ \text{kg})(\ln \frac{1}{2})}{0.070\ \text{kg/s}}$$

$$= 5.0\ \text{s.} \qquad \text{(Answer)}$$

Because $T = 0.34$ s, this is about 15 periods of oscillation.

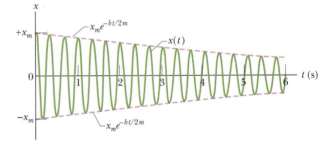

Fig. 16-16 The displacement function $x(t)$ for the damped oscillator of Fig. 16-15, with values given in Sample Problem 16-7. The amplitude, which is $x_m e^{-bt/2m}$, decreases exponentially with time.

(c) How long does it take for the mechanical energy to drop to one-half its initial value?

SOLUTION: Here the **Key Idea** is that, from Eq. 16-42, the mechanical energy at time t is $\frac{1}{2}kx_m^2\, e^{-bt/m}$. It has the value $\frac{1}{2}kx_m^2$ at $t = 0$. Thus, we must find the value of t for which

$$\tfrac{1}{2}kx_m^2\, e^{-bt/m} = \tfrac{1}{2}(\tfrac{1}{2}kx_m^2).$$

If we divide both sides of this equation by $\frac{1}{2}kx_m^2$ and solve for t as we did above, we find

$$t = \frac{-m \ln \frac{1}{2}}{b} = \frac{-(0.25\ \text{kg})(\ln \frac{1}{2})}{0.070\ \text{kg/s}} = 2.5\ \text{s.} \quad \text{(Answer)}$$

This is exactly half the time we calculated in (b), or about 7.5 periods of oscillation. Figure 16-16 was drawn to illustrate this sample problem.

16-9   Forced Oscillations and Resonance

A person swinging in a swing without anyone pushing it is an example of *free oscillation.* However, if someone pushes the swing periodically, the swing has *forced,* or *driven, oscillations.* *Two* angular frequencies are associated with a system undergoing driven oscillations: (1) the *natural* angular frequency ω of the system, which is the angular frequency at which it would oscillate if it were suddenly disturbed and then left to oscillate freely, and (2) the angular frequency ω_d of the external driving force causing the driven oscillations.

We can use Fig. 16-15 to represent an idealized forced simple harmonic oscillator if we allow the structure marked "rigid support" to move up and down at a variable angular frequency ω_d. Such a forced oscillator oscillates at the angular frequency ω_d of the driving force, and its displacement $x(t)$ is given by

$$x(t) = x_m \cos(\omega_d t + \phi), \qquad (16\text{-}43)$$

where x_m is the amplitude of the oscillations.

How large the displacement amplitude x_m is depends on a complicated function of ω_d and ω. The velocity amplitude v_m of the oscillations is easier to describe: it is greatest when

$$\omega_d = \omega \qquad \text{(resonance)}, \qquad (16\text{-}44)$$

a condition called **resonance.** Equation 16-44 is also *approximately* the condition at which the displacement amplitude x_m of the oscillations is greatest. Thus, if you push a swing at its natural angular frequency, the displacement and velocity amplitudes will increase to large values, a fact that children learn quickly by trial and error. If you push at other angular frequencies, either higher or lower, the displacement and velocity amplitudes will be smaller.

Figure 16-17 shows how the displacement amplitude of an oscillator depends on the angular frequency ω_d of the driving force, for three values of the damping coefficient b. Note that for all three the amplitude is approximately greatest when $\omega_d/\omega = 1$—that is, when the resonance condition of Eq. 16-44 is satisfied. The curves of Fig. 16-17 show that less damping gives a taller and narrower *resonance peak.*

All mechanical structures have one or more natural angular frequencies, and if a structure is subjected to a strong external driving force that matches one of these

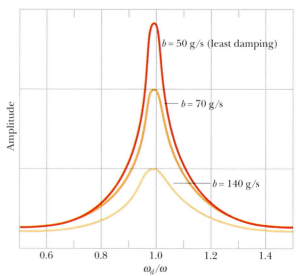

Fig. 16-17 The displacement amplitude x_m of a forced oscillator varies as the angular frequency ω_d of the driving force is varied. The amplitude is greatest *approximately* at $\omega_d/\omega = 1$, the resonance condition. The curves here correspond to three values of the damping constant b.

angular frequencies, the resulting oscillations of the structure may rupture it. Thus, for example, aircraft designers must make sure that none of the natural angular frequencies at which a wing can oscillate matches the angular frequency of the engines in flight. A wing that flaps violently at certain engine speeds would obviously be dangerous.

Mexico's earthquake in September 1985 was a major earthquake (8.1 on the Richter scale), but the seismic waves from it should have been too weak to cause extensive damage when they reached Mexico City about 400 km away. However, Mexico City is largely built on an ancient lake bed, where the soil is still soft with water. Although the amplitude of the seismic waves was weak in the firmer ground en route to Mexico City, their amplitude substantially increased in the loose soil of the city. Acceleration amplitudes of the waves were as much as $0.20g$, and the angular frequency was (surprisingly) concentrated around 3 rad/s. Not only was the ground severely oscillated, but many of the buildings with intermediate height had resonant angular frequencies of about 3 rad/s. Most of those intermediate-height buildings collapsed during the shaking, while shorter buildings (with higher resonant angular frequencies) and taller buildings (with lower resonant angular frequencies) remained standing.

REVIEW & SUMMARY

Frequency   The *frequency f* of periodic or oscillatory motion is the number of oscillations per second. In the SI system, it is measured in hertz:

$$1 \text{ hertz} = 1 \text{ Hz} = 1 \text{ oscillation per second} = 1 \text{ s}^{-1}. \quad (16\text{-}1)$$

Period   The *period T* is the time required for one complete oscillation, or **cycle.** It is related to the frequency by

$$T = \frac{1}{f}. \quad (16\text{-}2)$$

Simple Harmonic Motion   In *simple harmonic motion* (SHM), the displacement $x(t)$ of a particle from its equilibrium position is described by the equation

$$x = x_m \cos(\omega t + \phi) \quad \text{(displacement)}, \quad (16\text{-}3)$$

in which x_m is the **amplitude** of the displacement, the quantity $(\omega t + \phi)$ is the **phase** of the motion, and ϕ is the **phase constant.** The **angular frequency** ω is related to the period and frequency of the motion by

$$\omega = \frac{2\pi}{T} = 2\pi f \quad \text{(angular frequency)}. \quad (16\text{-}5)$$

Differentiating Eq. 16-3 leads to equations for the particle's velocity and acceleration during SHM as functions of time:

$$v = -\omega x_m \sin(\omega t + \phi) \quad \text{(velocity)} \quad (16\text{-}6)$$

and
$$a = -\omega^2 x_m \cos(\omega t + \phi) \quad \text{(acceleration)}. \quad (16\text{-}7)$$

In Eq. 16-6, the positive quantity ωx_m is the **velocity amplitude** v_m of the motion. In Eq. 16-7, the positive quantity $\omega^2 x_m$ is the **acceleration amplitude** a_m of the motion.

The Linear Oscillator   A particle with mass m that moves under the influence of a Hooke's law restoring force given by $F = -kx$ exhibits simple harmonic motion with

$$\omega = \sqrt{\frac{k}{m}} \quad \text{(angular frequency)} \quad (16\text{-}12)$$

and
$$T = 2\pi \sqrt{\frac{m}{k}} \quad \text{(period)}. \quad (16\text{-}13)$$

Such a system is called a **linear simple harmonic oscillator.**

Energy   A particle in simple harmonic motion has, at any time, kinetic energy $K = \frac{1}{2}mv^2$ and potential energy $U = \frac{1}{2}kx^2$. If no friction is present, the mechanical energy $E = K + U$ remains constant even though K and U change.

Pendulums   Examples of devices that undergo simple harmonic motion are the **torsion pendulum** of Fig. 16-7, the **simple pendulum** of Fig. 16-9, and the **physical pendulum** of Fig. 16-10. Their periods of oscillation for small oscillations are, respectively,

$$T = 2\pi\sqrt{I/\kappa}, \quad (16\text{-}23)$$

$$T = 2\pi\sqrt{L/g}, \quad (16\text{-}28)$$

and
$$T = 2\pi\sqrt{I/mgh}. \quad (16\text{-}29)$$

Simple Harmonic Motion and Uniform Circular Motion   Simple harmonic motion is the projection of uniform circular motion onto the diameter of the circle in which the latter motion occurs. Figure 16-14 shows that all parameters of circular motion (position, velocity, and acceleration) project to the corresponding values for simple harmonic motion.

Damped Harmonic Motion   The mechanical energy E in a real oscillating system decreases during the oscillations because external forces, such as a drag force, inhibit the oscillations and transfer mechanical energy to thermal energy. The real oscillator and its motion are then said to be **damped.** If the **damping force** is given by $\vec{F}_d = -b\vec{v}$, where v is the velocity of the oscillator and b is a **damping constant,** then the displacement of the oscillator is given by

$$x(t) = x_m\, e^{-bt/2m} \cos(\omega' t + \phi), \tag{16-40}$$

where ω', the angular frequency of the damped oscillator, is given by

$$\omega' = \sqrt{\frac{k}{m} - \frac{b^2}{4m^2}}. \tag{16-41}$$

If the damping constant is small ($b \ll \sqrt{km}$), then $\omega' \approx \omega$, where ω is the angular frequency of the undamped oscillator. For small b, the mechanical energy E of the oscillator is given by

$$E(t) \approx \tfrac{1}{2}kx_m^2\, e^{-bt/m}. \tag{16-42}$$

Forced Oscillations and Resonance   If an external driving force with angular frequency ω_d acts on an oscillating system with *natural* angular frequency ω, the system oscillates with angular frequency ω_d. The velocity amplitude v_m of the system is greatest when

$$\omega_d = \omega, \tag{16-44}$$

a condition called **resonance.** The amplitude x_m of the system is (approximately) greatest under the same condition.

QUESTIONS

1. Which of the following relationships between the acceleration a and the displacement x of a particle involve SHM: (a) $a = 0.5x$, (b) $a = 400x^2$, (c) $a = -20x$, (d) $a = -3x^2$?

2. Given $x = (2.0 \text{ m}) \cos(5t)$ for SHM and needing to find the velocity at $t = 2$ s, should you substitute for t and then differentiate with respect to t or vice versa?

3. The acceleration $a(t)$ of a particle undergoing SHM is graphed in Fig. 16-18. (a) Which of the labeled points corresponds to the particle at $-x_m$? (b) At point 4, is the velocity of the particle positive, negative, or zero? (c) At point 5, is the particle at $-x_m$, at $+x_m$, at 0, between $-x_m$ and 0, or between 0 and $+x_m$?

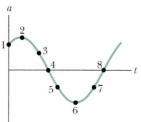

Fig. 16-18 Question 3.

4. Which of the following describe ϕ for the SHM of Fig. 16-19a:

(a) $-\pi < \phi < -\pi/2$,   (b) $\pi < \phi < 3\pi/2$,
(c) $-3\pi/2 < \phi < -\pi$?

(a)   (b)

Fig. 16-19 Questions 4 and 5.

5. The velocity $v(t)$ of a particle undergoing SHM is graphed in Fig. 16-19b. Is the particle momentarily stationary, headed toward $-x_m$, or headed toward $+x_m$ at (a) point A on the graph and (b) point B? Is the particle at $-x_m$, at $+x_m$, at 0, between $-x_m$ and 0, or between 0 and $+x_m$ when its velocity is represented by (c) point A and (d) point B? Is the speed of the particle increasing or decreasing at (e) point A and (f) point B?

6. Figure 16-20 gives, for three situations, the displacements $x(t)$ of a pair of simple harmonic oscillators (A and B) that are identical except for phase. For each pair, what phase shift (in radians and in degrees) is needed to shift the curve for A to coincide with the curve for B? Of the many possible answers, choose the shift with the smallest absolute magnitude.

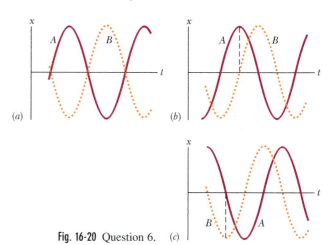

Fig. 16-20 Question 6.   (c)

7. Figures 16-21a and b show the positions of four linear oscillators with identical masses and spring constants, in snapshots at the same instant. What is the phase difference of the two linear oscil-

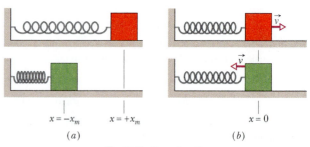

(a)   (b)

Fig. 16-21 Question 7.

lators in (a) Fig. 16-21*a* and (b) Fig. 16-21*b*? (c) What is the phase difference between the red oscillator in Fig. 16-21*a* and the green oscillator in Fig. 16-21*b*?

8. (a) Which curve in Fig. 16-22*a* gives the acceleration $a(t)$ versus displacement $x(t)$ of a simple harmonic oscillator? (b) Which curve in Fig. 16-22*b* gives the velocity $v(t)$ versus $x(t)$?

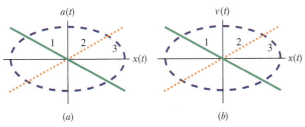

Fig. 16-22 Question 8.

9. In Fig. 16-23, a small block *A* sits on a large block *B* with a certain nonzero coefficient of static friction between the two blocks. Block *B*, which lies on a frictionless surface, is initially at $x = 0$, with the spring at its relaxed length; then we pull the block a distance d to the right and release it. As the spring–blocks system undergoes SHM, with amplitude x_m, block *A* is on the verge of slipping over *B*.

(a) Is the acceleration of block *A* constant or does it vary? (b) Is the magnitude of the frictional force accelerating *A* constant or does it vary? (c) Is *A* more likely to slip at $x = 0$ or at $x = \pm x_m$? (d) If the SHM began with an initial displacement that was greater than d, would slippage then be more likely or less likely? (Warm-up for Problem 16)

Fig. 16-23 Question 9.

10. In Fig. 16-24, a spring–block system is put into SHM in two experiments. In the first, the block is pulled from the equilibrium position through a displacement d_1 and then released. In the second, it is pulled from the equilibrium position through a greater displacement d_2 and then released. Are the (a) amplitude, (b) period, (c) frequency, (d) maximum kinetic energy, and (e) maximum potential energy in the second experiment greater than, less than, or the same as those in the first experiment?

Fig. 16-24 Question 10.

11. Figure 16-25 shows three physical pendulums consisting of identical uniform spheres of the same mass that are rigidly connected by identical rods of negligible mass. Each pendulum is vertical and can pivot about suspension point *O*. Rank the pendulums according to period of oscillation, greatest first.

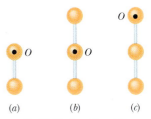

Fig. 16-25 Question 11.

12. Follow-up to Exercise 36: If the speed of the bullet were greater, would the following quantities of the resulting SHM then be greater, less, or the same: (a) amplitude, (b) period, (c) maximum potential energy?

13. You are to build the oscillation transfer device shown in Fig. 16-26. It consists of two spring–block systems hanging from a flexible rod. When the spring of system 1 is stretched and then released, the resulting SHM of system 1 at frequency f_1 oscillates the rod. The rod then provides a driving force on system 2, at the same frequency f_1. You can choose from four springs with spring constants k of 1600, 1500, 1400, and 1200 N/m, and four blocks with masses m of 800, 500, 400, and 200 kg. Mentally determine which spring should go with which block in each of the two systems to maximize the amplitude of oscillations in system 2.

Fig. 16-26 Question 13.

System 1   System 2

EXERCISES & PROBLEMS

SEC. 16-3 The Force Law for Simple Harmonic Motion

1E. An object undergoing simple harmonic motion takes 0.25 s to travel from one point of zero velocity to the next such point. The distance between those points is 36 cm. Calculate (a) the period, (b) the frequency, and (c) the amplitude of the motion.

2E. An oscillating block–spring system takes 0.75 s to begin repeating its motion. Find (a) the period, (b) the frequency in hertz, and (c) the angular frequency in radians per second.

3E. An oscillator consists of a block of mass 0.500 kg connected to a spring. When set into oscillation with amplitude 35.0 cm, the oscillator repeats its motion every 0.500 s. Find (a) the period, (b) the frequency, (c) the angular frequency, (d) the spring constant, (e) the maximum speed, and (f) the magnitude of the maximum force on the block from the spring. **ssm**

4E. What is the maximum acceleration of a platform that oscillates with an amplitude of 2.20 cm at a frequency of 6.60 Hz?

5E. A loudspeaker produces a musical sound by means of the oscillation of a diaphragm. If the amplitude of oscillation is limited to 1.0×10^{-3} mm, what frequencies will result in the magnitude of the diaphragm's acceleration exceeding g? ssm

6E. The scale of a spring balance that reads from 0 to 15.0 kg is 12.0 cm long. A package suspended from the balance is found to oscillate vertically with a frequency of 2.00 Hz. (a) What is the spring constant? (b) How much does the package weigh?

7E. A particle with a mass of 1.00×10^{-20} kg is oscillating with simple harmonic motion with a period of 1.00×10^{-5} s and a maximum speed of 1.00×10^3 m/s. Calculate (a) the angular frequency and (b) the maximum displacement of the particle. ssm

8E. A small body of mass 0.12 kg is undergoing simple harmonic motion of amplitude 8.5 cm and period 0.20 s. (a) What is the magnitude of the maximum force acting on it? (b) If the oscillations are produced by a spring, what is the spring constant?

9E. In an electric shaver, the blade moves back and forth over a distance of 2.0 mm in simple harmonic motion, with frequency 120 Hz. Find (a) the amplitude, (b) the maximum blade speed, and (c) the magnitude of the maximum blade acceleration. ssm

10E. A loudspeaker diaphragm is oscillating in simple harmonic motion with a frequency of 440 Hz and a maximum displacement of 0.75 mm. What are (a) the angular frequency, (b) the maximum speed, and (c) the magnitude of the maximum acceleration?

11E. An automobile can be considered to be mounted on four identical springs as far as vertical oscillations are concerned. The springs of a certain car are adjusted so that the oscillations have a frequency of 3.00 Hz. (a) What is the spring constant of each spring if the mass of the car is 1450 kg and the mass is evenly distributed over the springs? (b) What will be the oscillation frequency if five passengers, averaging 73.0 kg each, ride in the car? (Again, consider an even distribution of mass.)

12E. A body oscillates with simple harmonic motion according to the equation

$$x = (6.0 \text{ m}) \cos[(3\pi \text{ rad/s})t + \pi/3 \text{ rad}].$$

At $t = 2.0$ s, what are (a) the displacement, (b) the velocity, (c) the acceleration, and (d) the phase of the motion? Also, what are (e) the frequency and (f) the period of the motion?

13E. The piston in the cylinder head of a locomotive has a stroke (twice the amplitude) of 0.76 m. If the piston moves with simple harmonic motion with an angular frequency of 180 rev/min, what is its maximum speed? ssm

14P. Figure 16-27 shows an astronaut on a body-mass measuring device (BMMD). Designed for use on orbiting space vehicles, its purpose is to allow astronauts to measure their mass in the "weightless" conditions in Earth orbit. The BMMD is a spring-mounted chair; an astronaut measures his or her period of oscillation in the chair; the mass follows from the formula for the period of an oscillating block–spring system. (a) If M is the mass of the astronaut and m the effective mass of that part of the BMMD that also oscillates, show that

$$M = (k/4\pi^2)T^2 - m,$$

where T is the period of oscillation and k is the spring constant. (b) The spring constant was $k = 605.6$ N/m for the BMMD on

Fig. 16-27 Problem 14.

Skylab Mission Two; the period of oscillation of the empty chair was 0.90149 s. Calculate the effective mass of the chair. (c) With an astronaut in the chair, the period of oscillation became 2.08832 s. Calculate the mass of the astronaut.

15P. At a certain harbor, the tides cause the ocean surface to rise and fall a distance d (from highest level to lowest level) in simple harmonic motion, with a period of 12.5 h. How long does it take for the water to fall a distance $d/4$ from its highest level?

16P. In Fig. 16-28, two blocks ($m = 1.0$ kg and $M = 10$ kg) and a spring ($k = 200$ N/m) are arranged on a horizontal, frictionless surface. The coefficient of static friction between the two blocks is 0.40. What amplitude of simple harmonic motion of the spring–blocks system puts the smaller block on the verge of slipping over the larger block?

Fig. 16-28 Problem 16.

17P. A block is on a horizontal surface (a shake table) that is moving back and forth horizontally with simple harmonic motion of frequency 2.0 Hz. The coefficient of static friction between block and surface is 0.50. How great can the amplitude of the SHM be if the block is not to slip along the surface? ssm www

18P. A block rides on a piston that is moving vertically with simple harmonic motion. (a) If the SHM has period 1.0 s, at what amplitude of motion will the block and piston separate? (b) If the piston has an amplitude of 5.0 cm, what is the maximum frequency for which the block and piston will be in contact continuously?

19P. An oscillator consists of a block attached to a spring ($k = 400$ N/m). At some time t, the position (measured from the system's equilibrium location), velocity, and acceleration of the block are $x = 0.100$ m, $v = -13.6$ m/s, and $a = -123$ m/s². Calculate (a) the frequency of oscillation, (b) the mass of the block, and (c) the amplitude of the motion. ilw

20P. A simple harmonic oscillator consists of a block of mass 2.00 kg attached to a spring of spring constant 100 N/m. When $t = 1.00$ s, the position and velocity of the block are $x = 0.129$ m and $v = 3.415$ m/s. (a) What is the amplitude of the oscillations? What were the (b) position and (c) velocity of the block at $t = 0$ s?

21P. A massless spring hangs from the ceiling with a small object attached to its lower end. The object is initially held at rest in a position y_i such that the spring is at its rest length. The object is then released from y_i and oscillates up and down, with its lowest position being 10 cm below y_i. (a) What is the frequency of the oscillation? (b) What is the speed of the object when it is 8.0 cm below the initial position? (c) An object of mass 300 g is attached to the first object, after which the system oscillates with half the original frequency. What is the mass of the first object? (d) Relative to y_i, where is the new equilibrium (rest) position with both objects attached to the spring? **ssm**

22P. Two particles execute simple harmonic motion of the same amplitude and frequency along close parallel lines. They pass each other moving in opposite directions each time their displacement is half their amplitude. What is their phase difference?

23P. Two particles oscillate in simple harmonic motion along a common straight-line segment of length A. Each particle has a period of 1.5 s, but they differ in phase by $\pi/6$ rad. (a) How far apart are they (in terms of A) 0.50 s after the lagging particle leaves one end of the path? (b) Are they then moving in the same direction, toward each other, or away from each other? **ssm**

24P. In Fig. 16-29, two identical springs of spring constant k are attached to a block of mass m and to fixed supports. Show that the block's frequency of oscillation on the frictionless surface is

$$f = \frac{1}{2\pi}\sqrt{\frac{2k}{m}}.$$

25P. Suppose that the two springs in Fig. 16-29 have different spring constants k_1 and k_2. Show that the frequency f of oscillation of the block is then given by

$$f = \sqrt{f_1^2 + f_2^2},$$

where f_1 and f_2 are the frequencies at which the block would oscillate if connected only to spring 1 or only to spring 2. **ilw**

Fig. 16-29 Problems 24 and 25.

26P. The end of one of the prongs of a tuning fork that executes simple harmonic motion of frequency 1000 Hz has an amplitude of 0.40 mm. Find (a) the magnitude of the maximum acceleration and (b) the maximum speed of the end of the prong. Find (c) the magnitude of the acceleration and (d) the speed of the end of the prong when the end has a displacement of 0.20 mm.

27P. In Fig. 16-30, two springs are joined and connected to a block of mass m. The surface is frictionless. If the springs both have spring constant k, show that

$$f = \frac{1}{2\pi}\sqrt{\frac{k}{2m}}$$

gives the block's frequency of oscillation. **ssm**

Fig. 16-30 Problem 27.

28P. In Fig. 16-31, a block weighing 14.0 N, which slides without friction on a 40.0° incline, is connected to the top of the incline by a massless spring of unstretched length 0.450 m and spring constant 120 N/m. (a) How far from the top of the incline does the block stop? (b) If the block is pulled slightly down the incline and released, what is the period of the resulting oscillations?

Fig. 16-31 Problem 28.

29P. A uniform spring with unstretched length L and spring constant k is cut into two pieces of unstretched lengths L_1 and L_2, with $L_1 = nL_2$. What are the corresponding spring constants (a) k_1 and (b) k_2 in terms of n and k? If a block is attached to the original spring, as in Fig. 16-5, it oscillates with frequency f. If the spring is replaced with the piece L_1 or L_2, the corresponding frequency is f_1 or f_2. Find (c) f_1 and (d) f_2 in terms of f. **ssm** **www**

30P. In Fig. 16-32, three 10 000 kg ore cars are held at rest on a 30° incline on a mine railway using a cable that is parallel to the incline. The cable stretches 15 cm just before the coupling between the two lower cars breaks, detaching the lowest car. Assuming that the cable obeys Hooke's law, find (a) the frequency and (b) the amplitude of the resulting oscillations of the remaining two cars.

Fig. 16-32 Problem 30.

SEC. 16-4 Energy in Simple Harmonic Motion

31E. Find the mechanical energy of a block–spring system having a spring constant of 1.3 N/cm and an oscillation amplitude of 2.4 cm. **ssm**

32E. An oscillating block–spring system has a mechanical energy of 1.00 J, an amplitude of 10.0 cm, and a maximum speed of 1.20 m/s. Find (a) the spring constant, (b) the mass of the block and (c) the frequency of oscillation.

33E. A 5.00 kg object on a horizontal frictionless surface is attached to a spring with spring constant 1000 N/m. The object is displaced from equilibrium 50.0 cm horizontally and given an initial velocity of 10.0 m/s back toward the equilibrium position. (a) What is the frequency of the motion? What are (b) the initial potential energy of the block–spring system, (c) the initial kinetic energy, and (d) the amplitude of the oscillation? **ilw**

34E. A (hypothetical) large slingshot is stretched 1.50 m to launch a 130 g projectile with speed sufficient to escape from Earth (11.2 km/s). Assume the elastic bands of the slingshot obey Hooke's law. (a) What is the spring constant of the device, if all the elastic potential energy is converted to kinetic energy? (b) Assume that an average person can exert a force of 220 N. How many people are required to stretch the elastic bands?

35E. A vertical spring stretches 9.6 cm when a 1.3 kg block is hung from its end. (a) Calculate the spring constant. This block is then displaced an additional 5.0 cm downward and released from rest. Find (b) the period, (c) the frequency, (d) the amplitude, and (e) the maximum speed of the resulting SHM. ssm

36E. A block of mass M, at rest on a horizontal frictionless table, is attached to a rigid support by a spring of constant k. A bullet of mass m and velocity $\vec{v}$ strikes the block as shown in Fig. 16-33. The bullet is embedded in the

Fig. 16-33 Exercise 36.

block. Determine (a) the speed of the block immediately after the collision and (b) the amplitude of the resulting simple harmonic motion.

37E. When the displacement in SHM is one-half the amplitude x_m, what fraction of the total energy is (a) kinetic energy and (b) potential energy? (c) At what displacement, in terms of the amplitude, is the energy of the system half kinetic energy and half potential energy? ssm

38P. A 10 g particle is undergoing simple harmonic motion with an amplitude of 2.0×10^{-3} m and a maximum acceleration of magnitude 8.0×10^3 m/s². The phase constant is $-\pi/3$ rad. (a) Write an equation for the force on the particle as a function of time. (b) What is the period of the motion? (c) What is the maximum speed of the particle? (d) What is the total mechanical energy of this simple harmonic oscillator?

39P*. A 4.0 kg block is suspended from a spring with a spring constant of 500 N/m. A 50 g bullet is fired into the block from directly below with a speed of 150 m/s and becomes embedded in the block. (a) Find the amplitude of the resulting simple harmonic motion. (b) What fraction of the original kinetic energy of the bullet is transferred to mechanical energy of the harmonic oscillator? ssm www

SEC. 16-5   An Angular Simple Harmonic Oscillator

40E. A flat uniform circular disk has a mass of 3.00 kg and a radius of 70.0 cm. It is suspended in a horizontal plane by a vertical wire attached to its center. If the disk is rotated 2.50 rad about the wire, a torque of 0.0600 N·m is required to maintain that orientation. Calculate (a) the rotational inertia of the disk about the wire, (b) the torsion constant, and (c) the angular frequency of this torsion pendulum when it is set oscillating.

41P. The balance wheel of a watch oscillates with an angular amplitude of π rad and a period of 0.500 s. Find (a) the maximum angular speed of the wheel, (b) the angular speed of the wheel when its displacement is $\pi/2$ rad, and (c) the magnitude of the angular acceleration of the wheel when its displacement is $\pi/4$ rad. ssm

SEC. 16-6   Pendulums

42E. In Fig. 16-34, a 2500 kg demolition ball swings from the end of a crane. The length of the swinging segment of cable is 17 m. (a) Find the period of the swinging, assuming that the system can be treated as a simple pendulum. (b) Does the period depend on the ball's mass?

Fig. 16-34 Exercise 42.

43E. What is the length of a simple pendulum that marks seconds by completing a full swing from left to right and then back again every 2.0 s? ssm

44E. A performer seated on a trapeze is swinging back and forth with a period of 8.85 s. If she stands up, thus raising the center of mass of the *trapeze + performer* system by 35.0 cm, what will be the new period of the system? Treat *trapeze + performer* as a simple pendulum.

45E. A physical pendulum consists of a meter stick that is pivoted at a small hole drilled through the stick a distance d from the 50 cm mark. The period of oscillation is 2.5 s. Find d. ilw

46E. In Fig. 16-35, a physical pendulum consists of a uniform solid disk (of mass M and radius R) supported in a vertical plane by a pivot located a distance d from the center of the disk. The disk is displaced by a small angle and released. Find an expression for the period of the resulting simple harmonic motion.

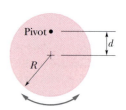

Fig. 16-35 Exercise 46.

47E. A pendulum is formed by pivoting a long thin rod of length L and mass m about a point on the rod that is a distance d above the center of the rod. (a) Find the period of this pendulum in terms of d, L, m, and g, assuming small-amplitude swinging. What happens to the period if (b) d is decreased, (c) L is increased, or (d) m is increased? ssm

48E. A uniform circular disk whose radius R is 12.5 cm is suspended as a physical pendulum from a point on its rim. (a) What is its period? (b) At what radial distance $r < R$ is there a pivot point that gives the same period?

49E. The pendulum in Fig. 16-36 consists of a uniform disk with

Fig. 16-36 Exercise 49.

radius 10.0 cm and mass 500 g attached to a uniform rod with length 500 mm and mass 270 g. (a) Calculate the rotational inertia of the pendulum about the pivot point. (b) What is the distance between the pivot point and the center of mass of the pendulum? (c) Calculate the period of oscillation. ssm

50E. (a) If the physical pendulum of Sample Problem 16-5 is inverted and suspended at point P, what is its period of oscillation? (b) Is the period now greater than, less than, or equal to its previous value?

51E. In Sample Problem 16-5, we saw that a physical pendulum has a center of oscillation at distance $2L/3$ from its point of suspension. Show that the distance between the point of suspension and the center of oscillation for a physical pendulum of any form is I/mh, where I and h have the meanings assigned to them in Eq. 16-29, and m is the mass of the pendulum.

52P. A stick with length L oscillates as a physical pendulum, pivoted about point O in Fig. 16-37. (a) Derive an expression for the period of the pendulum in terms of L and x, the distance from the pivot point to the center of mass of the pendulum. (b) For what value of x/L is the period a minimum? (c) Show that if $L = 1.00$ m and $g = 9.80$ m/s^2, this minimum is 1.53 s.

Fig. 16-37 Problem 52.

53P. In the overhead view of Fig. 16-38, a long uniform rod of length L and mass m is free to rotate in a horizontal plane about a vertical axis through its center. A spring with force constant k is connected horizontally between one end of the rod and a fixed wall. When the rod is in equilibrium, it is parallel to the wall. What is the period of the small oscillations that result when the rod is rotated slightly and released? ssm ilw www

Fig. 16-38 Problem 53.

54P. A simple pendulum of length L and mass m is suspended in a car that is traveling with constant speed v around a circle of radius R. If the pendulum undergoes small oscillations in a radial direction about its equilibrium position, what will be its frequency of oscillation?

55P. What is the frequency of a simple pendulum 2.0 m long (a) in a room, (b) in an elevator accelerating upward at a rate of 2.0 m/s^2, and (c) in free fall? ssm

56P. For a simple pendulum, find the angular amplitude θ_m at which the restoring torque required for simple harmonic motion deviates from the actual restoring torque by 1.0%. (See "Trigonometric Expansions" in Appendix E.)

57P. The bob on a simple pendulum of length R moves in an arc of a circle. (a) By considering that the radial acceleration of the

bob as it moves through its equilibrium position is that for uniform circular motion (v^2/R), show that the tension in the string at that position is $mg(1 + \theta_m^2)$ if the angular amplitude θ_m is small. (See "Trigonometric Expansions" in Appendix E.) (b) Is the tension at other positions of the bob greater, smaller, or the same?

58P. A wheel is free to rotate about its fixed axle. A spring is attached to one of its spokes a distance r from the axle, as shown in Fig. 16-39. (a) Assuming that the wheel is a hoop of mass m and radius R, obtain the angular frequency of small oscillations of this system in terms of m, R, r, and the spring constant k. How does the result change if (b) $r = R$ and (c) $r = 0$?

Fig. 16-39 Problem 58.

SEC. 16-8 Damped Simple Harmonic Motion

59E. In Sample Problem 16-7, what is the ratio of the amplitude of the damped oscillations to the initial amplitude when 20 full oscillations have elapsed? ssm

60E. The amplitude of a lightly damped oscillator decreases by 3.0% during each cycle. What fraction of the mechanical energy of the oscillator is lost in each full oscillation?

61E. For the system shown in Fig. 16-15, the block has a mass of 1.50 kg and the spring constant is 8.00 N/m. The damping force is given by $-b(dx/dt)$, where $b = 230$ g/s. Suppose that the block is initially pulled down a distance 12.0 cm and released. (a) Calculate the time required for the amplitude of the resulting oscillations to fall to one-third of its initial value. (b) How many oscillations are made by the block in this time? ssm

62P. Assume that you are examining the oscillation characteristics of the suspension system of a 2000 kg automobile. The suspension "sags" 10 cm when the entire automobile is placed on it. Also, the amplitude of oscillation decreases by 50% during one complete oscillation. Estimate the values of (a) the spring constant k and (b) the damping constant b for the spring and shock absorber system of one wheel, assuming each wheel supports 500 kg.

SEC. 16-9 Forced Oscillations and Resonance

63E. For Eq. 16-43, suppose the amplitude x_m is given by

$$x_m = \frac{F_m}{[m^2(\omega_d^2 - \omega^2)^2 + b^2\omega_d^2]^{1/2}},$$

where F_m is the (constant) amplitude of the external oscillating force exerted on the spring by the rigid support in Fig. 16-15. At resonance, what are (a) the amplitude and (b) the velocity amplitude of the oscillating object?

64P. A 1000 kg car carrying four 82 kg people travels over a rough "washboard" dirt road with corrugations 4.0 m apart, which cause the car to bounce on its spring suspension. The car bounces with maximum amplitude when its speed is 16 km/h. The car now stops, and the four people get out. By how much does the car body rise on its suspension owing to this decrease in mass?

NEW PROBLEMS

N1. What is the phase constant for the harmonic oscillator with the position function $x(t)$ given in Fig. 16N-1 if $x(t)$ has the form $x = x_m \cos(\omega t + \phi)$?

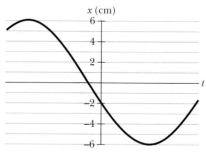

Fig. 16N-1 Problem N1.

N2. What is the phase constant for the harmonic oscillator with the velocity function $v(t)$ given in Fig. 16N-2 if the position function $x(t)$ has the form $x = x_m \cos(\omega t + \phi)$?

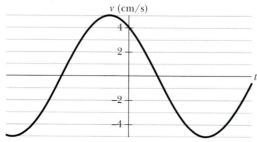

Fig. 16N-2 Problem N2.

N3. What is the phase constant for the harmonic oscillator with the acceleration function $a(t)$ given in Fig. 16N-3 if the position function $x(t)$ has the form $x = x_m \cos(\omega t + \phi)$?

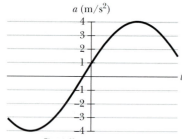

Fig. 16N-3 Problem N3.

N4. If the phase angle for a block–spring system in SHM is $\pi/6$ rad and the block's position is given by $x = x_m \cos(\omega t + \phi)$, what is the ratio of the kinetic energy to the potential energy at time $t = 0$?

N5. Hanging from a horizontal beam are nine simple pendulums of the following lengths: 0.10, 0.30, 0.40, 0.80, 1.2, 2.8, 3.5, 5.0, and 6.2 m. Suppose the beam undergoes horizontal oscillations with angular frequencies in the range from 2.00 rad/s to 4.00 rad/s. Which of the pendulums will be (strongly) set in motion? This is a model, albeit crude and very simplistic, of what occurred with the buildings in Mexico City in 1985 (see the opening photograph).

N6. Figure 16N-4a is a partial graph of the position function $x(t)$ for a simple harmonic oscillator with an angular frequency of 1.20 rad/s. Figure 16N-4b is a partial graph of the corresponding velocity function $v(t)$. What is the phase constant of the SHM if the position function $x(t)$ has the form $x = x_m \cos(\omega t + \phi)$?

(a)

(b)

Fig. 16N-4 Problem N6.

N7. Figure 16N-5 shows the kinetic energy K of a simple harmonic oscillator versus its position x. What is the spring constant?

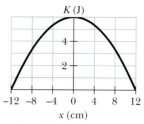

Fig. 16N-5 Problem N7.

N8. In Fig. 16N-6, a square block, with a mass of 3.00 kg and edge lengths $d = 6.00$ cm, is mounted on an axle through its center. A spring of spring constant $k = 1200$ N/m connects the block's upper corner with a rigid wall. Initially the spring is at its rest length. If the block is rotated by 3° and then released, what is the period of the resulting SHM?

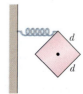

Fig. 16N-6 Problem N8.

N9. Figure 16N-7 gives the one-dimensional potential energy well for a 2.0 kg particle (the function $U(x)$ has the form of bx^2). If the particle passes through the equilibrium position with a velocity of magnitude 85 cm/s, will it be turned back before it reaches $x = 15$ cm? If so, at what position and if not, what is the speed of the particle at $x = 15$ cm?

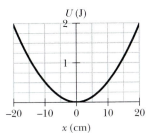

Fig. 16N-7 Problem N9.

N10. Figure 16N-8 gives the position of a 20 g block oscillating in SHM on the end of a spring. What are (a) the maximum kinetic energy of the block and (b) the number of times per second that maximum is reached? (*Hint:* Measuring a slope will probably not be very accurate. Can you think of another approach?)

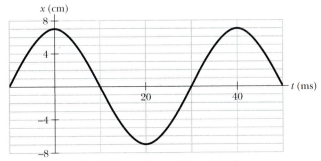

Fig. 16N-8 Problems N10 and N13.

N11. Suppose that a simple pendulum consists of a small 60 g bob at the end of a cord of negligible mass. If the angle θ between the cord and the vertical is given by

$$\theta = (0.080 \text{ rad}) \cos[(4.43 \text{ rad/s}) t + \phi],$$

what are (a) the pendulum's length and (b) its maximum kinetic energy?

N12. A rectangular block, with face lengths a and b, is to be suspended on a thin horizontal rod running through a narrow hole in the block. The block is then to be set swinging about the rod like a pendulum, through small angles so that it is in SHM. Figure 16N-9 shows one possible position of the hole, at distance r from the block's center, along a line connecting the center with a corner. (a) Plot the period of the pendulum versus distance r along that line such that the minimum in the curve is apparent. (b) For what value of r does that minimum occur? There is actually a line of points around the block's center for which the period of swinging has the same minimum value. (c) What shape does that line make and what are the dimensions of the shape?

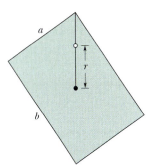

Fig. 16N-9 Problem N12.

N13. Simple harmonic motion is the projection of uniform circular motion on a diameter of that circle. That is, every SHM has a corresponding uniform circular motion. Figure 16N-8 gives the position $x(t)$ of a block oscillating in SHM on the end of a spring. What is (a) the speed and (b) the magnitude of the radial acceleration of a particle in the corresponding uniform circular motion?

N14. In Fig. 16N-10a, a metal plate is mounted on an axle through its center of mass. A spring of spring constant $k = 2000$ N/m connects a rigid wall with one point on the rim, at distance $r = 2.5$ cm from the plate's center of mass. Initially the spring is at its rest length. If the plate is rotated by an angle $\theta = 7°$ and then released, it then rotates about the axle in SHM, with its angular position given by Fig. 16N-10b. What is the rotational inertia of the plate about its center of mass?

(a)

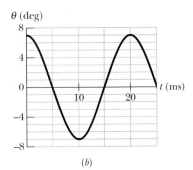

(b)

Fig. 16N-10 Problem N14.

N15. (a) In Eq. 16-39, find the ratio of the maximum damping force $(-b \, dx/dt)$ to the maximum spring force $(-kx)$ during the first oscillation for a damped simple harmonic oscillator, using the data of Sample Problem 16-7. (b) Does this ratio change appreciably during later oscillations?

N16. A torsion pendulum consists of a metal disk with a wire running through its center and soldered in place. The wire is mounted vertically on clamps and pulled taut. Figure 16N-11a gives the magnitude τ of the torque needed to rotate the disk about its center (and thus twist the wire) versus the rotation angle θ. The disk is rotated to $\theta = 0.200$ rad and then released. Figure 16N-11b shows the resulting oscillation of angular position θ versus time t. (a) What is the rotational inertia of the disk about its center? (b) What is the maximum angular speed $d\theta/dt$ of the disk? (*Caution:* Do not confuse the (constant) angular frequency of the SHM with the (varying) angular speed of the rotating disk, even though they usually have the same symbol ω. *Hint:* The potential energy U of a torsion pendulum is equal to $\frac{1}{2}\kappa\theta^2$, analogous to $U = \frac{1}{2}kx^2$ for a spring.)

(a)

(b)

Fig. 16N-11 Problem N16.

N17. Figure 16N-12 shows the kinetic energy K of a simple pendulum versus its angle θ from the vertical. The pendulum bob has mass 0.200 kg. What is the length of the pendulum?

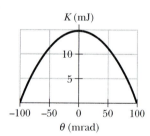

Fig. 16N-12 Problem N17.

N18. A block moves in SHM on the end of a spring, with its position given by $x = x_m \cos(\omega t + \phi)$. If $\phi = \pi/5$ rad, then at $t = 0$ what percentage of the total mechanical energy is potential energy?

N19. A 2.0 kg block is attached to the end of a spring with a spring constant of 350 N/m and forced to oscillate by an applied force $F = (15\ \text{N}) \sin(\omega_d t)$, where $\omega_d = 35$ rad/s. The damping constant is $b = 15$ kg/s. At $t = 0$, the block is at rest with the spring at its rest length. (a) Use numerical integration to plot the displacement of the block for the first 1.0 s of its motion. Use the motion near the end of the 1.0 s interval to estimate the amplitude, period, and angular frequency. Repeat the calculation for (b) $\omega_d = \sqrt{k/m}$ and (c) $\omega_d = 20$ rad/s.

N20. Figure 16N-13 shows block 1 of mass 0.200 kg sliding to the right over a frictionless elevated surface at a speed of 8.00 m/s. The block undergoes an elastic collision with stationary block 2, which is attached to a spring of spring constant 1208.5 N/m. (Assume that the spring does not affect the collision.) After the collision, block 2 oscillates in SHM with a period of 0.140 s, and block 1 slides off the opposite end of the elevated surface, landing a distance d from the base of that surface after falling height $h = 4.90$ m. What is the value of d?

Fig. 16N-13 Problem N20.

N21. A 2.5 kg disk, 42 cm in diameter, is supported by a massless rod, 76 cm long, which is pivoted at its end, as in Fig. 16N-14. (a) The massless torsion spring is initially not connected. What then is the period of oscillation? (b) The torsion spring is now connected so that, in equilibrium, the rod hangs vertically. What should be the torsion constant of the spring so that the new period of oscillation is 0.50 s shorter than before?

Fig. 16N-14 Problem N21.

N22. In Fig. 16N-15, block 2 of mass 2.0 kg oscillates on the end of a spring in SHM. It has a period of 20 ms, and its position is given by

$$x = (1.0\ \text{cm}) \cos(\omega t + \pi/2).$$

Block 1 of mass 4.0 kg slides toward block 2 with a velocity of magnitude 6.0 m/s, directed along the spring's length. The two blocks undergo a completely inelastic collision at time $t = 5.0$ ms. (The duration of the collision is much less than the period of motion.) What is the amplitude of the SHM after the collision?

Fig. 16N-15 Problem N22.

N23. In Fig. 16N-16, a book is suspended at one corner so that it can swing like a pendulum parallel to its plane. The edge lengths along the book face are 25 cm and 20 cm. If the angle through which it swings is only a few degrees, what is the period of the motion?

Fig. 16N-16 Problem N23.

N24. The angle θ of a certain simple pendulum (Fig. 16-9b) is given by $\theta = \theta_m \cos[(4.44 \text{ rad/s})t + \phi]$. If at time $t = 0$, $\theta = 0.040$ rad and $d\theta/dt = -0.200$ rad/s, what are (a) the phase constant ϕ and (b) the maximum angle θ_m? (*Hint:* Be careful not to confuse the rate $d\theta/dt$ at which the angle changes with the angular frequency ω of the pendulum's SHM.)

N25. A 55 g block oscillates in SHM on the end of a spring with a spring constant of 1500 N/m according to $x = x_m \cos(\omega t + \phi)$. How long does the block take to move from position $+0.800x_m$ to (a) position $+0.600x_m$ and (b) position $-0.800x_m$?

N26. A common device for entertaining a toddler is a *jump seat* that hangs from the horizontal portion of a doorframe via elastic cords (Fig. 16N-17). Assume that only one cord is on each side in spite of the more realistic arrangement shown. When a child is placed in the seat, they both descend by a distance d_s as the cords stretch (treat them as springs). Then the seat is pulled down an extra distance d_m and released, so that the child oscillates vertically, like a block on the end of a spring. Suppose you are the safety engineer for the manufacturer of the seat. You do not want the magnitude of the child's acceleration to exceed 0.20g for fear of hurting the child's neck. If $d_m = 10$ cm, what value of d_s corresponds to that acceleration magnitude?

Fig. 16N-17 Problem N26.

N27. You see two carts connected by a spring and oscillating on a horizontal air track. You happen to know that the spring constant of the spring is $k = 50.0$ N/m. You use Newton's second law to determine that the period of oscillation is the same for each cart and is related to the masses of the carts by

$$T = 2\pi \sqrt{\frac{m_1 m_2}{k(m_1 + m_2)}}.$$

Using a position detector, you determine that the positions of the carts as functions of time can be expressed as $x_1(t) = 2.70[1 - \cos(18.0t)]$ and $x_2(t) = 10.70 + 1.29 \cos(18.0t)$, where the coordinates are in centimeters and the time is in seconds. (a) Use the principle of conservation of momentum to find the masses of the carts. (b) Generate a table of the values of x_1 and x_2 for every 1 s from $t = 0$ to $t = 35$ s. For each of these times, also calculate the position of the center of mass, the total momentum of the two-cart system, and the force of the spring on each cart. Verify that the center of mass does not move, that the total momentum is conserved, and that the forces on the carts from the spring have the same magnitude but opposite directions. (c) Use the table to find the equilibrium length of the spring.

17 Waves—I

When a beetle moves along the sand within a few tens of centimeters of this sand scorpion, the scorpion immediately turns toward the beetle and dashes to it (for lunch). The scorpion can do this without seeing (it is nocturnal) or hearing the beetle.

How can the scorpion so precisely locate its prey?

The answer is in this chapter.

17-1 Waves and Particles

Two ways to get in touch with a friend in a distant city are to write a letter and to use the telephone.

The first choice (the letter) involves the concept of "particle": A material object moves from one point to another, carrying with it information and energy. Most of the preceding chapters deal with particles or with systems of particles.

The second choice (the telephone) involves the concept of "wave," the subject of this chapter and the next. In a wave, information and energy move from one point to another but no material object makes that journey. In your telephone call, a sound wave carries your message from your vocal cords to the telephone. There, an electromagnetic wave takes over, passing along a copper wire or an optical fiber or through the atmosphere, possibly by way of a communications satellite. At the receiving end there is another sound wave, from a telephone to your friend's ear. Although the message is passed, nothing that you have touched reaches your friend. Leonardo da Vinci understood about waves when he wrote of water waves: "It often happens that the wave flees the place of its creation, while the water does not; like the waves made in a field of grain by the wind, where we see the waves running across the field while the grain remains in place."

Particle and *wave* are the two great concepts in classical physics, in the sense that we seem able to associate almost every branch of the subject with one or the other. The two concepts are quite different. The word *particle* suggests a tiny concentration of matter capable of transmitting energy. The word *wave* suggests just the opposite—namely, a broad distribution of energy, filling the space through which it passes. The job at hand is to put aside particles for a while and to learn something about waves.

17-2 Types of Waves

Waves are of three main types:

1. *Mechanical waves.* These waves are most familiar because we encounter them almost constantly; common examples include water waves, sound waves, and seismic waves. All these waves have certain central features: They are governed by Newton's laws, and they can exist only within a material medium, such as water, air, and rock.

2. *Electromagnetic waves.* These waves are less familiar, but you use them constantly; common examples include visible and ultraviolet light, radio and television waves, microwaves, x rays, and radar waves. These waves require no material medium to exist. Light waves from stars, for example, travel through the vacuum of space to reach us. All electromagnetic waves travel through a vacuum at the same speed c, given by

$$c = 299\ 792\ 458\ \text{m/s} \quad \text{(speed of light)}. \tag{17-1}$$

3. *Matter waves.* Although these waves are commonly used in modern technology, their type is probably very unfamiliar to you. These waves are associated with electrons, protons, and other fundamental particles, and even atoms and molecules. Because we commonly think of these things as constituting matter, such waves are called matter waves.

Much of what we discuss in this chapter applies to waves of all kinds. However, for specific examples we shall refer to mechanical waves.

Fig. 17-1 (*a*) A single pulse is sent along a stretched string. A typical string element (marked with a dot) moves up once and then down as the pulse passes. The element's motion is perpendicular to the wave's direction of travel, so the pulse is a *transverse wave*. (*b*) A sinusoidal wave is sent along the string. A typical string element moves up and down continuously as the wave passes. This too is a transverse wave.

Fig. 17-2 A sound wave is set up in an air-filled pipe by moving a piston back and forth. Because the oscillations of an element of the air (represented by the black dot) are parallel to the direction in which the wave travels, the wave is a *longitudinal wave.*

17-3   Transverse and Longitudinal Waves

A wave sent along a stretched, taut string is the simplest mechanical wave. If you give one end of a stretched string a single up-and-down jerk, a wave in the form of a single *pulse* travels along the string, as in Fig. 17-1*a*. This pulse and its motion can occur because the string is under tension. When you pull your end of the string upward, it begins to pull upward on the adjacent section of the string via tension between the two sections. As the adjacent section moves upward, it begins to pull the next section upward, and so on. Meanwhile, you have pulled down on your end of the string. As each section moves upward in turn, it begins to be pulled back downward by neighboring sections that are already on the way down. The net result is that a distortion in the string's shape (the pulse) moves along the string at some velocity $\vec{v}$.

If you move your hand up and down in continuous simple harmonic motion, a continuous wave travels along the string at velocity $\vec{v}$. Because the motion of your hand is a sinusoidal function of time, the wave has a sinusoidal shape at any given instant, as in Fig. 17-1*b*; that is, the wave has the shape of a sine curve or a cosine curve.

We consider here only an "ideal" string, in which no frictionlike forces within the string cause the wave to die out as it travels along the string. In addition, we assume that the string is so long that we need not consider a wave rebounding from the far end.

One way to study the waves of Fig. 17-1 is to monitor the **wave forms** (shapes of the waves) as they move to the right. Alternatively, we could monitor the motion of an element of the string as the element oscillates up and down while a wave passes through it. We would find that the displacement of every such oscillating string element is *perpendicular* to the direction of travel of the wave, as indicated in Fig. 17-1. This motion is said to be **transverse,** and the wave is said to be a **transverse wave.**

Figure 17-2 shows how a sound wave can be produced by a piston in a long, air-filled pipe. If you suddenly move the piston rightward and then leftward, you can send a pulse of sound along the pipe. The rightward motion of the piston moves the elements of air next to it rightward, changing the air pressure there. The increased air pressure then pushes rightward on the elements of air somewhat farther along the pipe. Moving the piston leftward reduces the air pressure next to it. Once they have moved rightward, the nearest elements, and then farther elements, move back leftward. Thus, the motion of the air and the change in air pressure travel rightward along the pipe as a pulse.

If you push and pull on the piston in simple harmonic motion, as is being done in Fig. 17-2, a sinusoidal wave travels along the pipe. Because the motion of the elements of air is parallel to the direction of the wave's travel, the motion is said to be **longitudinal,** and the wave is said to be a **longitudinal wave.** In this chapter we concentrate on transverse waves, and string waves in particular; in Chapter 18 we shall concentrate on longitudinal waves, and sound waves in particular.

Both a transverse wave and a longitudinal wave are said to be **traveling waves** because they both travel from one point to another, as from one end of the string to the other end in Fig. 17-1 or from one end of the pipe to the other end in Fig. 17-2. Note that it is the wave that moves from end to end, not the material (string or air) through which the wave moves.

The sand scorpion shown in the photograph opening this chapter uses waves of both transverse and longitudinal motion to locate its prey. When a beetle even slightly disturbs the sand, it sends pulses along the sand's surface (Fig. 17-3). One

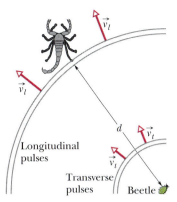

Fig. 17-3 A beetle's motion sends fast longitudinal pulses and slower transverse pulses along the sand's surface. The sand scorpion first intercepts the longitudinal pulses; here, it is the rearmost right leg that senses the pulses earliest.

set of pulses is longitudinal, traveling with speed $v_l = 150$ m/s. A second set is transverse, traveling with speed $v_t = 50$ m/s.

The scorpion, with its eight legs spread roughly in a circle about 5 cm in diameter, intercepts the faster longitudinal pulses first and learns the direction of the beetle; it is in the direction of whichever leg is disturbed earliest by the pulses. The scorpion then senses the time interval Δt between that first interception and the interception of the slower transverse waves and uses it to determine the distance d to the beetle. This distance is given by

$$\Delta t = \frac{d}{v_t} - \frac{d}{v_l},$$

and it turns out to be

$$d = (75 \text{ m/s}) \, \Delta t.$$

For example, if $\Delta t = 4.0$ ms, then $d = 30$ cm, which gives the scorpion a perfect fix on the beetle.

17-4 Wavelength and Frequency

To completely describe a wave on a string (and the motion of any element along its length), we need a function that gives the shape of the wave. This means that we need a relation in the form $y = h(x, t)$, in which y is the transverse displacement of any string element as a function h of the time t and the position x of the element along the string. In general, a sinusoidal shape like the wave in Fig. 17-1b can be described with h being either a sine function or a cosine function; both give the same general shape for the wave. In this chapter we use the sine function.

Imagine a sinusoidal wave like that of Fig. 17-1b traveling in the positive direction of an x axis. As the wave sweeps through succeeding elements (that is, very short sections) of the string, the elements oscillate parallel to the y axis. At time t, the displacement y of the element located at position x is given by

$$y(x, t) = y_m \sin(kx - \omega t). \tag{17-2}$$

Because this equation is written in terms of position x, it can be used to find the displacements of all the elements of the string as a function of time. Thus, it can tell us the shape of the wave at any given time and how that shape changes as the wave moves along the string. The names of the quantities in Eq. 17-2 are displayed in Fig. 17-4 and defined next.

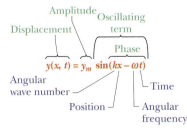

Fig. 17-4 The names of the quantities in Eq. 17-2, for a transverse sinusoidal wave.

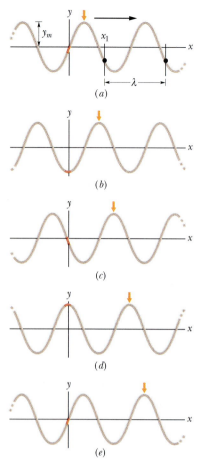

Fig. 17-5 Five "snapshots" of a string wave traveling in the positive direction of an x axis. The amplitude y_m is indicated. A typical wavelength λ, measured from an arbitrary position x_1, is also indicated.

Before we discuss them, however, let us examine Fig. 17-5, which shows five "snapshots" of a sinusoidal wave traveling in the positive direction of an x axis. The movement of the wave is indicated by the rightward progress of the short arrow pointing to a high point of the wave. From snapshot to snapshot, the short arrow moves to the right with the wave shape, but the string moves *only* parallel to the y axis. To see that, let us follow the motion of the red-dyed string element at $x = 0$. In the first snapshot (Fig. 17-5a), it is at displacement $y = 0$. In the next snapshot, it is at its extreme downward displacement because a *valley* (or extreme low point) of the wave is passing through it. It then moves back up through $y = 0$. In the fourth snapshot, it is at its extreme upward displacement because a *peak* (or extreme high point) of the wave is passing through it. In the fifth snapshot, it is again at $y = 0$, having completed one full oscillation.

Amplitude and Phase

The **amplitude** y_m of a wave, such as that in Fig. 17-5, is the magnitude of the maximum displacement of the elements from their equilibrium positions as the wave passes through them. (The subscript m stands for maximum.) Because y_m is a magnitude, it is always a positive quantity, even if it is measured downward in Fig. 17-5a, instead of upward as drawn.

The **phase** of the wave is the *argument* $kx - \omega t$ of the sine in Eq. 17-2. As the wave sweeps through a string element at a particular position x, the phase changes linearly with time t. This means that the sine also changes, oscillating between $+1$ and -1. Its extreme positive value $(+1)$ corresponds to a peak of the wave moving through the element; then, the value of y at position x is y_m. Its extreme negative value (-1) corresponds to a valley of the wave moving through the element; then, the value of y at position x is $-y_m$. Thus, the sine function and the time-dependent phase of a wave correspond to the oscillation of a string element, and the amplitude of the wave determines the extremes of the element's displacement.

Wavelength and Angular Wave Number

The **wavelength** λ of a wave is the distance (parallel to the direction of the wave's travel) between repetitions of the shape of the wave (or *wave shape*). A typical wavelength is marked in Fig. 17-5a, which is a snapshot of the wave at time $t = 0$. At that time, Eq. 17-2 gives, for the description of the wave shape,

$$y(x, 0) = y_m \sin kx. \tag{17-3}$$

By definition, the displacement y is the same at both ends of this wavelength—that is, at $x = x_1$ and $x = x_1 + \lambda$. Thus, by Eq. 17-3,

$$y_m \sin kx_1 = y_m \sin k(x_1 + \lambda)$$
$$= y_m \sin(kx_1 + k\lambda). \tag{17-4}$$

A sine function begins to repeat itself when its angle (or argument) is increased by 2π rad, so in Eq. 17-4 we must have $k\lambda = 2\pi$, or

$$k = \frac{2\pi}{\lambda} \qquad \text{(angular wave number).} \tag{17-5}$$

We call k the **angular wave number** of the wave; its SI unit is the radian per meter, or the inverse meter. (Note that the symbol k here does *not* represent a spring constant as previously.)

Notice that the wave in Fig. 17-5 moves to the right by $\frac{1}{4}\lambda$ from one snapshot to the next. Thus, by the fifth snapshot, it has moved to the right by 1λ.

Fig. 17-6 A graph of the displacement of the string element at $x = 0$ as a function of time, as the sinusoidal wave of Fig. 17-5 passes through it. The amplitude y_m is indicated. A typical period T, measured from an arbitrary time t_1, is also indicated.

Period, Angular Frequency, and Frequency

Figure 17-6 shows a graph of the displacement y of Eq. 17-2 versus time t at a certain position along the string, taken to be $x = 0$. If you were to monitor the string, you would see that the single element of the string at that position moves up and down in simple harmonic motion given by Eq. 17-2 with $x = 0$:

$$y(0, t) = y_m \sin(-\omega t)$$
$$= -y_m \sin \omega t \qquad (x = 0). \qquad (17\text{-}6)$$

Here we have made use of the fact that $\sin(-\alpha) = -\sin \alpha$, where α is any angle. Figure 17-6 is a graph of this equation; it *does not* show the shape of the wave.

We define the **period** of oscillation T of a wave to be the time any string element takes to move through one full oscillation. A typical period is marked on the graph of Fig. 17-6. Applying Eq. 17-6 to both ends of this time interval and equating the results yield

$$-y_m \sin \omega t_1 = -y_m \sin \omega(t_1 + T)$$
$$= -y_m \sin(\omega t_1 + \omega T). \qquad (17\text{-}7)$$

This can be true only if $\omega T = 2\pi$, or if

$$\omega = \frac{2\pi}{T} \qquad \text{(angular frequency)}. \qquad (17\text{-}8)$$

We call ω the **angular frequency** of the wave; its SI unit is the radian per second.

Look back at the five snapshots of a traveling wave in Fig. 17-5. The time between snapshots is $\frac{1}{4}T$. Thus, by the fifth snapshot, every string element has made one full oscillation.

The **frequency** f of a wave is defined as $1/T$ and is related to the angular frequency ω by

$$f = \frac{1}{T} = \frac{\omega}{2\pi} \qquad \text{(frequency)}. \qquad (17\text{-}9)$$

Like the frequency of simple harmonic motion in Chapter 16, this frequency f is a number of oscillations per unit time — here, the number made by a string element as the wave moves through it. As in Chapter 16, f is usually measured in hertz or its multiples, such as kilohertz.

✔ **CHECKPOINT 1:** The figure is a composite of three snapshots, each of a wave traveling along a particular string. The phases for the waves are given by (a) $2x - 4t$, (b) $4x - 8t$, and (c) $8x - 16t$. Which phase corresponds to which wave in the figure?

17-5 The Speed of a Traveling Wave

Figure 17-7 shows two snapshots of the wave of Eq. 17-2, taken a small time interval Δt apart. The wave is traveling in the positive direction of x (to the right in Fig. 17-7), the entire wave pattern moving a distance Δx in that direction during the interval Δt. The ratio $\Delta x/\Delta t$ (or, in the differential limit, dx/dt) is the **wave speed** v. How can we find its value?

As the wave in Fig. 17-7 moves, each point of the moving wave form, such as point A marked on a peak, retains its displacement y. (Points on the string do not retain their displacement, but points on the wave *form* do.) If point A retains its

Fig. 17-7 Two snapshots of the wave of Fig. 17-5, at time $t = 0$ and then at time $t = \Delta t$. As the wave moves to the right at velocity $\vec{v}$, the entire curve shifts a distance Δx during Δt. Point A "rides" with the wave form but the string elements move only up and down.

displacement as it moves, the phase in Eq. 17-2 giving it that displacement must remain a constant:

$$kx - \omega t = \text{a constant.} \tag{17-10}$$

Note that although this argument is constant, both x and t are changing. In fact, as t increases, x must also, to keep the argument constant. This confirms that the wave pattern is moving in the positive direction of x.

To find the wave speed v, we take the derivative of Eq. 17-10, getting

$$k\frac{dx}{dt} - \omega = 0$$

or
$$\frac{dx}{dt} = v = \frac{\omega}{k}. \tag{17-11}$$

Using Eq. 17-5 ($k = 2\pi/\lambda$) and Eq. 17-8 ($\omega = 2\pi/T$), we can rewrite the wave speed as

$$v = \frac{\omega}{k} = \frac{\lambda}{T} = \lambda f \qquad \text{(wave speed).} \tag{17-12}$$

The equation $v = \lambda/T$ tells us that the wave speed is one wavelength per period; the wave moves a distance of one wavelength in one period of oscillation.

Equation 17-2 describes a wave moving in the positive direction of x. We can find the equation of a wave traveling in the opposite direction by replacing t in Eq. 17-2 with $-t$. This corresponds to the condition

$$kx + \omega t = \text{a constant,} \tag{17-13}$$

which (compare Eq. 17-10) requires that x *decrease* with time. Thus, a wave traveling in the negative direction of x is described by the equation

$$y(x, t) = y_m \sin(kx + \omega t). \tag{17-14}$$

If you analyze the wave of Eq. 17-14 as we have just done for the wave of Eq. 17-2, you will find for its velocity

$$\frac{dx}{dt} = -\frac{\omega}{k}. \tag{17-15}$$

The minus sign (compare Eq. 17-11) verifies that the wave is indeed moving in the negative direction of x and justifies our switching the sign of the time variable.

Consider now a wave of arbitrary shape, given by

$$y(x, t) = h(kx \pm \omega t), \tag{17-16}$$

where h represents *any* function, the sine function being one possibility. Our previous analysis shows that all waves in which the variables x and t enter in the combination $kx \pm \omega t$ are traveling waves. Furthermore, all traveling waves *must* be of the form of Eq. 17-16. Thus, $y(x, t) = \sqrt{ax + bt}$ represents a possible (though perhaps physically a little bizarre) traveling wave. The function $y(x, t) = \sin(ax^2 - bt)$, on the other hand, does *not* represent a traveling wave.

Sample Problem 17-1

A wave traveling along a string is described by

$$y(x, t) = 0.00327 \sin(72.1x - 2.72t), \tag{17-17}$$

in which the numerical constants are in SI units (0.00327 m, 72.1 rad/m, and 2.72 rad/s).

(a) What is the amplitude of this wave?

SOLUTION: The Key Idea is that Eq. 17-17 is of the same form as Eq. 17-2,

$$y = y_m \sin(kx - \omega t), \qquad (17\text{-}18)$$

so we have a sinusoidal wave. By comparing the two equations, we see that the amplitude is

$$y_m = 0.00327 \text{ m} = 3.27 \text{ mm}. \qquad \text{(Answer)}$$

(b) What are the wavelength, period, and frequency of this wave?

SOLUTION: By comparing Eqs. 17-17 and 17-18, we see that the angular wave number and angular frequency are

$$k = 72.1 \text{ rad/m} \quad \text{and} \quad \omega = 2.72 \text{ rad/s}.$$

We then relate wavelength λ to k via Eq. 17-5:

$$\lambda = \frac{2\pi}{k} = \frac{2\pi \text{ rad}}{72.1 \text{ rad/m}}$$
$$= 0.0871 \text{ m} = 8.71 \text{ cm}. \qquad \text{(Answer)}$$

Next, we relate T to ω with Eq. 17-8:

$$T = \frac{2\pi}{\omega} = \frac{2\pi \text{ rad}}{2.72 \text{ rad/s}} = 2.31 \text{ s}, \qquad \text{(Answer)}$$

and from Eq. 17-9 we have

$$f = \frac{1}{T} = \frac{1}{2.31 \text{ s}} = 0.433 \text{ Hz}. \qquad \text{(Answer)}$$

(c) What is the velocity of this wave?

SOLUTION: The speed of the wave is given by Eq. 17-12:

$$v = \frac{\omega}{k} = \frac{2.72 \text{ rad/s}}{72.1 \text{ rad/m}} = 0.0377 \text{ m/s}$$
$$= 3.77 \text{ cm/s}. \qquad \text{(Answer)}$$

Because the phase in Eq. 17-17 contains the position variable x, the wave is moving along the x axis. Also, because the wave equation is written in the form of Eq. 17-2, the *minus* sign in front of the ωt term indicates that the wave is moving in the *positive* direction of the x axis. (Note that the quantities calculated in (b) and (c) are independent of the amplitude of the wave.)

(d) What is the displacement y at $x = 22.5$ cm and $t = 18.9$ s?

SOLUTION: The Key Idea here is that Eq. 17-17 gives the displacement as a function of position x and time t. Substituting the given values into the equation yields

$$y = 0.00327 \sin(72.1 \times 0.225 - 2.72 \times 18.9)$$
$$= (0.00327 \text{ m}) \sin(-35.1855 \text{ rad})$$
$$= (0.00327 \text{ m})(0.588)$$
$$= 0.00192 \text{ m} = 1.92 \text{ mm}. \qquad \text{(Answer)}$$

Thus, the displacement is positive. (Be sure to change your calculator mode to radians before evaluating the sine.)

Sample Problem 17-2

In Sample Problem 17-1d, we showed that at $t = 18.9$ s the transverse displacement y of the element of the string at $x = 0.255$ m due to the wave of Eq. 17-17 is 1.92 mm.

(a) What is u, the transverse velocity of the same element of the string, at that time? (This speed, which is associated with the transverse oscillation of an element of the string, is in the y direction. Do not confuse it with v, the constant velocity at which the *wave form* travels along the x axis.)

SOLUTION: The Key Idea here is that the transverse velocity u is the rate at which the displacement y of the element is changing. In general, that displacement is given by

$$y(x, t) = y_m \sin(kx - \omega t). \qquad (17\text{-}19)$$

For an element at a certain location x, we find the rate of change of y by taking the derivative of Eq. 17-19 with respect to t while treating x as a constant. A derivative taken while one (or more) of the variables is treated as a constant is called a *partial derivative* and is represented by the symbol $\partial/\partial x$ rather than d/dx. Here we have

$$u = \frac{\partial y}{\partial t} = -\omega y_m \cos(kx - \omega t). \qquad (17\text{-}20)$$

Next, substituting numerical values from Sample Problem 17-1, we obtain

$$u = (-2.72 \text{ rad/s})(3.27 \text{ mm}) \cos(-35.1855 \text{ rad})$$
$$= 7.20 \text{ mm/s}. \qquad \text{(Answer)}$$

Thus, at $t = 18.9$ s, the element of string at $x = 22.5$ cm is moving in the positive direction of y, with a velocity of 7.20 mm/s.

(b) What is the transverse acceleration a_y of the same element at that time?

SOLUTION: The Key Idea here is that the transverse acceleration a_y is the rate at which the transverse velocity of the element is changing. From Eq. 17-20, again treating x as a constant but allowing t to vary, we find

$$a_y = \frac{\partial u}{\partial t} = -\omega^2 y_m \sin(kx - \omega t).$$

Comparison with Eq. 17-19 shows that we can write this as

$$a_y = -\omega^2 y.$$

We see that the transverse acceleration of an oscillating string element is proportional to its transverse displacement but opposite in sign. This is completely consistent with the action of the element itself—namely, that it is moving transversely in simple harmonic motion. Substituting numerical values yields

$$a_y = -(2.72 \text{ rad/s})^2(1.92 \text{ mm})$$
$$= -14.2 \text{ mm/s}^2. \qquad \text{(Answer)}$$

Thus, at $t = 18.9$ s, the element of string at $x = 22.5$ cm is displaced from its equilibrium position by 1.92 mm in the positive y direction and has an acceleration of magnitude 14.2 mm/s^2 in the negative y direction.

✔**CHECKPOINT 2:**   Here are the equations of three waves:
(1) $y(x, t) = 2 \sin(4x - 2t)$,   (2) $y(x, t) = \sin(3x - 4t)$,   (3) $y(x, t) = 2 \sin(3x - 3t)$.
Rank the waves according to their (a) wave speed and (b) maximum transverse speed, greatest first.

PROBLEM-SOLVING TACTICS

Tactic 1: *Evaluating Large Phases*

Sometimes, as in Sample Problems 17-1d and 17-2, an angle much greater than 2π rad (or 360°) crops up and you are asked to find its sine or cosine. Adding or subtracting an integral multiple of 2π rad to such an angle does not change the value of any of its trigonometric functions. In Sample Problem 17-1d, for example, the angle is -35.1855 rad. Adding $(6)(2\pi \text{ rad})$ to this angle yields

$$-35.1855 \text{ rad} + (6)(2\pi \text{ rad}) = 2.51361 \text{ rad},$$

an angle of less than 2π rad that has the same trigonometric functions as -35.1855 rad (Fig. 17-8). As an example, the sine of both 2.51361 rad and -35.1855 rad is 0.588.

Your calculator will reduce such large angles for you automatically. *Caution:* Do not round off large angles if you intend to take their sines or cosines. In taking the sine of a very large angle, you are throwing away most of the angle and taking the sine of what is left over. If, for example, you were to round -35.1855 rad to -35 rad (a change of 0.5% and normally a reasonable step), you would be changing the sine of the angle by 27%. Also, if you change a large angle from degrees to radians, be sure to use an exact conversion factor (such as $180° = \pi$ rad) rather than an approximate one (such as $57.3° \approx 1$ rad).

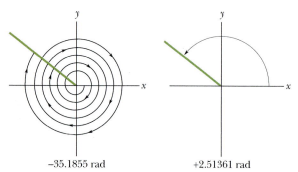

−35.1855 rad   +2.51361 rad

Fig. 17-8 These two angles are different but all their trigonometric functions are identical.

17-6 Wave Speed on a Stretched String

The speed of a wave is related to the wave's wavelength and frequency by Eq. 17-12, but *it is set by the properties of the medium*. If a wave is to travel through a medium such as water, air, steel, or a stretched string, it must cause the particles of that medium to oscillate as it passes. For that to happen, the medium must possess both mass (so that there can be kinetic energy) and elasticity (so that there can be potential energy). Thus, the medium's mass and elasticity properties determine how fast the wave can travel in the medium. Conversely, it should be possible to calculate the speed of the wave through the medium in terms of these properties. We do so now for a stretched string, in two ways.

Dimensional Analysis

In dimensional analysis we carefully examine the dimensions of all the physical quantities that enter into a given situation to determine the quantities they produce. In this case, we examine mass and elasticity to find a speed v, which has the dimension of length divided by time, or LT^{-1}.

For the mass, we use the mass of a string element, which is represented by the mass m of the string divided by the length l of the string. We call this ratio the *linear density* μ of the string. Thus, $\mu = m/l$, its dimension being mass divided by length, ML^{-1}.

You cannot send a wave along a string unless the string is under tension, which means that it has been stretched and pulled taut by forces at its two ends. The tension τ in the string is equal to the common magnitude of those two forces. As a wave travels along the string, it displaces elements of the string by causing additional stretching, with adjacent sections of string pulling on each other because of the tension. Thus, we can associate the tension in the string with the stretching (elasticity) of the string. The tension and the stretching forces it produces have the dimension of a force—namely, MLT^{-2} (from $F = ma$).

The goal here is to combine μ (dimension ML^{-1}) and τ (dimension MLT^{-2}) in such a way as to generate v (dimension LT^{-1}). A little juggling of various combinations suggests

$$v = C\sqrt{\frac{\tau}{\mu}}, \tag{17-21}$$

in which C is a dimensionless constant that cannot be determined with dimensional analysis. In our second approach to determining wave speed, you will see that Eq. 17-21 is indeed correct and that $C = 1$.

Derivation from Newton's Second Law

Instead of the sinusoidal wave of Fig. 17-1b, let us consider a single symmetrical pulse such as that of Fig. 17-9, moving from left to right along a string with speed v. For convenience, we choose a reference frame in which the pulse remains stationary; that is, we run along with the pulse, keeping it constantly in view. In this frame, the string appears to move past us, from right to left in Fig. 17-9, with speed v.

Consider a small string element within the pulse, of length Δl, forming an arc of a circle of radius R and subtending an angle 2θ at the center of that circle. A force $\vec{\tau}$ with a magnitude equal to the tension in the string pulls tangentially on this element at each end. The horizontal components of these forces cancel, but the vertical components add to form a radial restoring force $\vec{F}$. In magnitude,

$$F = 2(\tau \sin \theta) \approx \tau (2\theta) = \tau \frac{\Delta l}{R} \qquad \text{(force)}, \tag{17-22}$$

where we have approximated $\sin \theta$ as θ for the small angles θ in Fig. 17-9. From that figure, we have also used $2\theta = \Delta l/R$.

The mass of the element is given by

$$\Delta m = \mu \, \Delta l \qquad \text{(mass)}, \tag{17-23}$$

where μ is the string's linear density.

At the moment shown in Fig. 17-9, the string element Δl is moving in an arc of a circle. Thus, it has a centripetal acceleration toward the center of that circle, given by

$$a = \frac{v^2}{R} \qquad \text{(acceleration)}. \tag{17-24}$$

Equations 17-22, 17-23, and 17-24 contain the elements of Newton's second law. Combining them in the form

$$\text{force} = \text{mass} \times \text{acceleration}$$

gives

$$\frac{\tau \, \Delta l}{R} = (\mu \, \Delta l) \frac{v^2}{R}.$$

Solving this equation for the speed v yields

$$v = \sqrt{\frac{\tau}{\mu}} \qquad \text{(speed)}, \tag{17-25}$$

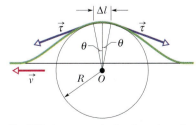

Fig. 17-9 A symmetrical pulse, viewed from a reference frame in which the pulse is stationary and the string appears to move right to left with speed v. We find speed v by applying Newton's second law to a string element of length Δl, located at the top of the pulse.

in exact agreement with Eq. 17-21 if the constant C in that equation is given the value unity. Equation 17-25 gives the speed of the pulse in Fig. 17-9 and the speed of *any* other wave on the same string under the same tension.

Equation 17-25 tells us:

> ▶ The speed of a wave along a stretched ideal string depends only on the tension and linear density of the string and not on the frequency of the wave.

The *frequency* of the wave is fixed entirely by whatever generates the wave (for example, the person in Fig. 17-1*b*). The *wavelength* of the wave is then fixed by Eq. 17-12 in the form $\lambda = v/f$.

✓**CHECKPOINT 3:**   You send a traveling wave along a particular string by oscillating one end. If you increase the frequency of the oscillations, do (a) the speed of the wave and (b) the wavelength of the wave increase, decrease, or remain the same? If, instead, you increase the tension in the string, do (c) the speed of the wave and (d) the wavelength of the wave increase, decrease, or remain the same?

Sample Problem 17-3

In Fig. 17-10, two strings have been tied together with a knot and then stretched between two rigid supports. The strings have linear densities $\mu_1 = 1.4 \times 10^{-4}$ kg/m and $\mu_2 = 2.8 \times 10^{-4}$ kg/m. Their lengths are $L_1 = 3.0$ m and $L_2 = 2.0$ m, and string 1 is under a tension of 400 N. Simultaneously, on each string a pulse is sent from the rigid support end, toward the knot. Which pulse reaches the knot first?

SOLUTION: We need several **Key Ideas** here:

1. The time t taken by a pulse to travel a length L is $t = L/v$, where v is the constant speed of the pulse.

2. The speed of a pulse on a stretched string depends on the string's

Fig. 17-10 Sample Problem 17-3. Two strings, of lengths L_1 and L_2, tied together with a knot and stretched between two rigid supports.

tension τ and linear density μ, and is given by Eq. 17-25 $(v = \sqrt{\tau/\mu})$.

3. Because the two strings are stretched together, they must both be under the same tension $\tau (= 400$ N).

Putting these three ideas together gives us, as the time for the pulse on string 1 to reach the knot,

$$t_1 = \frac{L_1}{v_1} = L_1 \sqrt{\frac{\mu_1}{\tau}} = (3.0 \text{ m}) \sqrt{\frac{1.4 \times 10^{-4} \text{ kg/m}}{400 \text{ N}}}$$
$$= 1.77 \times 10^{-3} \text{ s.}$$

Similarly, the data for the pulse on string 2 give us

$$t_2 = L_2 \sqrt{\frac{\mu_2}{\tau}} = 1.67 \times 10^{-3} \text{ s.}$$

Thus, the pulse on string 2 reaches the knot first.

Now look back at **Key Idea** 2. The linear density of string 2 is greater than that of string 1, so the pulse on string 2 must be slower than that on string 1. Could we have guessed the answer from that fact alone? No, because from the first **Key Idea** we see that the distance traveled by a pulse also matters.

17-7 Energy and Power of a Traveling String Wave

When we set up a wave on a stretched string, we provide energy for the motion of the string. As the wave moves away from us, it transports that energy as both kinetic energy and elastic potential energy. Let us consider each form in turn.

Kinetic Energy

An element of the string of mass dm, oscillating transversely in simple harmonic motion as the wave passes through it, has kinetic energy associated with its transverse

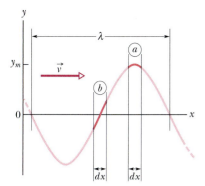

Fig. 17-11 A snapshot of a traveling wave on a string at time $t = 0$. String element a is at displacement $y = y_m$, and string element b is at displacement $y = 0$. The kinetic energy of the string element at each position depends on the transverse velocity of the element. The potential energy depends on the amount by which the string element is stretched as the wave passes through it.

velocity $\vec{u}$. When the element is rushing through its $y = 0$ position (element b in Fig. 17-11), its transverse velocity—and thus its kinetic energy—is a maximum. When the element is at its extreme position $y = y_m$ (as is element a), its transverse velocity—and thus its kinetic energy—is zero.

Elastic Potential Energy

To send a sinusoidal wave along a previously straight string, the wave must necessarily stretch the string. As a string element of length dx oscillates transversely, its length must increase and decrease in a periodic way if the string element is to fit the sinusoidal wave form. Elastic potential energy is associated with these length changes, just as for a spring.

When the string element is at its $y = y_m$ position (element a in Fig. 17-11), its length has its normal undisturbed value dx, so its elastic potential energy is zero. However, when the element is rushing through its $y = 0$ position, it is stretched to its maximum extent, and its elastic potential energy then is a maximum.

Energy Transport

The oscillating string element thus has both its maximum kinetic energy and its maximum elastic potential energy at $y = 0$. In the snapshot of Fig. 17-11, the regions of the string at maximum displacement have no energy, and the regions at zero displacement have maximum energy. As the wave travels along the string, forces due to the tension in the string continuously do work to transfer energy from regions with energy to regions with no energy.

Suppose we set up a wave on a string stretched along a horizontal x axis so that Eq. 17-2 describes the string's displacement. We might send a wave along the string by continuously oscillating one end of the string, as in Fig. 17-1b. In doing so, we continuously provide energy for the motion and stretching of the string—as the string sections oscillate perpendicularly to the x axis, they have kinetic energy and elastic potential energy. As the wave moves into sections that were previously at rest, energy is transferred into those new sections. Thus, we say that the wave *transports* the energy along the string.

The Rate of Energy Transmission

The kinetic energy dK associated with a string element of mass dm is given by

$$dK = \tfrac{1}{2}\, dm\, u^2, \tag{17-26}$$

where u is the transverse speed of the oscillating string element. To find u, we differentiate Eq. 17-2 with respect to time while holding x constant:

$$u = \frac{\partial y}{\partial t} = -\omega y_m \cos(kx - \omega t). \tag{17-27}$$

Using this relation and putting $dm = \mu\, dx$, we rewrite Eq. 17-26 as

$$dK = \tfrac{1}{2}(\mu\, dx)(-\omega y_m)^2 \cos^2(kx - \omega t). \tag{17-28}$$

Dividing Eq. 17-28 by dt gives the rate at which the kinetic energy of a string element changes, and thus the rate at which kinetic energy is carried along by the wave. The ratio dx/dt that then appears on the right of Eq. 17-28 is the wave speed v, so we obtain

$$\frac{dK}{dt} = \tfrac{1}{2}\mu v \omega^2 y_m^2 \cos^2(kx - \omega t). \tag{17-29}$$

The *average* rate at which kinetic energy is transported is

$$\left(\frac{dK}{dt}\right)_{avg} = \tfrac{1}{2}\mu v\omega^2 y_m^2 [\cos^2(kx - \omega t)]_{avg}$$

$$= \tfrac{1}{4}\mu v\omega^2 y_m^2. \qquad (17\text{-}30)$$

Here we have taken the average over an integer number of wavelengths and have used the fact that the average value of the square of a cosine function over an integer number of periods is $\tfrac{1}{2}$.

Elastic potential energy is also carried along with the wave, and at the same average rate given by Eq. 17-30. Although we shall not examine the proof, you should recall that, in an oscillating system such as a pendulum or a spring–block system, the average kinetic energy and the average potential energy are indeed equal.

The **average power,** which is the average rate at which energy of both kinds is transmitted by the wave, is then

$$P_{avg} = 2\left(\frac{dK}{dt}\right)_{avg} \qquad (17\text{-}31)$$

or, from Eq. 17-30,

$$P_{avg} = \tfrac{1}{2}\mu v\omega^2 y_m^2 \qquad \text{(average power).} \qquad (17\text{-}32)$$

The factors μ and v in this equation depend on the material and tension of the string. The factors ω and y_m depend on the process that generates the wave. The dependence of the average power of a wave on the square of its amplitude and also on the square of its angular frequency is a general result, true for waves of all types.

Sample Problem 17-4

A stretched string has linear density $\mu = 525$ g/m and is under tension $\tau = 45$ N. We send a sinusoidal wave with frequency $f = 120$ Hz and amplitude $y_m = 8.5$ mm along the string from one end. At what average rate does the wave transport energy?

SOLUTION: The **Key Idea** here is that the average rate of energy transport is the average power P_{avg} as given by Eq. 17-32. To use that equation, however, we first must calculate the angular frequency ω and the wave speed v. From Eq. 17-9,

$$\omega = 2\pi f = (2\pi)(120 \text{ Hz}) = 754 \text{ rad/s.}$$

From Eq. 17-25 we have

$$v = \sqrt{\frac{\tau}{\mu}} = \sqrt{\frac{45 \text{ N}}{0.525 \text{ kg/m}}} = 9.26 \text{ m/s.}$$

Equation 17-32 then yields

$$P_{avg} = \tfrac{1}{2}\mu v\omega^2 y_m^2$$
$$= (\tfrac{1}{2})(0.525 \text{ kg/m})(9.26 \text{ m/s})(754 \text{ rad/s})^2 (0.0085 \text{ m})^2$$
$$\approx 100 \text{ W.} \qquad \text{(Answer)}$$

✔**CHECKPOINT 4:** For the string and wave of this sample problem, we can adjust three parameters: the tension in the string, the frequency of the wave, and the amplitude of the wave. Does the average rate at which the wave transports energy along the string increase, decrease, or remain the same if we increase (a) the tension, (b) the frequency, and (c) the amplitude?

17-8 The Principle of Superposition for Waves

It often happens that two or more waves pass simultaneously through the same region. When we listen to a concert, for example, sound waves from many instruments fall simultaneously on our eardrums. The electrons in the antennas of our radio and television receivers are set in motion by the net effect of many electromagnetic waves from many different broadcasting centers. The water of a lake or harbor may be churned up by waves in the wakes of many boats.

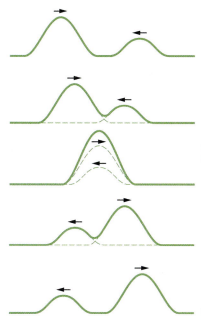

Fig. 17-12 A series of snapshots that show two pulses traveling in opposite directions along a stretched string. The superposition principle applies as the pulses move through each other.

Suppose that two waves travel simultaneously along the same stretched string. Let $y_1(x, t)$ and $y_2(x, t)$ be the displacements that the string would experience if each wave traveled alone. The displacement of the string when the waves overlap is then the algebraic sum

$$y'(x, t) = y_1(x, t) + y_2(x, t). \tag{17-33}$$

This summation of displacements along the string means that

▶ Overlapping waves algebraically add to produce a **resultant wave** (or **net wave**).

This is another example of the **principle of superposition,** which says that when several effects occur simultaneously, their net effect is the sum of the individual effects.

Figure 17-12 shows a sequence of snapshots of two pulses traveling in opposite directions on the same stretched string. When the pulses overlap, the resultant pulse is their sum. Moreover, each pulse moves through the other, as if the other were not present:

▶ Overlapping waves do not in any way alter the travel of each other.

17-9 Interference of Waves

Suppose we send two sinusoidal waves of the same wavelength and amplitude in the same direction along a stretched string. The superposition principle applies. What resultant wave does it predict for the string?

The resultant wave depends on the extent to which the waves are *in phase* (in step) with respect to each other—that is, how much one wave form is shifted from the other wave form. If the waves are exactly in phase (so that the peaks and valleys of one are exactly aligned with those of the other), they combine to double the displacement of either wave acting alone. If they are exactly out of phase (the peaks of one are exactly aligned with the valleys of the other), they combine to cancel everywhere, and the string remains straight. We call this phenomenon of combining waves **interference,** and the waves are said to **interfere.** (These terms refer only to the displacements of the waves; the travel of the waves is unaffected.)

Let one wave traveling along a stretched string be given by

$$y_1(x, t) = y_m \sin(kx - \omega t) \tag{17-34}$$

and another, shifted from the first, by

$$y_2(x, t) = y_m \sin(kx - \omega t + \phi). \tag{17-35}$$

These waves have the same angular frequency ω (and thus the same frequency f), the same angular wave number k (and thus the same wavelength λ), and the same amplitude y_m. They both travel in the positive direction of the x axis, with the same speed, given by Eq. 17-25. They differ only by a constant angle ϕ, which we call the **phase constant.** These waves are said to be *out of phase* by ϕ or to have a *phase difference* of ϕ, or one wave is said to be *phase-shifted* from the other by ϕ.

From the principle of superposition (Eq. 17-33), the resultant wave is the algebraic sum of the two interfering waves and has displacement

$$y'(x, t) = y_1(x, t) + y_2(x, t)$$
$$= y_m \sin(kx - \omega t) + y_m \sin(kx - \omega t + \phi). \tag{17-36}$$

Displacement

$$y'(x,t) = \underbrace{[2y_m \cos \tfrac{1}{2}\phi]}_{\text{Amplitude}} \underbrace{\sin(kx - \omega t + \tfrac{1}{2}\phi)}_{\substack{\text{Oscillating} \\ \text{term}}}$$

Fig. 17-13 The resultant wave of Eq. 17-38, due to the interference of two sinusoidal transverse waves, is also a sinusoidal transverse wave, with an amplitude and an oscillating term.

In Appendix E we see that we can write the sum of the sines of two angles α and β as

$$\sin \alpha + \sin \beta = 2 \sin \tfrac{1}{2}(\alpha + \beta) \cos \tfrac{1}{2}(\alpha - \beta). \qquad (17\text{-}37)$$

Applying this relation to Eq. 17-36 leads to

$$y'(x, t) = [2y_m \cos \tfrac{1}{2}\phi] \sin(kx - \omega t + \tfrac{1}{2}\phi). \qquad (17\text{-}38)$$

As Fig. 17-13 shows, the resultant wave is also a sinusoidal wave traveling in the direction of increasing x. It is the only wave you would actually see on the string (you would *not* see the two interfering waves of Eqs. 17-34 and 17-35).

> If two sinusoidal waves of the same amplitude and wavelength travel in the *same* direction along a stretched string, they interfere to produce a resultant sinusoidal wave traveling in that direction.

The resultant wave differs from the interfering waves in two respects: (1) its phase constant is $\tfrac{1}{2}\phi$, and (2) its amplitude y'_m is the quantity in the brackets in Eq. 17-38:

$$y'_m = 2y_m \cos \tfrac{1}{2}\phi \qquad \text{(amplitude).} \qquad (17\text{-}39)$$

If $\phi = 0$ rad (or 0°), the two interfering waves are exactly in phase, as in Fig. 17-14a. Then Eq. 17-38 reduces to

$$y'(x, t) = 2y_m \sin(kx - \omega t) \qquad (\phi = 0). \qquad (17\text{-}40)$$

This resultant wave is plotted in Fig. 17-14d. Note from both that figure and Eq. 17-40 that the amplitude of the resultant wave is twice the amplitude of either interfering wave. That is the greatest amplitude the resultant wave can have, because the cosine term in Eqs. 17-38 and 17-39 has its greatest value (unity) when $\phi = 0$. Interference that produces the greatest possible amplitude is called *fully constructive interference.*

If $\phi = \pi$ rad (or 180°), the interfering waves are exactly out of phase as in Fig. 17-14b. Then $\cos \tfrac{1}{2}\phi$ becomes $\cos \pi/2 = 0$, and the amplitude of the resultant

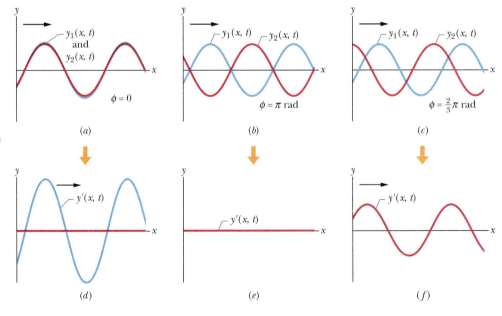

Fig. 17-14 Two identical sinusoidal waves, $y_1(x, t)$ and $y_2(x, t)$, travel along a string in the positive direction of an x axis. They interfere to give a resultant wave $y'(x, t)$. The resultant wave is what is actually seen on the string. The phase difference ϕ between the two interfering waves is (*a*) 0 rad or 0°, (*b*) π rad or 180°, and (*c*) $\tfrac{2}{3}\pi$ rad or 120°. The corresponding resultant waves are shown in (*d*), (*e*), and (*f*).

TABLE 17-1   Phase Differences and Resulting Interference Types[a]

Phase Difference, in			Amplitude of Resultant Wave	Type of Interference
Degrees	Radians	Wavelengths		
0	0	0	$2y_m$	Fully constructive
120	$\frac{2}{3}\pi$	0.33	y_m	Intermediate
180	π	0.50	0	Fully destructive
240	$\frac{4}{3}\pi$	0.67	y_m	Intermediate
360	2π	1.00	$2y_m$	Fully constructive
865	15.1	2.40	$0.60y_m$	Intermediate

[a]The phase difference is between two otherwise identical waves, with amplitude y_m, moving in the same direction.

wave as given by Eq. 17-39 is zero. We then have, for all values of x and t,

$$y'(x, t) = 0 \qquad (\phi = \pi \text{ rad}). \qquad (17\text{-}41)$$

The resultant wave is plotted in Fig. 17-14e. Although we sent two waves along the string, we see no motion of the string. This type of interference is called *fully destructive interference.*

Because a sinusoidal wave repeats its shape every 2π rad, a phase difference $\phi = 2\pi$ rad (or 360°) corresponds to a shift of one wave relative to the other wave by a distance equivalent to one wavelength. Thus, phase differences can be described in terms of wavelengths as well as angles. For example, in Fig. 17-14b the waves may be said to be 0.50 wavelength out of phase. Table 17-1 shows some other examples of phase differences and the interference they produce. Note that when interference is neither fully constructive nor fully destructive, it is called *intermediate interference.* The amplitude of the resultant wave is then intermediate between 0 and $2y_m$. For example, from Table 17-1, if the interfering waves have a phase difference of 120° ($\phi = \frac{2}{3}\pi$ rad = 0.33 wavelength), then the resultant wave has an amplitude of y_m, the same as the interfering waves (see Figs. 17-14c and f).

Two waves with the same wavelength are in phase if their phase difference is zero or any integer number of wavelengths. Thus, the integer part of any phase difference *expressed in wavelengths* may be discarded. For example, a phase difference of 0.40 wavelength is equivalent in every way to one of 2.40 wavelengths, and so the simpler of the two numbers can be used in computations.

Sample Problem 17-5

Two identical sinusoidal waves, moving in the same direction along a stretched string, interfere with each other. The amplitude y_m of each wave is 9.8 mm, and the phase difference ϕ between them is 100°.

(a) What is the amplitude y'_m of the resultant wave due to the interference of these two waves, and what type of interference occurs?

SOLUTION: The Key Idea here is that these are identical sinusoidal waves traveling in the *same direction* along a string, so they interfere to produce a sinusoidal traveling wave. Because they are identical, they have the *same amplitude.* Thus, the amplitude y'_m of the

resultant wave is given by Eq. 17-39:

$$y'_m = 2y_m \cos \tfrac{1}{2}\phi = (2)(9.8 \text{ mm}) \cos(100°/2)$$
$$= 13 \text{ mm}. \qquad \text{(Answer)}$$

We can tell that the interference is *intermediate* in two ways. The phase difference is between 0 and 180° and, correspondingly, amplitude y'_m is between 0 and $2y_m$ (= 19.6 mm).

(b) What phase difference, in radians and wavelengths, will give the resultant wave an amplitude of 4.9 mm?

SOLUTION: The same Key Idea applies here as in part (a), but now we are given y'_m and seek ϕ. From Eq. 17-39,

$$y'_m = 2y_m \cos \tfrac{1}{2}\phi.$$

We now have

$$4.9 \text{ mm} = (2)(9.8 \text{ mm}) \cos \tfrac{1}{2}\phi,$$

which gives us (with a calculator in the radian mode)

$$\phi = 2 \cos^{-1} \frac{4.9 \text{ mm}}{(2)(9.8 \text{ mm})}$$

$$= \pm 2.636 \text{ rad} \approx \pm 2.6 \text{ rad}. \qquad \text{(Answer)}$$

There are two solutions because we can obtain the same resultant wave by letting the first wave *lead* (travel ahead of) or *lag* (travel behind) the second wave by 2.6 rad. In wavelengths, the phase difference is

$$\frac{\phi}{2\pi \text{ rad/wavelength}} = \frac{\pm 2.636 \text{ rad}}{2\pi \text{ rad/wavelength}}$$

$$= \pm 0.42 \text{ wavelength}. \qquad \text{(Answer)}$$

✓**CHECKPOINT 5:** Here are four other possible phase differences between the two waves of this sample problem, expressed in wavelengths: 0.20, 0.45, 0.60, and 0.80. Rank them according to the amplitude of the resultant wave, greatest first.

17-10 Phasors

We can represent a string wave (or any other type of wave) vectorially with a **phasor.** In essence, a phasor is a vector that has a magnitude equal to the amplitude of the wave and that rotates around an origin; the angular speed of the phasor is equal to the angular frequency ω of the wave. For example, the wave

$$y_1(x, t) = y_{m1} \sin(kx - \omega t) \qquad (17\text{-}42)$$

is represented by the phasor shown in Fig. 17-15a. The magnitude of the phasor is the amplitude y_{m1} of the wave. As the phasor rotates around the origin at angular speed ω, its projection y_1 on the vertical axis varies sinusoidally, from a maximum of y_{m1} through zero to a minimum of $-y_{m1}$ and then back to y_{m1}. This variation corresponds to the sinusoidal variation in the displacement y_1 of any point along the string as the wave passes through it.

When two waves travel along the same string in the same direction, we can represent them and their resultant wave in a *phasor diagram.* The phasors in Fig. 17-15b represent the wave of Eq. 17-42 and a second wave given by

$$y_2(x, t) = y_{m2} \sin(kx - \omega t + \phi). \qquad (17\text{-}43)$$

This second wave is phase shifted from the first wave by phase constant ϕ. Because the phasors rotate at the same angular speed ω, the angle between the two phasors is always ϕ. If ϕ is a *positive* quantity, then the phasor for wave 2 *lags* the phasor for wave 1 as they rotate, as drawn in Fig. 17-15b. If ϕ is a negative quantity, then the phasor for wave 2 *leads* the phasor for wave 1.

Because waves y_1 and y_2 have the same angular wave number k and angular frequency ω, we know from Eq. 17-38 that their resultant is of the form

$$y'(x, t) = y'_m \sin(kx - \omega t + \beta), \qquad (17\text{-}44)$$

where y'_m is the amplitude of the resultant wave and β is its phase constant. To find the values of y'_m and β, we would have to sum the two combining waves, as we did to obtain Eq. 17-38.

To do this on a phasor diagram, we vectorially add the two phasors at any instant during their rotation, as in Fig. 17-15c where phasor y_{m2} has been shifted to the

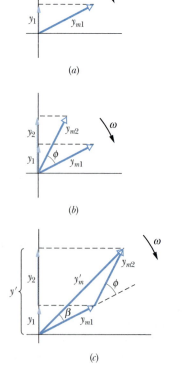

(a)

(b)

(c)

Fig. 17-15 (a) A phasor of magnitude y_{m1} rotating about an origin at angular speed ω represents a sinusoidal wave. The phasor's projection y_1 on the vertical axis represents the displacement of a point through which the wave passes. (b) A second phasor, also of angular speed ω but of magnitude y_{m2} and rotating at a constant angle ϕ from the first phasor, represents a second wave, with a phase constant ϕ. (c) The resultant wave of the two waves is represented by the vector sum y'_m of the two phasors. The projection y' on the vertical axis represents the displacement of a point as that resultant wave passes through it.

head of phasor y_{m1}. The magnitude of the vector sum equals the amplitude y'_m in Eq. 17-44. The angle between the vector sum and the phasor for y_1 equals the phase constant β in Eq. 17-44.

Note that, in contrast to the method of Section 17-9:

> We can use phasors to combine waves *even if their amplitudes are different.*

Sample Problem 17-6

Two sinusoidal waves $y_1(x, t)$ and $y_2(x, t)$ have the same wavelength and travel together in the same direction along a string. Their amplitudes are $y_{m1} = 4.0$ mm and $y_{m2} = 3.0$ mm, and their phase constants are 0 and $\pi/3$ rad, respectively. What are the amplitude y'_m and phase constant β of the resultant wave? Write the resultant wave in the form of Eq. 17-44.

SOLUTION: One **Key Idea** here is that the two waves have a number of properties in common: Because they travel along the same string, they must have the same speed v, as set by the tension and linear density of the string according to Eq. 17-25. With the same wavelength λ, they must have the same angular wave number $k (= 2\pi/\lambda)$. Also, with the same wave number k and speed v, they must have the same angular frequency $\omega (= kv)$.

A second **Key Idea** is that the waves (call them waves 1 and 2) can be represented by phasors rotating at the same angular speed ω about an origin. Because the phase constant for wave 2 is *greater* than that for wave 1 by $\pi/3$, phasor 2 must *lag* phasor 1 by $\pi/3$ rad in their clockwise rotation, as shown in Fig. 17-16a. The resultant wave due to the interference of waves 1 and 2 can then be represented by a phasor that is the vector sum of phasors 1 and 2.

To simplify the vector summation, we drew phasors 1 and 2 in Fig. 17-16a at the instant when phasor 1 lies along the horizontal axis. We then drew lagging phasor 2 at positive angle $\pi/3$ rad. In Fig. 17-16b we shifted phasor 2 so its tail is at the head of phasor 1. Then we can draw the phasor y'_m of the resultant wave from the tail of phasor 1 to the head of phasor 2. The phase constant β is the angle it makes with phasor 1.

To find values for y'_m and β, we can sum phasors 1 and 2 directly on a vector-capable calculator, by adding a vector of magnitude 4.0 and angle 0 rad to a vector of magnitude 3.0 and angle $\pi/3$ rad, or we can add the vectors by components. For the hori-

Fig. 17-16 Sample Problem 17-6. (*a*) Two phasors of magnitudes y_{m1} and y_{m2} and with phase difference $\pi/3$. (*b*) Vector addition of these phasors at any instant during their rotation gives the magnitude y'_m of the phasor for the resultant wave.

zontal components we have

$$y'_{mh} = y_{m1} \cos 0 + y_{m2} \cos \pi/3$$
$$= 4.0 \text{ mm} + (3.0 \text{ mm}) \cos \pi/3 = 5.50 \text{ mm}.$$

For the vertical components we have

$$y'_{mv} = y_{m1} \sin 0 + y_{m2} \sin \pi/3$$
$$= 0 + (3.0 \text{ mm}) \sin \pi/3 = 2.60 \text{ mm}.$$

Thus, the resultant wave has an amplitude of

$$y'_m = \sqrt{(5.50 \text{ mm})^2 + (2.60 \text{ mm})^2}$$
$$= 6.1 \text{ mm} \qquad \text{(Answer)}$$

and a phase constant of

$$\beta = \tan^{-1} \frac{2.60 \text{ mm}}{5.50 \text{ mm}} = 0.44 \text{ rad.} \qquad \text{(Answer)}$$

From Fig. 17-16b, phase constant β is a *positive* angle relative to phasor 1. Thus, the resultant wave *lags* wave 1 in their travel by phase constant $\beta = +0.44$ rad. From Eq. 17-44, we can write the resultant wave as

$$y'(x, t) = (6.1 \text{ mm}) \sin(kx - \omega t + 0.44 \text{ rad}). \quad \text{(Answer)}$$

17-11 Standing Waves

In the preceding two sections, we discussed two sinusoidal waves of the same wavelength and amplitude traveling *in the same direction* along a stretched string. What if they travel in opposite directions? We can again find the resultant wave by applying the superposition principle.

Figure 17-17 suggests the situation graphically. It shows the two combining waves, one traveling to the left in Fig. 17-17a, the other to the right in Fig. 17-17b. Figure 17-17c shows their sum, obtained by applying the superposition principle graphically. The outstanding feature of the resultant wave is that there are places

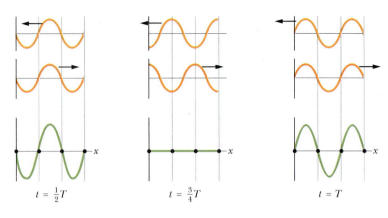

(a)

(b)

(c)

$t = 0$ $t = \frac{1}{4}T$ $t = \frac{1}{2}T$ $t = \frac{3}{4}T$ $t = T$

Fig. 17-17 (a) Five snapshots of a wave traveling to the left, at the times t indicated below part (c) (T is the period of oscillation). (b) Five snapshots of a wave identical to that in (a) but traveling to the right, at the same times t. (c) Corresponding snapshots for the superposition of the two waves on the same string. At $t = 0$, $\frac{1}{2}T$, and T, fully constructive interference occurs because of the alignment of peaks with peaks and valleys with valleys. At $t = \frac{1}{4}T$ and $\frac{3}{4}T$, fully destructive interference occurs because of the alignment of peaks with valleys. Some points (the nodes, marked with dots) never oscillate; some points (the antinodes) oscillate the most.

along the string, called **nodes**, where the string never moves. Four such nodes are marked by dots in Fig. 17-17c. Halfway between adjacent nodes are **antinodes,** where the amplitude of the resultant wave is a maximum. Wave patterns such as that of Fig. 17-17c are called **standing waves** because the wave patterns do not move left or right; the locations of the maxima and minima do not change.

> If two sinusoidal waves of the same amplitude and wavelength travel in *opposite* directions along a stretched string, their interference with each other produces a standing wave.

To analyze a standing wave, we represent the two combining waves with the equations

$$y_1(x, t) = y_m \sin(kx - \omega t) \tag{17-45}$$

and

$$y_2(x, t) = y_m \sin(kx + \omega t). \tag{17-46}$$

The principle of superposition gives, for the combined wave,

$$y'(x, t) = y_1(x, t) + y_2(x, t) = y_m \sin(kx - \omega t) + y_m \sin(kx + \omega t).$$

Applying the trigonometric relation of Eq. 17-37 leads to

$$y'(x, t) = [2y_m \sin kx] \cos \omega t, \tag{17-47}$$

which is displayed in Fig. 17-18. This equation does not describe a traveling wave because it is not of the form of Eq. 17-16. Instead, it describes a standing wave.

The quantity $2y_m \sin kx$ in the brackets of Eq. 17-47 can be viewed as the amplitude of oscillation of the string element that is located at position x. However, since an amplitude is always positive and $\sin kx$ can be negative, we take the absolute value of the quantity $2y_m \sin kx$ to be the amplitude at x.

In a traveling sinusoidal wave, the amplitude of the wave is the same for all string elements. That is not true for a standing wave, in which the amplitude *varies with position*. In the standing wave of Eq. 17-47, for example, the amplitude is zero for values of kx that give $\sin kx = 0$. Those values are

$$kx = n\pi, \quad \text{for } n = 0, 1, 2, \ldots. \tag{17-48}$$

Substituting $k = 2\pi/\lambda$ in this equation and rearranging, we get

$$x = n\frac{\lambda}{2}, \quad \text{for } n = 0, 1, 2, \ldots \quad \text{(nodes)}, \tag{17-49}$$

Displacement

$$y'(x,t) = [\underbrace{2y_m \sin kx}_{\substack{\text{Amplitude} \\ \text{at position } x}}]\underbrace{\cos \omega t}_{\substack{\text{Oscillating} \\ \text{term}}}$$

Fig. 17-18 The resultant wave of Eq. 17-47 is a standing wave and is due to the interference of two sinusoidal waves of the same amplitude and wavelength that travel in opposite directions.

as the positions of zero amplitude—the nodes—for the standing wave of Eq. 17-47. Note that adjacent nodes are separated by $\lambda/2$, half a wavelength.

Fig. 17-19 (*a*) A pulse incident from the right is reflected at the left end of the string, which is tied to a wall. Note that the reflected pulse is inverted from the incident pulse. (*b*) Here the left end of the string is tied to a ring that can slide without friction up and down the rod. Now the pulse is not inverted by the reflection.

(*a*)   (*b*)

The amplitude of the standing wave of Eq. 17-47 has a maximum value of $2y_m$, which occurs for values of kx that give $|\sin kx| = 1$. Those values are

$$kx = \tfrac{1}{2}\pi, \tfrac{3}{2}\pi, \tfrac{5}{2}\pi, \ldots$$
$$= (n + \tfrac{1}{2})\pi, \qquad \text{for } n = 0, 1, 2, \ldots. \qquad (17\text{-}50)$$

Substituting $k = 2\pi/\lambda$ in Eq. 17-50 and rearranging, we get

$$x = \left(n + \frac{1}{2}\right)\frac{\lambda}{2}, \qquad \text{for } n = 0, 1, 2, \ldots \qquad \text{(antinodes),} \qquad (17\text{-}51)$$

as the positions of maximum amplitude—the antinodes—of the standing wave of Eq. 17-47. The antinodes are separated by $\lambda/2$ and are located halfway between pairs of nodes.

Reflections at a Boundary

We can set up a standing wave in a stretched string by allowing a traveling wave to be reflected from the far end of the string so that it travels back through itself. The incident (original) wave and the reflected wave can then be described by Eqs. 17-45 and 17-46, respectively, and they can combine to form a pattern of standing waves.

In Fig. 17-19, we use a single pulse to show how such reflections take place. In Fig. 17-19*a*, the string is fixed at its left end. When the pulse arrives at that end, it exerts an upward force on the support (the wall). By Newton's third law, the support exerts an opposite force of equal magnitude on the string. This second force generates a pulse at the support, which travels back along the string in the direction opposite that of the incident pulse. In a "hard" reflection of this kind, there must be a node at the support because the string is fixed there. The reflected and incident pulses must have opposite signs, so as to cancel each other at that point.

In Fig. 17-19*b*, the left end of the string is fastened to a light ring that is free to slide without friction along a rod. When the incident pulse arrives, the ring moves up the rod. As the ring moves, it pulls on the string, stretching the string and producing a reflected pulse with the same sign and amplitude as the incident pulse. Thus, in such a "soft" reflection, the incident and reflected pulses reinforce each other, creating an antinode at the end of the string; the maximum displacement of the ring is twice the amplitude of either of these pulses.

✔**CHECKPOINT 6:** Two waves with the same amplitude and wavelength interfere in three different situations to produce resultant waves with the following equations:
(1) $y'(x, t) = 4 \sin(5x - 4t)$
(2) $y'(x, t) = 4 \sin(5x) \cos(4t)$
(3) $y'(x, t) = 4 \sin(5x + 4t)$
In which situation are the two combining waves traveling (a) toward positive x, (b) toward negative x, and (c) in opposite directions?

17-12 Standing Waves and Resonance

Consider a string, such as a guitar string, that is stretched between two clamps. Suppose we send a continuous sinusoidal wave of a certain frequency along the string, say, toward the right. When the wave reaches the right end, it reflects and begins to travel back to the left. That left-going wave then overlaps the wave that is still traveling to the right. When the left-going wave reaches the left end, it reflects again and the newly reflected wave begins to travel to the right, overlapping the left-

Fig. 17-20 Stroboscopic photographs reveal (imperfect) standing wave patterns on a string being made to oscillate by a vibrator at the left end. The patterns occur at certain frequencies of oscillation.

going and right-going waves. In short, we very soon have many overlapping traveling waves, which interfere with one another.

For certain frequencies, the interference produces a standing wave pattern (or **oscillation mode**) with nodes and large antinodes like those in Fig. 17-20. Such a standing wave is said to be produced by **resonance,** and the string is said to *resonate* at these certain frequencies, called **resonant frequencies.** If the string is oscillated at some frequency other than a resonant frequency, a standing wave is not set up. Then the interference of the right-going and left-going traveling waves results in only small (perhaps imperceptible) oscillations of the string.

Let a string be stretched between two clamps separated by a fixed distance L. To find expressions for the resonant frequencies of the string, we note that a node must exist at each of its ends, because each end is fixed and cannot oscillate. The simplest pattern that meets this key requirement is that in Fig. 17-21a, which shows the string at both its extreme displacements (one solid and one dashed, together forming a single "loop"). There is only one antinode, which is at the center of the string. Note that half a wavelength spans the length L, which we take to be the string's length. Thus, for this pattern, $\lambda/2 = L$. This condition tells us that if the left-going and right-going traveling waves are to set up this pattern by their interference, they must have the wavelength $\lambda = 2L$.

A second simple pattern meeting the requirement of nodes at the fixed ends is shown in Fig. 17-21b. This pattern has three nodes and two antinodes and is said to be a two-loop pattern. For the left-going and right-going waves to set it up, they must have a wavelength $\lambda = L$. A third pattern is shown in Fig. 17-21c. It has four nodes, three antinodes, and three loops, and the wavelength is $\lambda = \frac{2}{3}L$. We could continue this progression by drawing increasingly more complicated patterns. In each step of the progression, the pattern would have one more node and one more antinode than the preceding step, and an additional $\lambda/2$ would be fitted into the distance L.

Thus, a standing wave can be set up on a string of length L by a wave with a wavelength equal to one of the values

(a)
$$L = \frac{\lambda}{2}$$

(b)
$$L = \lambda = \frac{2\lambda}{2}$$

(c)
$$L = \frac{3\lambda}{2}$$

Fig. 17-21 A string, stretched between two clamps, is made to oscillate in standing wave patterns. (a) The simplest possible pattern consists of one *loop,* which refers to the composite shape formed by the string in its extreme displacements (the solid and dashed lines). (b) The next simplest pattern has two loops. (c) The next has three loops.

$$\lambda = \frac{2L}{n}, \qquad \text{for } n = 1, 2, 3, \ldots . \qquad (17\text{-}52)$$

The resonant frequencies that correspond to these wavelengths follow from Eq. 17-12:

$$f = \frac{v}{\lambda} = n\frac{v}{2L}, \qquad \text{for } n = 1, 2, 3, \ldots . \qquad (17\text{-}53)$$

Here v is the speed of traveling waves on the string.

Equation 17-53 tells us that the resonant frequencies are integer multiples of the lowest resonant frequency, $f = v/2L$, which corresponds to $n = 1$. The oscillation mode with that lowest frequency is called the *fundamental mode* or the *first harmonic.* The *second harmonic* is the oscillation mode with $n = 2$, the *third harmonic* is that with $n = 3$, and so on. The frequencies associated with these modes are often

Fig. 17-22 One of many possible standing wave patterns for a kettledrum head, made visible by dark powder sprinkled on the drumhead. As the head is set into oscillation at a single frequency by a mechanical vibrator at the upper left of the photograph, the powder collects at the nodes, which are circles and straight lines in this two-dimensional example.

labeled f_1, f_2, f_3, and so on. The collection of all possible oscillation modes is called the **harmonic series,** and n is called the **harmonic number** of the nth harmonic.

The phenomenon of resonance is common to all oscillating systems and can occur in two and three dimensions. For example, Fig. 17-22 shows a two-dimensional standing wave pattern on the oscillating head of a kettledrum.

✔**CHECKPOINT 7:** In the following series of resonant frequencies, one frequency (lower than 400 Hz) is missing: 150, 225, 300, 375 Hz. (a) What is the missing frequency? (b) What is the frequency of the seventh harmonic?

Sample Problem 17-7

In Fig. 17-23, a string, tied to a sinusoidal vibrator at P and running over a support at Q, is stretched by a block of mass m. The separation L between P and Q is 1.2 m, the linear density of the string is 1.6 g/m, and the frequency f of the vibrator is fixed at 120 Hz. The amplitude of the motion at P is small enough for that point to be considered a node. A node also exists at Q.

(a) What mass m allows the vibrator to set up the fourth harmonic on the string?

SOLUTION: One **Key Idea** here is that the string will resonate at only certain frequencies, determined by the wave speed v on the string and the length L of the string. From Eq. 17-53, these resonant frequencies are

$$f = n\frac{v}{2L}, \qquad \text{for } n = 1, 2, 3, \ldots . \qquad (17\text{-}54)$$

To set up the fourth harmonic (for which $n = 4$), we need to adjust the right side of this equation, with $n = 4$, so that the left side equals the frequency of the vibrator (120 Hz).

We cannot adjust L in Eq. 17-54; it is set. However, a second **Key Idea** is that we *can* adjust v, because it depends on how much mass m we hang on the string. According to Eq. 17-25, wave speed $v = \sqrt{\tau/\mu}$. Here the tension τ in the string is equal to the weight mg of the block. Thus,

$$v = \sqrt{\frac{\tau}{\mu}} = \sqrt{\frac{mg}{\mu}}. \qquad (17\text{-}55)$$

Fig. 17-23 Sample Problem 17-7. A string under tension connected to a vibrator. For a fixed vibrator frequency, standing wave patterns will occur for certain values of the string tension.

Substituting v from Eq. 17-55 into Eq. 17-54, setting $n = 4$ for the fourth harmonic, and solving for m give us

$$m = \frac{4L^2f^2\mu}{n^2g} \qquad (17\text{-}56)$$

$$= \frac{(4)(1.2 \text{ m})^2(120 \text{ Hz})^2(0.0016 \text{ kg/m})}{(4)^2(9.8 \text{ m/s}^2)}$$

$$= 0.846 \text{ kg} \approx 0.85 \text{ kg}. \qquad \text{(Answer)}$$

(b) What standing wave mode is set up if $m = 1.00$ kg?

SOLUTION: If we insert this value of m into Eq. 17-56 and solve for n, we find that $n = 3.7$. A **Key Idea** here is that n must be an integer, so $n = 3.7$ is impossible. Thus, with $m = 1.00$ kg, the vibrator cannot set up a standing wave on the string, and any oscillation of the string will be small, perhaps even imperceptible.

Tactic 2: *Harmonics on a String*
When you need to obtain information about a certain harmonic on a stretched string of given length L, first draw that harmonic (as in Fig. 17-21). If you are asked about, say, the fifth harmonic, you need to draw five loops between the fixed support points. That would mean that five loops, each of length $\lambda/2$, occupy the length

L of the string. Thus, $5(\lambda/2) = L$, and $\lambda = 2L/5$. You can then use Eq. 17-12 ($f = v/\lambda$) to find the frequency of the harmonic.

Keep in mind that the wavelength of a harmonic is set only by the length L of the string, but the frequency depends also on the wave speed v, which is set by the tension and the linear density of the string via Eq. 17-25.

REVIEW & SUMMARY

Transverse and Longitudinal Waves   Mechanical waves can exist only in material media and are governed by Newton's laws. **Transverse** mechanical waves, like those on a stretched string, are waves in which the particles of the medium oscillate perpendicular to the wave's direction of travel. Waves in which the particles of the medium oscillate parallel to the wave's direction of travel are **longitudinal** waves.

Sinusoidal Waves   A sinusoidal wave moving in the positive x direction has the mathematical form

$$y(x, t) = y_m \sin(kx - \omega t), \qquad (17\text{-}2)$$

where y_m is the **amplitude** of the wave, k is the **angular wave number,** ω is the **angular frequency,** and $kx - \omega t$ is the **phase.** The **wavelength** λ is related to k by

$$k = \frac{2\pi}{\lambda}. \qquad (17\text{-}5)$$

The **period** T and **frequency** f of the wave are related to ω by

$$\frac{\omega}{2\pi} = f = \frac{1}{T}. \qquad (17\text{-}9)$$

Finally, the **wave speed** v is related to these other parameters by

$$v = \frac{\omega}{k} = \frac{\lambda}{T} = \lambda f. \qquad (17\text{-}12)$$

Equation of a Traveling Wave   Any function of the form

$$y(x, t) = h(kx \pm \omega t) \qquad (17\text{-}16)$$

can represent a **traveling wave** with a wave speed given by Eq. 17-12 and a wave shape given by the mathematical form of h. The plus sign denotes a wave traveling in the negative x direction, and the minus sign a wave traveling in the positive x direction.

Wave Speed on Stretched String   The speed of a wave on a stretched string is set by properties of the string. The speed on a string with tension τ and linear density μ is

$$v = \sqrt{\frac{\tau}{\mu}}. \qquad (17\text{-}25)$$

Power   The **average power,** or average rate at which energy is transmitted by a sinusoidal wave on a stretched string, is given by

$$P_{\text{avg}} = \tfrac{1}{2}\mu v \omega^2 y_m^2. \qquad (17\text{-}32)$$

Superposition of Waves   When two or more waves traverse the same medium, the displacement of any particle of the medium is the sum of the displacements that the individual waves would give it.

Interference of Waves   Two sinusoidal waves on the same string exhibit **interference,** adding or canceling according to the principle of superposition. If the two are traveling in the same direction and have the same amplitude y_m and frequency (hence the same wavelength) but differ in phase by a **phase constant** ϕ, the result is a single wave with this same frequency:

$$y'(x, t) = [2y_m \cos \tfrac{1}{2}\phi] \sin(kx - \omega t + \tfrac{1}{2}\phi). \qquad (17\text{-}38)$$

If $\phi = 0$, the waves are exactly in phase and their interference is fully constructive; if $\phi = \pi$ rad, they are exactly out of phase and their interference is fully destructive.

Phasors   A wave $y(x, t)$ can be represented with a *phasor*. This is a vector that has a magnitude equal to the amplitude y_m of the wave and that rotates about an origin with an angular speed equal to the angular frequency ω of the wave. The projection of the rotating phasor on a vertical axis gives the displacement y of a point along the wave's travel.

Standing Waves   The interference of two identical sinusoidal waves moving in opposite directions produces **standing waves.** For a string with fixed ends, the standing wave is given by

$$y'(x, t) = [2y_m \sin kx] \cos \omega t. \qquad (17\text{-}47)$$

Standing waves are characterized by fixed locations of zero displacement called **nodes** and fixed locations of maximum displacement called **antinodes.**

Resonance   Standing waves on a string can be set up by reflection of traveling waves from the ends of the string. If an end is fixed, it must be the position of a node. This limits the frequencies at which standing waves will occur on a given string. Each possible frequency is a **resonant frequency,** and the corresponding standing wave pattern is an **oscillation mode.** For a stretched string of length L with fixed ends, the resonant frequencies are

$$f = \frac{v}{\lambda} = n\frac{v}{2L}, \qquad \text{for } n = 1, 2, 3, \dots. \qquad (17\text{-}53)$$

The oscillation mode corresponding to $n = 1$ is called the *fundamental mode* or the *first harmonic*; the mode corresponding to $n = 2$ is the *second harmonic*; and so on.

QUESTIONS

1. What is the wavelength of the (strange) wave in Fig. 17-24, where each segment of the wave has length d?

Fig. 17-24 Question 1.

2. Figure 17-25a gives a snapshot of a wave traveling in the direction of positive x along a string under tension. Four string elements are indicated by the lettered points. For each of those elements, determine whether, at the instant of the snapshot, the element is moving upward or downward or is momentarily at rest. (*Hint:* Imagine the wave as it moves through the four string elements.)

Figure 17-25b gives the displacement of a string element

located at, say, $x = 0$ as a function of time. At the lettered times, is the element moving upward or downward or is it momentarily at rest?

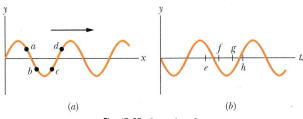

Fig. 17-25 Question 2.

3. In Fig. 17-26, five points are indicated on a snapshot of a sinusoidal wave. What is the phase difference between point 1 and (a) point 2, (b) point 3, (c) point 4, and (d) point 5? Answer in radians and in terms of the wavelength of the wave. The snapshot shows a point of zero displacement at $x = 0$. In terms of the period T of the wave, when will (e) a peak and (f) the next point of zero displacement reach $x = 0$?

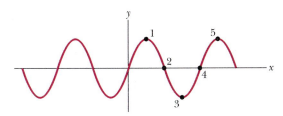

Fig. 17-26 Question 3.

4. The following four waves are sent along strings with the same linear densities (x is in meters and t is in seconds). Rank the waves according to (a) their wave speed and (b) the tension in the strings along which they travel, greatest first:

(1) $y_1 = (3 \text{ mm}) \sin(x - 3t)$,   (3) $y_3 = (1 \text{ mm}) \sin(4x - t)$,

(2) $y_2 = (6 \text{ mm}) \sin(2x - t)$,   (4) $y_4 = (2 \text{ mm}) \sin(x - 2t)$.

5. In Fig. 17-27, wave 1 consists of a rectangular peak of height 4 units and width d, and a rectangular valley of depth 2 units and width d. The wave travels rightward along an x axis. Choices 2, 3, and 4 are similar waves, with the same heights, depths, and widths, that will travel leftward along that axis and through wave 1. With

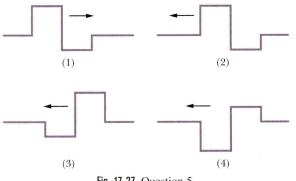

Fig. 17-27 Question 5.

which choice will the interference give, for an instant, (a) the deepest valley, (b) a flat line, and (c) a level peak $2d$ wide?

6. If you start with two sinusoidal waves of the same amplitude traveling in phase on a string and then somehow phase-shift one of them by 5.4 wavelengths, what type of interference will occur on the string?

7. The amplitudes and phase differences for four pairs of waves of equal wavelengths are (a) 2 mm, 6 mm, and π rad; (b) 3 mm, 5 mm, and π rad; (c) 7 mm, 9 mm, and π rad; (d) 2 mm, 2 mm, and 0 rad. Each pair travels in the same direction along the same string. Without written calculation, rank the four pairs according to the amplitude of their resultant wave, greatest first. (*Hint:* Construct phasor diagrams.)

8. If you set up the seventh harmonic on a string, (a) how many nodes are present, and (b) is there a node, antinode, or some intermediate state at the midpoint? If you next set up the sixth harmonic, (c) is its resonant wavelength longer or shorter than that for the seventh harmonic, and (d) is the resonant frequency higher or lower?

9. Strings A and B have identical lengths and linear densities, but string B is under greater tension than string A. Figure 17-28 shows four situations, (a) through (d), in which standing wave patterns exist on the two strings. In which situations is there the possibility that strings A and B are oscillating at the same resonant frequency?

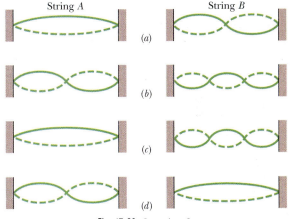

Fig. 17-28 Question 9.

10. (a) If a standing wave on a string is given by

$$y'(t) = (3 \text{ mm}) \sin(5x) \cos(4t),$$

is there a node or an antinode of the oscillations of the string at $x = 0$? (b) If the standing wave is given by

$$y'(t) = (3 \text{ mm}) \sin(5x + \pi/2) \cos(4t),$$

is there a node or an antinode at $x = 0$?

11. (a) In Sample Problem 17-7 and Fig. 17-23, if we gradually increase the mass of the block (the frequency remains fixed), new resonant modes appear. Do the harmonic numbers of the new resonant modes increase or decrease from one to the next? (b) Is the shift from one resonant mode to the next gradual, or does each resonant mode disappear well before the next one appears?

EXERCISES & PROBLEMS

ssm Solution is in the Student Solutions Manual.
www Solution is available on the World Wide Web at:
        http://www.wiley.com/college/hrw
ilw Solution is available on the Interactive LearningWare.

SEC. 17-5   The Speed of a Traveling Wave

1E. A wave has an angular frequency of 110 rad/s and a wavelength of 1.80 m. Calculate (a) the angular wave number and (b) the speed of the wave.

2E. The speed of electromagnetic waves (which include visible light, radio, and x rays) in vacuum is 3.0×10^8 m/s. (a) Wavelengths of visible light waves range from about 400 nm in the violet to about 700 nm in the red. What is the range of frequencies of these waves? (b) The range of frequencies for shortwave radio (for example, FM radio and VHF television) is 1.5 to 300 MHz. What is the corresponding wavelength range? (c) X ray wavelengths range from about 5.0 nm to about 1.0×10^{-2} nm. What is the frequency range for x rays?

3E. A sinusoidal wave travels along a string. The time for a particular point to move from maximum displacement to zero is 0.170 s. What are the (a) period and (b) frequency? (c) The wavelength is 1.40 m; what is the wave speed? ssm

4E. Write the equation for a sinusoidal wave traveling in the negative direction along an x axis and having an amplitude of 0.010 m, a frequency of 550 Hz, and a speed of 330 m/s.

5E. Show that

$$y = y_m \sin k(x - vt), \qquad y = y_m \sin 2\pi\left(\frac{x}{\lambda} - ft\right),$$

$$y = y_m \sin \omega\left(\frac{x}{v} - t\right), \qquad y = y_m \sin 2\pi\left(\frac{x}{\lambda} - \frac{t}{T}\right)$$

are all equivalent to $y = y_m \sin(kx - \omega t)$. ssm

6P. The equation of a transverse wave traveling along a very long string is $y = 6.0 \sin(0.020\pi x + 4.0\pi t)$, where x and y are expressed in centimeters and t is in seconds. Determine (a) the amplitude, (b) the wavelength, (c) the frequency, (d) the speed, (e) the direction of propagation of the wave, and (f) the maximum transverse speed of a particle in the string. (g) What is the transverse displacement at $x = 3.5$ cm when $t = 0.26$ s?

7P. (a) Write an equation describing a sinusoidal transverse wave traveling on a cord in the $+x$ direction with a wavelength of 10 cm, a frequency of 400 Hz, and an amplitude of 2.0 cm. (b) What is the maximum speed of a point on the cord? (c) What is the speed of the wave? ssm

8P. A transverse sinusoidal wave of wavelength 20 cm is moving along a string in the positive x direction. The transverse displacement of the string particle at $x = 0$ as a function of time is shown in Fig. 17-29. (a) Make a rough sketch of one wavelength

Fig. 17-29 Problem 8.

of the wave (the portion between $x = 0$ and $x = 20$ cm) at time $t = 0$. (b) What is the speed of the wave? (c) Write the equation for the wave with all the constants evaluated. (d) What is the transverse velocity of the particle at $x = 0$ at $t = 5.0$ s?

9P. A sinusoidal wave of frequency 500 Hz has a speed of 350 m/s. (a) How far apart are two points that differ in phase by $\pi/3$ rad? (b) What is the phase difference between two displacements at a certain point at times 1.00 ms apart? ilw

SEC. 17-6   Wave Speed on a Stretched String

10E. The heaviest and lightest strings on a certain violin have linear densities of 3.0 and 0.29 g/m. What is the ratio of the diameter of the heaviest string to that of the lightest string, assuming that the strings are of the same material?

11E. What is the speed of a transverse wave in a rope of length 2.00 m and mass 60.0 g under a tension of 500 N? ssm

12E. The tension in a wire clamped at both ends is doubled without appreciably changing the wire's length between the clamps. What is the ratio of the new to the old wave speed for transverse waves traveling along this wire?

13E. The linear density of a string is 1.6×10^{-4} kg/m. A transverse wave on the string is described by the equation

$$y = (0.021 \text{ m}) \sin[(2.0 \text{ m}^{-1})x + (30 \text{ s}^{-1})t].$$

What is (a) the wave speed and (b) the tension in the string? ssm

14E. The equation of a transverse wave on a string is

$$y = (2.0 \text{ mm}) \sin[(20 \text{ m}^{-1})x - (600 \text{ s}^{-1})t].$$

The tension in the string is 15 N. (a) What is the wave speed? (b) Find the linear density of this string in grams per meter.

15P. A stretched string has a mass per unit length of 5.0 g/cm and a tension of 10 N. A sinusoidal wave on this string has an amplitude of 0.12 mm and a frequency of 100 Hz and is traveling in the negative direction of x. Write an equation for this wave. ssm

16P. What is the fastest transverse wave that can be sent along a steel wire? For safety reasons, the maximum tensile stress to which steel wires should be subjected is 7.0×10^8 N/m^2. The density of steel is 7800 kg/m^3. Show that your answer does not depend on the diameter of the wire.

17P. A sinusoidal transverse wave of amplitude y_m and wavelength λ travels on a stretched cord. (a) Find the ratio of the maximum particle speed (the speed with which a single particle in the cord moves transverse to the wave) to the wave speed. (b) If a wave having a certain wavelength and amplitude is sent along a cord, would this speed ratio depend on the material of which the cord is made, such as wire or nylon? ssm

18P. A sinusoidal wave is traveling on a string with speed 40 cm/s. The displacement of the particles of the string at $x = 10$ cm is found to vary with time according to the equation $y = (5.0 \text{ cm}) \sin[1.0 - (4.0 \text{ s}^{-1})t]$. The linear density of the string is 4.0 g/cm. What are (a) the frequency and (b) the wavelength of the wave? (c) Write the general equation giving the transverse displacement of the particles of the string as a function of position and time. (d) Calculate the tension in the string.

19P. A sinusoidal transverse wave is traveling along a string in the negative direction of an x axis. Figure 17-30 shows a plot of the displacement as a function of position at time $t = 0$; the y intercept is 4.0 cm. The string tension is 3.6 N, and its linear density is 25 g/m. Find (a) the amplitude, (b) the wavelength, (c) the wave speed, and (d) the period of the wave. (e) Find the maximum transverse speed of a particle in the string. (f) Write an equation describing the traveling wave. ssm ilw

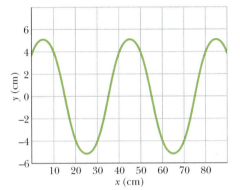

Fig. 17-30
Problem 19.

20P. In Fig. 17-31a, string 1 has a linear density of 3.00 g/m, and string 2 has a linear density of 5.00 g/m. They are under tension owing to the hanging block of mass $M = 500$ g. Calculate the wave speed on (a) string 1 and (b) string 2. (*Hint:* When a string loops halfway around a pulley, it pulls on the pulley with a net force that is twice the tension in the string.) Next the block is divided into two blocks (with $M_1 + M_2 = M$) and the apparatus is rearranged as shown in Fig. 17-31b. Find (c) M_1 and (d) M_2 such that the wave speeds in the two strings are equal.

(a)

(b)

Fig. 17-31 Problem 20.

21P. A wire 10.0 m long and having a mass of 100 g is stretched under a tension of 250 N. If two pulses, separated in time by 30.0 ms, are generated, one at each end of the wire, where will the pulses first meet? ssm ilw

22P. The type of rubber band used inside some baseballs and golf balls obeys Hooke's law over a wide range of elongation of the band. A segment of this material has an unstretched length ℓ and a mass m. When a force F is applied, the band stretches an additional length $\Delta\ell$. (a) What is the speed (in terms of m, $\Delta\ell$, and the spring constant k) of transverse waves on this stretched rubber band? (b) Using your answer to (a), show that the time required for a transverse pulse to travel the length of the rubber band is proportional to $1/\sqrt{\Delta\ell}$ if $\Delta\ell \ll \ell$ and is constant if $\Delta\ell \gg \ell$.

23P*. A uniform rope of mass m and length L hangs from a ceiling. (a) Show that the speed of a transverse wave on the rope is a function of y, the distance from the lower end, and is given by $v = \sqrt{gy}$. (b) Show that the time a transverse wave takes to travel the length of the rope is given by $t = 2\sqrt{L/g}$. ssm

SEC. 17-7 Energy and Power of a Traveling String Wave

24E. A string along which waves can travel is 2.70 m long and has a mass of 260 g. The tension in the string is 36.0 N. What must be the frequency of traveling waves of amplitude 7.70 mm for the average power to be 85.0 W?

25P. A transverse sinusoidal wave is generated at one end of a long, horizontal string by a bar that moves up and down through a distance of 1.00 cm. The motion is continuous and is repeated regularly 120 times per second. The string has linear density 120 g/m and is kept under a tension of 90.0 N. Find the maximum value of (a) the transverse speed u and (b) the transverse component of the tension τ. (*Hint:* That component is $\tau \sin \theta$, where θ is the angle the string makes with the horizontal. You will need to relate angle θ to dy/dx.)

(c) Show that the two maximum values calculated above occur at the same phase values for the wave. What is the transverse displacement y of the string at these phases? (d) What is the maximum rate of energy transfer along the string? (e) What is the transverse displacement y when this maximum transfer occurs? (f) What is the minimum rate of energy transfer along the string? (g) What is the transverse displacement y when this minimum transfer occurs? ssm www

SEC. 17-9 Interference of Waves

26E. What phase difference between two otherwise identical traveling waves, moving in the same direction along a stretched string, will result in the combined wave having an amplitude 1.50 times that of the common amplitude of the two combining waves? Express your answer in (a) degrees, (b) radians, and (c) wavelengths.

27E. Two identical traveling waves, moving in the same direction, are out of phase by $\pi/2$ rad. What is the amplitude of the resultant wave in terms of the common amplitude y_m of the two combining waves? ssm

28P. Two sinusoidal waves, identical except for phase, travel in the same direction along a string and interfere to produce a resultant wave given by $y'(x, t) = (3.0 \text{ mm}) \sin(20x - 4.0t + 0.820 \text{ rad})$, with x in meters and t in seconds. What are (a) the wavelength λ of the two waves, (b) the phase difference between them, and (c) their amplitude y_m?

SEC. 17-10 Phasors

29E. Determine the amplitude of the resultant wave when two sinusoidal string waves having the same frequency and traveling in the same direction on the same string are combined, if their amplitudes are 3.0 cm and 4.0 cm and they have phase constants of 0 and $\pi/2$ rad, respectively. ssm

30P. Two sinusoidal waves of the same period, with amplitudes of 5.0 and 7.0 mm, travel in the same direction along a stretched string; they produce a resultant wave with an amplitude of 9.0 mm. The phase constant of the 5.0 mm wave is 0. What is the phase constant of the 7.0 mm wave?

31P. Three sinusoidal waves of the same frequency travel along a string in the positive direction of an x axis. Their amplitudes are y_1, $y_1/2$, and $y_1/3$, and their phase constants are 0, $\pi/2$, and π, respectively. What are (a) the amplitude and (b) the phase constant of the resultant wave? (c) Plot the wave form of the resultant wave at $t = 0$, and discuss its behavior as t increases. **ssm** **www**

SEC. 17-12   Standing Waves and Resonance

32E. A string under tension τ_i oscillates in the third harmonic at frequency f_3, and the waves on the string have wavelength λ_3. If the tension is increased to $\tau_f = 4\tau_i$ and the string is again made to oscillate in the third harmonic, what then are (a) the frequency of oscillation in terms of f_3 and (b) the wavelength of the waves in terms of λ_3?

33E. A nylon guitar string has a linear density of 7.2 g/m and is under a tension of 150 N. The fixed supports are 90 cm apart.

Fig. 17-32 Exercise 33.

The string is oscillating in the standing wave pattern shown in Fig. 17-32. Calculate the (a) speed, (b) wavelength, and (c) frequency of the traveling waves whose superposition gives this standing wave. **ilw**

34E. Two sinusoidal waves with identical wavelengths and amplitudes travel in opposite directions along a string with a speed of 10 cm/s. If the time interval between instants when the string is flat is 0.50 s, what is the wavelength of the waves?

35E. A string fixed at both ends is 8.40 m long and has a mass of 0.120 kg. It is subjected to a tension of 96.0 N and set oscillating. (a) What is the speed of the waves on the string? (b) What is the longest possible wavelength for a standing wave? (c) Give the frequency of that wave. **ssm**

36E. A 125 cm length of string has a mass of 2.00 g. It is stretched with a tension of 7.00 N between fixed supports. (a) What is the wave speed for this string? (b) What is the lowest resonant frequency of this string?

37E. What are the three lowest frequencies for standing waves on a wire 10.0 m long having a mass of 100 g, which is stretched under a tension of 250 N? **ssm**

38E. String A is stretched between two clamps separated by distance L. String B, with the same linear density and under the same tension as string A, is stretched between two clamps separated by distance 4L. Consider the first eight harmonics of string B. Which, if any, has a resonant frequency that matches a resonant frequency of string A?

39P. A string that is stretched between fixed supports separated by 75.0 cm has resonant frequencies of 420 and 315 Hz, with no intermediate resonant frequencies. What are (a) the lowest resonant frequency and (b) the wave speed? **ssm** **ilw** **www**

40P. In Fig. 17-33, two pulses travel along a string in opposite

Fig. 17-33 Problem 40.

directions. The wave speed v is 2.0 m/s and the pulses are 6.0 cm apart at $t = 0$. (a) Sketch the wave patterns when t is equal to 5.0, 10, 15, 20, and 25 ms. (b) In what form (or type) is the energy of the pulses at $t = 15$ ms?

41P. A string oscillates according to the equation

$$y' = (0.50 \text{ cm}) \sin\left[\left(\frac{\pi}{3} \text{ cm}^{-1}\right)x\right] \cos[(40\pi \text{ s}^{-1})t].$$

What are (a) the amplitude and (b) the speed of the two waves (identical except for direction of travel) whose superposition gives this oscillation? (c) What is the distance between nodes? (d) What is the speed of a particle of the string at the position $x = 1.5$ cm when $t = \frac{9}{8}$ s? **ssm**

42P. A standing wave results from the sum of two transverse traveling waves given by

$$y_1 = 0.050 \cos(\pi x - 4\pi t)$$

and

$$y_2 = 0.050 \cos(\pi x + 4\pi t),$$

where x, y_1, and y_2 are in meters and t is in seconds. (a) What is the smallest positive value of x that corresponds to a node? (b) At what times during the interval $0 \le t \le 0.50$ s will the particle at $x = 0$ have zero velocity?

43P. A string 3.0 m long is oscillating as a three-loop standing wave with an amplitude of 1.0 cm. The wave speed is 100 m/s. (a) What is the frequency? (b) Write equations for two waves that, when combined, will result in this standing wave. **ssm**

44P. In an experiment on standing waves, a string 90 cm long is attached to the prong of an electrically driven tuning fork that oscillates perpendicular to the length of the string at a frequency of 60 Hz. The mass of the string is 0.044 kg. What tension must the string be under (weights are attached to the other end) if it is to oscillate in four loops?

45P. Oscillation of a 600 Hz tuning fork sets up standing waves in a string clamped at both ends. The wave speed for the string is 400 m/s. The standing wave has four loops and an amplitude of 2.0 mm. (a) What is the length of the string? (b) Write an equation for the displacement of the string as a function of position and time. **ssm**

46P. A rope, under a tension of 200 N and fixed at both ends, oscillates in a second-harmonic standing wave pattern. The displacement of the rope is given by

$$y = (0.10 \text{ m})(\sin \pi x/2) \sin 12\pi t,$$

where $x = 0$ at one end of the rope, x is in meters, and t is in seconds. What are (a) the length of the rope, (b) the speed of the waves on the rope, and (c) the mass of the rope? (d) If the rope oscillates in a third-harmonic standing wave pattern, what will be the period of oscillation?

47P. A generator at one end of a very long string creates a wave given by

$$y = (6.0 \text{ cm}) \cos \frac{\pi}{2} [(2.0 \text{ m}^{-1})x + (8.0 \text{ s}^{-1})t],$$

and one at the other end creates the wave

$$y = (6.0 \text{ cm}) \cos \frac{\pi}{2} [(2.0 \text{ m}^{-1})x - (8.0 \text{ s}^{-1})t].$$

Calculate (a) the frequency, (b) the wavelength, and (c) the speed of each wave. At what x values are (d) the nodes and (e) the antinodes? ssm

48P. A standing wave pattern on a string is described by

$$y(x, t) = 0.040 \sin 5\pi x \cos 40\pi t,$$

where x and y are in meters and t is in seconds. (a) Determine the location of all nodes for $0 \le x \le 0.40$ m. (b) What is the period of the oscillatory motion of any (nonnode) point on the string? What are (c) the speed and (d) the amplitude of the two traveling waves that interfere to produce this wave? (e) At what times for $0 \le t \le 0.050$ s will all the points on the string have zero transverse velocity?

49P. Show that the maximum kinetic energy in each loop of a standing wave produced by two traveling waves of identical amplitudes is $2\pi^2 \mu y_m^2 f v$. ssm www

50P. For a certain transverse standing wave on a long string, an antinode is at $x = 0$ and a node is at $x = 0.10$ m. The displacement $y(t)$ of the string particle at $x = 0$ is shown in Fig. 17-34. When $t = 0.50$ s, what are the displacements of the string particles at (a) $x = 0.20$ m and (b) $x = 0.30$ m? At $x = 0.20$ m, what are the transverse velocities of the string particles at (c) $t = 0.50$ s and (d) $t = 1.0$ s? (e) Sketch the standing wave at $t = 0.50$ s for the range $x = 0$ to $x = 0.40$ m.

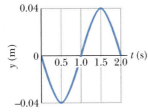

Fig. 17-34 Problem 50.

51P. In Fig. 17-35, an aluminum wire, of length $L_1 = 60.0$ cm, cross-sectional area 1.00×10^{-2} cm^2, and density 2.60 g/cm^3, is joined to a steel wire, of density 7.80 g/cm^3 and the same cross-sectional area. The compound wire, loaded with a block of mass $m = 10.0$ kg, is arranged so that the distance L_2 from the joint to the supporting pulley is 86.6 cm. Transverse waves are set up in the wire by using an external source of variable frequency; a node is located at the pulley. (a) Find the lowest frequency of excitation for which standing waves are observed such that the joint in the wire is one of the nodes. (b) How many nodes are observed at this frequency? ssm

Fig. 17-35 Problem 51.

Additional Problem

52. *Body armor*. When a high-speed projectile such as a bullet or bomb fragment strikes modern body armor, the fabric of the armor stops the projectile and prevents penetration by quickly spreading the projectile's energy over a large area. This spreading is done by longitudinal and transverse pulses that move *radially* from the impact point, where the projectile pushes a cone-shaped dent into the fabric. The longitudinal pulse, racing along the fibers of the fabric at speed v_l ahead of the denting, causes the fibers to thin and stretch, with material flowing radially inward into the dent. One such radial fiber is shown in Fig. 17-36a. Part of the projectile's energy goes into this motion and stretching. The transverse pulse, moving at a slower speed v_t, is due to the denting. As the projectile increases the dent's depth, the dent increases in radius, causing the material in the fibers to move in the same direction as the projectile (perpendicular to the transverse pulse's direction of travel). The rest of the projectile's energy goes into this motion. All the energy that does not eventually go into permanently deforming the fibers ends up as thermal energy.

Figure 17-36b is a graph of speed v versus time t for a bullet of mass 10.2 g fired from a .38 Special revolver directly into body armor. Take $v_l = 2000$ m/s, and assume that the half-angle θ of the conical dent is 60°. At the end of the collision, what are the radii of (a) the thinned region and (b) the dent (assuming that the person wearing the armor remains stationary)?

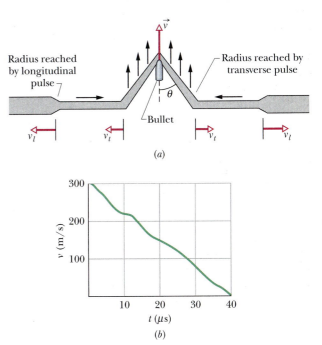

Fig. 17-36 Problem 52.

NEW PROBLEMS

N1. A sinusoidal wave moving along a string is shown twice in Fig. 17N-1, as crest A travels in the positive direction of an x axis by distance $d = 6.0$ cm in 4.0 ms. The tick marks along the axis are separated by 10 cm. Write an equation for the wave in the form $y(x, t) = y_m \sin(kx \pm \omega t)$, where $\pm$ indicates that the proper sign must be determined.

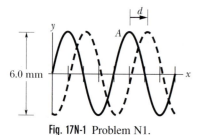

Fig. 17N-1 Problem N1.

N2. A sinusoidal wave travels along a string under tension. Figure 17N-2 gives the slopes along the string at time $t = 0$. What is the amplitude of the wave?

Fig. 17N-2 Problem N2.

N3. Figure 17N-3 shows the displacement y versus time t of the point on a string at $x = 0$, as a wave passes through it. The wave has the form $y(x, t) = y_m \sin(kx - \omega t + \phi)$. What is ϕ? (*Caution:* A calculator does not always give the proper inverse trig function, so check your answer by substituting it and an assumed value of ω into $y(x, t)$ and then plotting the function.)

Fig. 17N-3 Problem N3.

N4. Figure 17N-4 shows the transverse velocity u versus time t of the point on a string at $x = 0$, as a wave passes through it. The wave has the form $y(x, t) = y_m \sin(kx - \omega t + \phi)$. What is ϕ? (*Caution:* A calculator does not always give the proper inverse trig function, so check your answer by substituting it and an assumed value of ω into $y(x, t)$ and then plotting the function.)

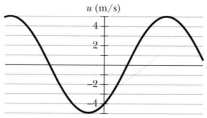

Fig. 17N-4 Problem N4.

N5. Figure 17N-5 shows the transverse acceleration a_y versus time t of the point on a string at $x = 0$, as a wave of the form $y(x, t) = y_m \sin(kx - \omega t + \phi)$ passes through it. What is ϕ? (*Caution:* A calculator does not always give the proper inverse trig function, so check your answer by substituting it and an assumed value of ω into $y(x, t)$ and then plotting the function.)

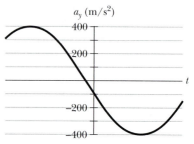

Fig. 17N-5 Problem N5.

N6. At time $t = 0$ and at position $x = 0$ m along a string, a traveling sinusoidal wave with an angular frequency of 440 rad/s has displacement $y = +4.5$ mm and transverse velocity $u = -0.75$ m/s. If the wave has the form $y(x, t) = y_m \sin(kx - \omega t + \phi)$, what is phase constant ϕ?

N7. The function $y(x, t) = (15$ cm$) \cos(\pi x - 15\pi t)$, with x in meters and t in seconds, describes a wave on a taut string. What is the transverse speed for a point on the string at an instant when that point has the displacement $y = +12$ cm?

N8. A sinusoidal wave is sent along a cord with a linear density of 2.0 g/m. As it travels, the kinetic energies of the mass elements along the cord vary. Figure 17N-6a gives the rate at which the kinetic energy dK/dt is changing at a particular instant, plotted as a function of distance x along the string. Figure 17N-6b is similar except that it gives the rate at which the kinetic energy of a particular mass element (at a particular location) changes with time t. What is the amplitude of the wave?

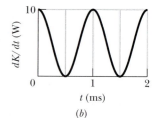

Fig. 17N-6 Problem N8.

N9. If $y(x, t) = (6.0$ mm$) \sin(kx + (600$ rad/s$)t + \phi)$ describes a wave traveling along a string, how much time does any given point on the string take to move between displacements $y = +2.0$ mm and $y = -2.0$ mm?

N10. Two sinusoidal waves of the same frequency are to be sent in the same direction along a taut string. One wave has an amplitude of 5.0 mm, the other 8.0 mm. (a) What phase difference ϕ_1 between the two waves results in the least possible amplitude of the resultant wave? (b) What is that least possible amplitude? (c) What phase difference ϕ_2 results in the greatest possible amplitude of the resultant wave? (d) What is that greatest possible amplitude? (e) What amplitude occurs if the phase angle is $(\phi_1 - \phi_2)/2$?

N11. In a demonstration, a 1.2 kg horizontal rope is fixed in place at its two ends ($x = 0$ and $x = 2.0$ m) and made to oscillate up and down in the fundamental mode, at frequency 5.0 Hz. At $t = 0$, the point at $x = 1.0$ m has zero displacement and is moving upward in the positive direction of a y axis with a transverse velocity of 5.0 m/s. What are (a) the amplitude of the motion of that point and (b) the tension in the rope? (c) Write the standing wave equation for the fundamental mode.

N12. Two sinusoidal waves, of the same frequency 120 Hz, are to be sent in the positive direction of an x axis that is directed along a cord under tension. They can be sent in phase, or one can be shifted from the other. Figure 17N-7 shows the amplitude y' of the resulting wave versus the distance of the shift (how far one wave is shifted from the other wave). Write an equation (complete with numerical values) for each wave for when they are in phase.

Fig. 17N-7 Problem N12.

N13. These two waves travel along the same string:

$$y_1(x, t) = (4.60 \text{ mm}) \sin(2\pi x - 400\pi t)$$
$$y_2(x, t) = (5.60 \text{ mm}) \sin(2\pi x - 400\pi t + 0.80\pi \text{ rad}).$$

What are (a) the amplitude and (b) the phase angle (relative to wave 1) of the resultant wave? (c) If a third wave of amplitude 5.00 mm is also to be sent along the string in the same direction as the first two waves, what should be its phase angle to maximize the amplitude of the new resultant wave?

N14. Four waves are to be sent along the same string, in the same direction:

$$y_1(x, t) = (4.00 \text{ mm}) \sin(2\pi x - 400\pi t)$$
$$y_2(x, t) = (4.00 \text{ mm}) \sin(2\pi x - 400\pi t + 0.7\pi)$$
$$y_3(x, t) = (4.00 \text{ mm}) \sin(2\pi x - 400\pi t + \pi)$$
$$y_4(x, t) = (4.00 \text{ mm}) \sin(2\pi x - 400\pi t + 1.7\pi).$$

What is the resultant wave?

N15. Four sinusoidal waves travel in the positive x direction along the same string. Their frequencies are in the ratio 1:2:3:4, and their amplitudes are in the ratio $1:\frac{1}{2}:\frac{1}{3}:\frac{1}{4}$, respectively. When $t = 0$, at $x = 0$, the first and third waves are 180° out of phase with the second and fourth. Plot the resultant wave form when $t = 0$, and discuss its behavior as t increases.

N16. The following two waves are sent in opposite directions on a string so as to create a standing wave, with vertical oscillations:

$$y_1(x, t) = (6.00 \text{ mm}) \sin(4.00\pi x - 400\pi t)$$
$$y_2(x, t) = (6.00 \text{ mm}) \sin(4.00\pi x + 400\pi t).$$

An antinode is located at point A. When that point moves from maximum upward displacement to maximum downward displacement, how far does each wave move along the string?

N17. A sinusoidal wave of angular frequency 1200 rad/s and amplitude 3.00 mm is sent along a cord with linear density 2.00 g/m and tension 1200 N. (a) What is the average rate at which energy is transported by the wave to the opposite end of the cord? (b) If, simultaneously, an identical wave travels along an adjacent, identical cord, what is the total average rate at which energy is transported to the opposite ends of the two cords by the waves? If, instead, those two waves are sent along the *same* cord simultaneously, what is the total average rate at which they transport energy to the opposite end of the cord when their phase difference is (c) 0, (d) 0.4π rad, and (e) π rad?

N18. Two sinusoidal waves with the same amplitude of 9.00 mm and the same wavelength travel together along a string that is stretched along an x axis. Their resultant wave is shown twice in Fig. 17N-8, as valley A travels in the negative direction of an x axis by distance $d = 56.0$ cm in 8.0 ms. The tick marks along the axis are separated by 10 cm. Write equations for the two interfering waves in the form $y(x, t) = y_m \sin(kx \pm \omega t + \phi)$, where $\pm$ indicates that the proper sign must be determined.

Fig. 17N-8 Problem N18.

N19. Standing waves are set up with the equipment in Sample Problem 17-7 (Fig. 17-23) except that a different string is used. A standing wave appears when the mass of the hanging block is 286.1 g or 447.0 g, but not for any intermediate mass. What is the linear density of the string?

N20. Two sinusoidal waves with the same amplitude and wavelength travel together along a string that is stretched along an x axis. Their resultant wave is shown twice in Fig. 17N-9, as the antinode A travels from an extreme upward displacement to an extreme downward displacement in 6.0 ms. The tick marks along the axis are separated by 10 cm. Write equations for the two interfering waves in the form $y(x, t) = y_m \sin(kx \pm \omega t)$, where $\pm$ indicates that the proper sign must be determined.

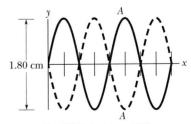
Fig. 17N-9 Problem N20.

N21. One of the harmonic frequencies for a particular string under tension is 325 Hz. The next higher harmonic frequency is 390 Hz. What harmonic frequency is next higher after the harmonic frequency 195 Hz?

18 Waves—II

This horseshoe bat not only can locate a moth flying in total darkness but can also determine the moth's relative speed, to home in on the insect.

How does the bat's detection system work, and how can a moth "jam" the system or otherwise reduce its effectiveness?

The answer is in this chapter.

Fig. 18-1 An image of a fetus searching for a thumb to suck; the image is made with ultrasound (which has a frequency above your hearing range).

18-1 Sound Waves

As we saw in Chapter 17, mechanical waves are waves that require a material medium to exist. There are two types of mechanical waves: *Transverse waves* involve oscillations perpendicular to the direction in which the wave travels; *longitudinal waves* involve oscillations parallel to the direction of wave travel.

In this book, a **sound wave** is defined roughly as any longitudinal wave. Seismic prospecting teams use such waves to probe Earth's crust for oil. Ships carry sound-ranging gear (sonar) to detect underwater obstacles. Submarines use sound waves to stalk other submarines, largely by listening for the characteristic noises produced by the propulsion system. Figure 18-1, a computer-processed image of a fetal head and arm, shows how sound waves can be used to explore the soft tissues of the human body. In this chapter we shall focus on sound waves that travel through the air and that are audible to people.

Figure 18-2 illustrates several ideas that we shall use in our discussions. Point S represents a tiny sound source, called a *point source*, that emits sound waves in all directions. The *wavefronts* and *rays* indicate the direction of travel and the spread of the sound waves. **Wavefronts** are surfaces over which the oscillations of the air due to the sound wave have the same value; such surfaces are represented by whole or partial circles in a two-dimensional drawing for a point source. **Rays** are directed lines perpendicular to the wavefronts that indicate the direction of travel of the wavefronts. The short double arrows superimposed on the rays of Fig. 18-2 indicate that the longitudinal oscillations of the air are parallel to the rays.

Near a point source like that of Fig. 18-2, the wavefronts are spherical and spread out in three dimensions, and there the waves are said to be *spherical*. As the wavefronts move outward and their radii become larger, their curvature decreases. Far from the source, we approximate the wavefronts as planes (or lines on two-dimensional drawings), and the waves are said to be *planar*.

18-2 The Speed of Sound

The speed of any mechanical wave, transverse or longitudinal, depends on both an inertial property of the medium (to store kinetic energy) and an elastic property of the medium (to store potential energy). Thus, we can generalize Eq. 17-25, which gives the speed of a transverse wave along a stretched string, by writing

$$v = \sqrt{\frac{\tau}{\mu}} = \sqrt{\frac{\text{elastic property}}{\text{inertial property}}}, \tag{18-1}$$

where (for transverse waves) τ is the tension in the string and μ is the string's linear density. If the medium is air and the wave is longitudinal, we can guess that the inertial property, corresponding to μ, is the volume density ρ of air. What shall we put for the elastic property?

In a stretched string, potential energy is associated with the periodic stretching of the string elements as the wave passes through them. As a sound wave passes through air, potential energy is associated with periodic compressions and expansions of small volume elements of the air. The property that determines the extent to which an element of a medium changes in volume when the pressure (force per unit area) on it changes is the **bulk modulus** B, defined (from Eq. 13-27) as

$$B = -\frac{\Delta p}{\Delta V/V} \qquad \text{(definition of bulk modulus)}. \tag{18-2}$$

Here $\Delta V/V$ is the fractional change in volume produced by a change in pressure Δp. As explained in Section 15-3, the SI unit for pressure is the newton per square meter,

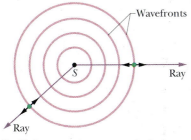

Fig. 18-2 A sound wave travels from a point source S through a three-dimensional medium. The wavefronts form spheres centered on S; the rays are radial to S. The short, double-headed arrows indicate that elements of the medium oscillate parallel to the rays.

TABLE 18-1   The Speed of Sound[a]

Medium	Speed (m/s)
Gases	
Air (0°C)	331
Air (20°C)	343
Helium	965
Hydrogen	1284
Liquids	
Water (0°C)	1402
Water (20°C)	1482
Seawater[b]	1522
Solids	
Aluminum	6420
Steel	5941
Granite	6000

[a]At 0°C and 1 atm pressure, except where noted.

[b]At 20°C and 3.5% salinity.

which is given a special name, the *pascal* (Pa). From Eq. 18-2 we see that the unit for B is also the pascal. The signs of Δp and ΔV are always opposite: When we increase the pressure on an element (Δp is positive), its volume decreases (ΔV is negative). We include a minus sign in Eq. 18-2 so that B is always a positive quantity. Now substituting B for τ and ρ for μ in Eq. 18-1 yields

$$v = \sqrt{\frac{B}{\rho}} \quad \text{(speed of sound)} \tag{18-3}$$

as the speed of sound in a medium with bulk modulus B and density ρ. This is actually the correct equation, as we shall derive shortly. Table 18-1 lists the speed of sound in various media.

The density of water is almost 1000 times greater than the density of air. If this were the only relevant factor, we would expect from Eq. 18-3 that the speed of sound in water would be considerably less than the speed of sound in air. However, Table 18-1 shows us that the reverse is true. We conclude (again from Eq. 18-3) that the bulk modulus of water must be more than 1000 times greater than that of air. This is indeed the case. Water is much more incompressible than air, which (see Eq. 18-2) is another way of saying that its bulk modulus is much greater.

Formal Derivation of Eq. 18-3

We now derive Eq. 18-3 by direct application of Newton's laws. Let a single pulse in which air is compressed travel (from right to left) with speed v through the air in a long tube, like that in Fig. 17-2. Let us run along with the pulse at that speed, so that the pulse appears to stand still in our reference frame. Figure 18-3a shows the situation as it is viewed from that frame. The pulse is standing still, and air is moving at speed v through it from left to right.

Let the pressure of the undisturbed air be p and the pressure inside the pulse be $p + \Delta p$, where Δp is positive owing to the compression. Consider a slice of air of thickness Δx and face area A, moving toward the pulse at speed v. As this element of air enters the pulse, its leading face encounters a region of higher pressure, which slows it to speed $v + \Delta v$, in which Δv is negative. This slowing is complete when the rear face reaches the pulse, which requires time interval

$$\Delta t = \frac{\Delta x}{v}. \tag{18-4}$$

Fig. 18-3 A compression pulse is sent down a long air-filled tube. The reference frame of the figure is chosen so that the pulse is at rest and the air moves from left to right. (a) A slice of air of width Δx moves toward the pulse with speed v. (b) The leading face of the slice enters the pulse. The forces acting on the leading and trailing faces (due to air pressure) are shown.

Let us apply Newton's second law to the element. During Δt, the average force on the element's trailing face is pA toward the right, and the average force on the leading face is $(p + \Delta p)A$ toward the left (Fig. 18-3b). Therefore, the average net force on the element during Δt is

$$F = pA - (p + \Delta p)A$$
$$= -\Delta p\, A \qquad \text{(net force).} \qquad (18\text{-}5)$$

The minus sign indicates that the net force on the fluid element is directed to the left in Fig. 18-3b. The volume of the element is $A\,\Delta x$, so with the aid of Eq. 18-4, we can write its mass as

$$\Delta m = \rho A\,\Delta x = \rho A v\,\Delta t \qquad \text{(mass).} \qquad (18\text{-}6)$$

Then the average acceleration of the element during Δt is

$$a = \frac{\Delta v}{\Delta t} \qquad \text{(acceleration).} \qquad (18\text{-}7)$$

From Newton's second law ($F = ma$), we have, from Eqs. 18-5, 18-6, and 18-7,

$$-\Delta p\, A = (\rho A v\,\Delta t)\frac{\Delta v}{\Delta t},$$

which we can write as

$$\rho v^2 = -\frac{\Delta p}{\Delta v / v}. \qquad (18\text{-}8)$$

The air that occupies a volume $V\,(= Av\,\Delta t)$ outside the pulse is compressed by an amount $\Delta V\,(= A\,\Delta v\,\Delta t)$ as it enters the pulse. Thus,

$$\frac{\Delta V}{V} = \frac{A\,\Delta v\,\Delta t}{Av\,\Delta t} = \frac{\Delta v}{v}. \qquad (18\text{-}9)$$

Substituting Eq. 18-9 and then Eq. 18-2 into Eq. 18-8 leads to

$$\rho v^2 = -\frac{\Delta p}{\Delta v / v} = -\frac{\Delta p}{\Delta V / V} = B.$$

Solving for v yields Eq. 18-3 for the speed of the air toward the right in Fig. 18-3, and thus for the actual speed of the pulse toward the left.

Sample Problem 18-1

One clue used by your brain to determine the direction of a source of sound is the time delay Δt between the arrival of the sound at the ear closer to the source and the arrival at the farther ear. Assume that the source is distant so that a wavefront from it is approximately planar when it reaches you, and let D represent the separation between your ears.

(a) Find an expression that gives Δt in terms of D and the angle θ between the direction of the source and the forward direction.

SOLUTION: The situation is shown (from an overhead view) in Fig. 18-4, where wavefronts approach you from a source that is located in front of you and to your right. The Key Idea here is that

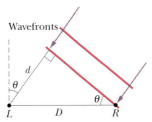

Fig. 18-4 Sample Problem 18-1. A wavefront travels a distance d ($= D \sin \theta$) farther to reach the left ear (L) than to reach the right ear (R).

the time delay Δt is due to the distance d that each wavefront must travel to reach your left ear (L) after it reaches your right ear (R). From Fig. 18-4, we find

$$\Delta t = \frac{d}{v} = \frac{D \sin \theta}{v}, \qquad \text{(Answer)} \quad (18\text{-}10)$$

where v is the speed of sound in air. Based on a lifetime of experience, your brain correlates each detected value of Δt (from zero to the maximum value) with a value of θ (from zero to 90°) for the direction of the sound source.

(b) Suppose that you are submerged in water at 20°C when a wavefront arrives from directly to your right. Based on the time-delay clue, at what angle θ from the forward direction does the source seem to be?

SOLUTION: The Key Idea here is that the speed of the sound is now the speed v_w in water, so in Eq. 18-10 we substitute v_w for v and

90° for θ, finding that

$$\Delta t_w = \frac{D \sin 90°}{v_w} = \frac{D}{v_w}. \qquad (18\text{-}11)$$

Since v_w is about four times v, delay Δt_w is about one-fourth the maximum time delay in air. Based on experience, your brain will process the water time delay as if it occurred in air. Thus, the sound source appears to be at an angle θ smaller than 90°. To find that apparent angle, we substitute the time delay D/v_w from Eq. 18-11 for Δt in Eq. 18-10, obtaining

$$\frac{D}{v_w} = \frac{D \sin \theta}{v}. \qquad (18\text{-}12)$$

Then, to solve for θ we substitute $v = 343$ m/s and $v_w = 1482$ m/s (from Table 18-1) into Eq. 18-12, finding

$$\sin \theta = \frac{v}{v_w} = \frac{343 \text{ m/s}}{1482 \text{ m/s}} = 0.231$$

and thus

$$\theta = 13°. \qquad \text{(Answer)}$$

18-3   Traveling Sound Waves

Here we examine the displacements and pressure variations associated with a sinusoidal sound wave traveling through air. Figure 18-5a displays such a wave traveling rightward through a long air-filled tube. Recall from Chapter 17 that we can produce such a wave by sinusoidally moving a piston at the left end of the tube (as in Fig. 17-2). The piston's rightward motion moves the element of air next to it and compresses that air; the piston's leftward motion allows the element of air to move back to the left and the pressure to decrease. As each element of air pushes on the next element in turn, the right–left motion of the air and the change in its pressure travel along the tube as a sound wave.

   Consider a thin element of air of thickness Δx, located at a position x along the tube. As the wave travels through x, the element of air oscillates left and right in simple harmonic motion about its equilibrium position (Fig. 18-5b). Thus, the oscillations of each air element due to the traveling sound wave are like those of a string element due to a transverse wave, except that the air element oscillates longitudinally rather than transversely. Because string elements oscillate parallel to the y axis, we write their displacements in the form $y(x, t)$. Similarly, because air ele-

Fig. 18-5 (a) A sound wave, traveling through a long air-filled tube with speed v, consists of a moving, periodic pattern of expansions and compressions of the air. The wave is shown at an arbitrary instant. (b) A horizontally expanded view of a short piece of the tube. As the wave passes, a fluid element of thickness Δx oscillates left and right in simple harmonic motion about its equilibrium position. At the instant shown in (b), the element happens to be displaced a distance s to the right of its equilibrium position. Its maximum displacement, either right or left, is s_m.

(a) $s(x, t) = s_m \cos(kx - \omega t)$

Amplitude — Oscillating term

(b) $\Delta p(x, t) = \Delta p_m \sin(kx - \omega t)$

Pressure variation

Fig. 18-6 (a) The displacement function and (b) the pressure-variation function of a traveling sound wave consist of an amplitude and an oscillating term.

ments oscillate parallel to the x axis, we could write their displacements in the form $x(x, t)$. However, we shall avoid that confusing notation and use $s(x, t)$ instead.

To show that the displacements $s(x, t)$ are sinusoidal functions of x and t, we can use either a sine function or a cosine function. In this chapter we use a cosine function, writing

$$s(x, t) = s_m \cos(kx - \omega t). \qquad (18\text{-}13)$$

Figure 18-6a shows the major parts of this equation. In it, s_m is the **displacement amplitude**—that is, the maximum displacement of the air element to either side of its equilibrium position (see Fig. 18-5b). The angular wave number k, angular frequency ω, frequency f, wavelength λ, speed v, and period T for a sound (longitudinal) wave are defined and interrelated exactly as for a transverse wave, except that λ is now the distance (again along the direction of travel) in which the pattern of compression and expansion due to the wave begins to repeat itself (see Fig. 18-5a). (We assume s_m is much less than λ.)

As the wave moves, the air pressure at any position x in Fig. 18-5a varies sinusoidally, as we prove next. To describe this variation we write

$$\Delta p(x, t) = \Delta p_m \sin(kx - \omega t). \qquad (18\text{-}14)$$

Figure 18-6b shows the major parts of this equation. A negative value of Δp in Eq. 18-14 corresponds to an expansion of the air, and a positive value to a compression. Here Δp_m is the **pressure amplitude,** which is the maximum increase or decrease in pressure due to the wave; Δp_m is normally very much less than the pressure p present when there is no wave. As we shall prove, the pressure amplitude Δp_m is related to the displacement amplitude s_m in Eq. 18-13 by

$$\Delta p_m = (v\rho\omega)s_m. \qquad (18\text{-}15)$$

Figure 18-7 shows plots of Eqs. 18-13 and 18-14 at $t = 0$; with time, the two curves would move rightward along the horizontal axes. Note that the displacement and pressure variation are $\pi/2$ rad (or 90°) out of phase. Thus, for example, the pressure variation Δp at any point along the wave is zero when the displacement there is a maximum.

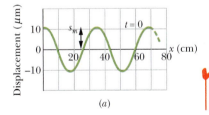

(a)

✔**CHECKPOINT 1:** When the oscillating fluid element in Fig. 18-5b is moving rightward through the point of zero displacement, is the pressure in the element at its equilibrium value, just beginning to increase, or just beginning to decrease?

Derivation of Eqs. 18-14 and 18-15

Figure 18-5b shows an oscillating element of air of cross-sectional area A and thickness Δx, with its center displaced from its equilibrium position by distance s.

From Eq. 18-2 we can write, for the pressure variation in the displaced element,

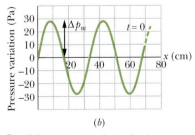

(b)

Fig. 18-7 (a) A plot of the displacement function (Eq. 18-13) for $t = 0$. (b) A similar plot of the pressure-variation function (Eq. 18-14). Both plots are for a 1000 Hz sound wave whose pressure amplitude is at the threshold of pain; see Sample Problem 18-2.

$$\Delta p = -B \frac{\Delta V}{V}. \qquad (18\text{-}16)$$

The quantity V in Eq. 18-16 is the volume of the element, given by

$$V = A \Delta x. \qquad (18\text{-}17)$$

The quantity ΔV in Eq. 18-16 is the change in volume that occurs when the element is displaced. This volume change comes about because the displacements of the two faces of the element are not quite the same, differing by some amount Δs. Thus, we

can write the change in volume as

$$\Delta V = A\,\Delta s. \qquad (18\text{-}18)$$

Substituting Eqs. 18-17 and 18-18 into Eq. 18-16 and passing to the differential limit yield

$$\Delta p = -B\frac{\Delta s}{\Delta x} = -B\frac{\partial s}{\partial x}. \qquad (18\text{-}19)$$

The symbols ∂ indicate that the derivative in Eq. 18-19 is a *partial derivative*, which tells us how s changes with x when the time t is fixed. From Eq. 18-13 we then have, treating t as a constant,

$$\frac{\partial s}{\partial x} = \frac{\partial}{\partial x}[s_m\cos(kx-\omega t)] = -ks_m\sin(kx-\omega t).$$

Substituting this quantity for the partial derivative in Eq. 18-19 yields

$$\Delta p = Bks_m\sin(kx-\omega t).$$

Setting $\Delta p_m = Bks_m$, this yields Eq. 18-14, which we set out to prove.

Using Eq. 18-3, we can now write

$$\Delta p_m = (Bk)s_m = (v^2\rho k)s_m.$$

Equation 18-15, which we also promised to prove, follows at once if we substitute ω/v for k from Eq. 17-12.

Sample Problem 18-2

The maximum pressure amplitude Δp_m that the human ear can tolerate in loud sounds is about 28 Pa (which is very much less than the normal air pressure of about 10^5 Pa). What is the displacement amplitude s_m for such a sound in air of density $\rho = 1.21\ \text{kg/m}^3$, at a frequency of 1000 Hz and a speed of 343 m/s?

SOLUTION: The Key Idea is that the displacement amplitude s_m of a sound wave is related to the pressure amplitude Δp_m of the wave according to Eq. 18-15. Solving that equation for s_m yields

$$s_m = \frac{\Delta p_m}{v\rho\omega} = \frac{\Delta p_m}{v\rho(2\pi f)}.$$

Substituting known data then gives us

$$s_m = \frac{28\ \text{Pa}}{(343\ \text{m/s})(1.21\ \text{kg/m}^3)(2\pi)(1000\ \text{Hz})}$$
$$= 1.1\times10^{-5}\ \text{m} = 11\ \mu\text{m}. \qquad \text{(Answer)}$$

That is only about one-seventh the thickness of this page. Obviously, the displacement amplitude of even the loudest sound that the ear can tolerate is very small.

The pressure amplitude Δp_m for the *faintest* detectable sound at 1000 Hz is 2.8×10^{-5} Pa. Proceeding as above leads to $s_m = 1.1\times10^{-11}$ m or 11 pm, which is about one-tenth the radius of a typical atom. The ear is indeed a sensitive detector of sound waves.

18-4 Interference

Like transverse waves, sound waves can undergo interference. Let us consider, in particular, the interference between two identical sound waves traveling in the same direction. Figure 18-8 shows how we can set up such a situation: Two point sources S_1 and S_2 emit sound waves that are in phase and of identical wavelength λ. Thus, the sources themselves are said to be in phase; that is, as the waves emerge from the sources, their displacements are always identical. We are interested in the waves that then travel through point P in Fig. 18-8. We assume that the distance to P is much greater than the distance between the sources so that we can approximate the waves as traveling in the same direction at P.

If the waves traveled along paths with identical lengths to reach point P, they would be in phase there. As with transverse waves, this means that they would undergo fully constructive interference there. However, in Fig. 18-8, path L_2 traveled

Fig. 18-8 Two point sources S_1 and S_2 emit spherical sound waves in phase. The rays indicate that the waves pass through a common point P.

by the wave from S_2 is longer than path L_1 traveled by the wave from S_1. The difference in path lengths means that the waves may not be in phase at point P. In other words, their phase difference ϕ at P depends on their **path length difference** $\Delta L = |L_2 - L_1|$.

To relate phase difference ϕ to path length difference ΔL, we recall (from Section 17-4) that a phase difference of 2π rad corresponds to one wavelength. Thus, we can write the proportion

$$\frac{\phi}{2\pi} = \frac{\Delta L}{\lambda}, \tag{18-20}$$

from which

$$\phi = \frac{\Delta L}{\lambda}\, 2\pi. \tag{18-21}$$

Fully constructive interference occurs when ϕ is zero, 2π, or any integer multiple of 2π. We can write this condition as

$$\phi = m(2\pi), \qquad \text{for } m = 0, 1, 2, \ldots \qquad \text{(fully constructive interference).} \tag{18-22}$$

From Eq. 18-21, this occurs when the ratio $\Delta L/\lambda$ is

$$\frac{\Delta L}{\lambda} = 0, 1, 2, \ldots \qquad \text{(fully constructive interference).} \tag{18-23}$$

For example, if the path length difference $\Delta L = |L_2 - L_1|$ in Fig. 18-8 is equal to 2λ, then $\Delta L/\lambda = 2$ and the waves undergo fully constructive interference at point P. The interference is fully constructive because the wave from S_2 is phase-shifted relative to the wave from S_1 by 2λ, putting the two waves *exactly in phase* at P.

Fully destructive interference occurs when ϕ is an odd multiple of π, a condition we can write as

$$\phi = (2m + 1)\pi, \qquad \text{for } m = 0, 1, 2, \ldots \qquad \text{(fully destructive interference).} \tag{18-24}$$

From Eq. 18-21, this occurs when the ratio $\Delta L/\lambda$ is

$$\frac{\Delta L}{\lambda} = 0.5, 1.5, 2.5, \ldots \qquad \text{(fully destructive interference).} \tag{18-25}$$

For example, if the path length difference $\Delta L = |L_2 - L_1|$ in Fig. 18-8 is equal to 2.5λ, then $\Delta L/\lambda = 2.5$ and the waves undergo fully destructive interference at point P. The interference is fully destructive because the wave from S_2 is phase-shifted relative to the wave from S_1 by 2.5 wavelengths, which puts the two waves *exactly out of phase* at P.

Of course, two waves could produce intermediate interference as, say, when $\Delta L/\lambda = 1.2$. This would be closer to fully constructive interference ($\Delta L/\lambda = 1.0$) than to fully destructive interference ($\Delta L/\lambda = 1.5$).

Sample Problem 18-3

In Fig. 18-9a, two point sources S_1 and S_2, which are in phase and separated by distance $D = 1.5\lambda$, emit identical sound waves of wavelength λ.

(a) What is the path length difference of the waves from S_1 and S_2 at point P_1, which lies on the perpendicular bisector of distance D, at a distance greater than D from the sources? What type of interference occurs at P_1?

SOLUTION: The Key Idea here is that, because the waves travel identical distances to reach P_1, their path length difference is

$$\Delta L = 0. \qquad \text{(Answer)}$$

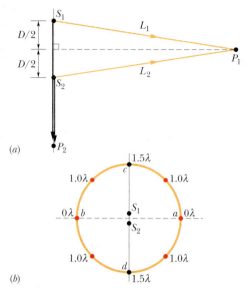

Fig. 18-9 Sample Problem 18-3. (a) Two point sources S_1 and S_2, separated by distance D, emit spherical sound waves in phase. The waves travel equal distances to reach point P_1. Point P_2 is on the line extending through S_1 and S_2. (b) The path length difference (in terms of wavelength) between the waves from S_1 and S_2, at eight points on a large circle around the sources.

From Eq. 18-23, this means that the waves undergo fully constructive interference at P_1.

(b) What are the path length difference and type of interference at point P_2 in Fig. 18-9a?

SOLUTION: Now the **Key Idea** is that the wave from S_1 travels the extra distance D (= 1.5λ) to reach P_2. Thus, the path length difference is

$$\Delta L = 1.5\lambda. \qquad \text{(Answer)}$$

From Eq. 18-25, this means that the waves are exactly out of phase at P_2 and undergo fully destructive interference there.

(c) Figure 18-9b shows a circle with a radius much greater than D, centered on the midpoint between sources S_1 and S_2. What is the number of points N around this circle at which the interference is fully constructive?

SOLUTION: Imagine that, starting at point a, we move clockwise along the circle to point d. One **Key Idea** here is that as we move to point d, the path length difference ΔL increases and so the type of interference changes. From (a), we know that the path length difference is $\Delta L = 0\lambda$ at point a. From (b), we know that $\Delta L = 1.5\lambda$ at point d. Thus, there must be one point along the circle between a and d at which $\Delta L = \lambda$, as indicated in Fig. 18-9b. From Eq. 18-23, fully constructive interference occurs at that point. Also, there can be no other point along the way from point a to point d at which fully constructive interference occurs, because there is no other integer than 1 between 0 and 1.5.

Another **Key Idea** here is to use symmetry to locate the other points of fully constructive interference along the rest of the circle. Symmetry about line cd gives us point b, at which $\Delta L = 0\lambda$. Also, there are three more points at which $\Delta L = \lambda$. In all we have

$$N = 6. \qquad \text{(Answer)}$$

✓CHECKPOINT 2: In this sample problem, if the distance D between sources S_1 and S_2 were, instead, equal to 4λ, what would be the path length difference and what type of interference would occur at (a) point P_1 and (b) point P_2?

18-5 Intensity and Sound Level

If you have ever tried to sleep while someone played loud music nearby, you are well aware that there is more to sound than frequency, wavelength, and speed. There is also intensity. The **intensity** I of a sound wave at a surface is the average rate per unit area at which energy is transferred by the wave through or onto the surface. We can write this as

$$I = \frac{P}{A}, \qquad (18\text{-}26)$$

where P is the time rate of energy transfer (the power) of the sound wave, and A is the area of the surface intercepting the sound. As we shall derive shortly, the intensity I is related to the displacement amplitude s_m of the sound wave by

$$I = \tfrac{1}{2}\rho v \omega^2 s_m^2. \qquad (18\text{-}27)$$

Variation of Intensity with Distance

How intensity varies with distance from a real sound source is often complex. Some real sources (like loudspeakers) may transmit sound only in particular directions,

Sound can cause the wall of a drinking glass to oscillate. If the sound produces a standing wave of oscillations and if the intensity of the sound is large enough, the glass will shatter.

and the environment usually produces echoes (reflected sound waves) that overlap the direct sound waves. In some situations, however, we can ignore echoes and assume that the sound source is a point source that emits the sound *isotropically*— that is, with equal intensity in all directions. The wavefronts spreading from such an isotropic point source S at a particular instant are shown in Fig. 18-10.

Let us assume that the mechanical energy of the sound waves is conserved as they spread from this source. Let us also center an imaginary sphere of radius r on the source, as shown in Fig. 18-10. All the energy emitted by the source must pass through the surface of the sphere. Thus, the time rate at which energy is transferred through the surface by the sound waves must equal the time rate at which energy is emitted by the source (that is, the power P_s of the source). From Eq. 18-26, the intensity I at the sphere must then be

$$I = \frac{P_s}{4\pi r^2},\qquad(18\text{-}28)$$

where $4\pi r^2$ is the area of the sphere. Equation 18-28 tells us that the intensity of sound from an isotropic point source decreases with the square of the distance r from the source.

✔**CHECKPOINT 3:**   The figure indicates three small patches 1, 2, and 3 that lie on the surfaces of two imaginary spheres; the spheres are centered on an isotropic point source S of sound. The rates at which energy is transmitted through the three patches by the sound waves are equal. Rank the patches according to (a) the intensity of the sound on them and (b) their area, greatest first.

The Decibel Scale

You saw in Sample Problem 18-2 that the displacement amplitude at the human ear ranges from about 10^{-5} m for the loudest tolerable sound to about 10^{-11} m for the faintest detectable sound, a ratio of 10^6. From Eq. 18-27 we see that the intensity of a sound varies as the *square* of its amplitude, so the ratio of intensities at these two limits of the human auditory system is 10^{12}. Humans can hear over an enormous range of intensities.

We deal with such an enormous range of values by using logarithms. Consider the relation

$$y = \log x,$$

in which x and y are variables. It is a property of this equation that if we *multiply x by 10*, then y increases by 1. To see this, we write

$$y' = \log(10x) = \log 10 + \log x = 1 + y.$$

Similarly, if we multiply x by 10^{12}, y increases by only 12.

Thus, instead of speaking of the intensity I of a sound wave, it is much more convenient to speak of its **sound level** β, defined as

$$\beta = (10 \text{ dB}) \log \frac{I}{I_0}.\qquad(18\text{-}29)$$

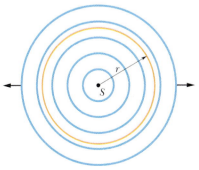

Fig. 18-10 A point source S emits sound waves uniformly in all directions. The waves pass through an imaginary sphere of radius r that is centered on S.

Here dB is the abbreviation for **decibel,** the unit of sound level, a name that was chosen to recognize the work of Alexander Graham Bell. I_0 in Eq. 18-29 is a standard reference intensity ($= 10^{-12}$ W/m^2), chosen because it is near the lower limit of the human range of hearing. For $I = I_0$, Eq. 18-29 gives $\beta = 10 \log 1 = 0$, so our

TABLE 18-2   Some Sound Levels (dB)

Hearing threshold	0
Rustle of leaves	10
Conversation	60
Rock concert	110
Pain threshold	120
Jet engine	130

standard reference level corresponds to zero decibels. Then β increases by 10 dB every time the sound intensity increases by an order of magnitude (a factor of 10). Thus, $\beta = 40$ corresponds to an intensity that is 10^4 times the standard reference level. Table 18-2 lists the sound levels for a variety of environments.

Derivation of Eq. 18-27

Consider, in Fig. 18-5a, a thin slice of air of thickness dx, area A, and mass dm, oscillating back and forth as the sound wave of Eq. 18-13 passes through it. The kinetic energy dK of the slice of air is

$$dK = \tfrac{1}{2}dm \, v_s^2. \tag{18-30}$$

Here v_s is not the speed of the wave but the speed of the oscillating element of air, obtained from Eq. 18-13 as

$$v_s = \frac{\partial s}{\partial t} = -\omega s_m \sin(kx - \omega t).$$

Using this relation and putting $dm = \rho A \, dx$ allow us to rewrite Eq. 18-30 as

$$dK = \tfrac{1}{2}(\rho A \, dx)(-\omega s_m)^2 \sin^2(kx - \omega t). \tag{18-31}$$

Dividing Eq. 18-31 by dt gives the rate at which kinetic energy moves along with the wave. As we saw in Chapter 17 for transverse waves, dx/dt is the wave speed v, so we have

$$\frac{dK}{dt} = \tfrac{1}{2}\rho A v \omega^2 s_m^2 \sin^2(kx - \omega t). \tag{18-32}$$

The *average* rate at which kinetic energy is transported is

$$\left(\frac{dK}{dt}\right)_{\text{avg}} = \tfrac{1}{2}\rho A v \omega^2 s_m^2[\sin^2(kx - \omega t)]_{\text{avg}}$$

$$= \tfrac{1}{4}\rho A v \omega^2 s_m^2. \tag{18-33}$$

To obtain this equation, we have used the fact that the average value of the square of a sine (or a cosine) function over one full oscillation is $\tfrac{1}{2}$.

We assume that *potential* energy is carried along with the wave at this same average rate. The wave intensity I, which is the average rate per unit area at which energy of both kinds is transmitted by the wave, is then, from Eq. 18-33,

$$I = \frac{2(dK/dt)_{\text{avg}}}{A} = \tfrac{1}{2}\rho v \omega^2 s_m^2,$$

which is Eq. 18-27, the equation we set out to derive.

Sample Problem 18-4

An electric spark jumps along a straight line of length $L = 10$ m, emitting a pulse of sound that travels radially outward from the spark. (The spark is said to be a *line source* of sound.) The power of the emission is $P_s = 1.6 \times 10^4$ W.

(a) What is the intensity I of the sound when it reaches a distance $r = 12$ m from the spark?

SOLUTION: Let us center an imaginary cylinder of radius $r = 12$ m and length $L = 10$ m (open at both ends) on the spark, as shown in Fig. 18-11. One Key Idea here is that the intensity I at the cylin-

drical surface is the ratio P/A of the time rate P at which sound energy passes through the surface to the surface area A. Another Key Idea is to assume that the principle of conservation of energy applies to the sound energy. This means that the rate P at which energy is transferred through the cylinder must equal the rate P_s at which energy is emitted by the source. Putting these ideas together and noting that the area of the cylindrical surface is $A = 2\pi r L$, we have

$$I = \frac{P}{A} = \frac{P_s}{2\pi r L}. \tag{18-34}$$

This tells us that the intensity of the sound from a line source

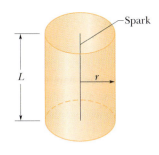

Fig. 18-11 Sample Problem 18-4. A spark along a straight line of length L emits sound waves radially outward. The waves pass through an imaginary cylinder of radius r and length L that is centered on the spark.

decreases with distance r (and not with the square of distance r as for a point source). Substituting the given data, we find

$$I = \frac{1.6 \times 10^4 \text{ W}}{2\pi(12 \text{ m})(10 \text{ m})} = 21.2 \text{ W/m}^2 \approx 21 \text{ W/m}^2. \quad \text{(Answer)}$$

(b) At what time rate P_d is sound energy intercepted by an acoustic detector of area $A_d = 2.0 \text{ cm}^2$, aimed at the spark and located a distance $r = 12 \text{ m}$ from the spark?

SOLUTION: Applying the first Key Idea of part (a), we know that the intensity of sound at the detector is the ratio of the energy transfer rate P_d there to the detector's area A_d:

$$I = \frac{P_d}{A_d}. \quad (18\text{-}35)$$

We can imagine that the detector lies on the cylindrical surface of (a). Then the sound intensity at the detector is the intensity $I \, (= 21.2 \text{ W/m}^2)$ at the cylindrical surface. Solving Eq. 18-35 for P_d gives us

$$P_d = (21.2 \text{ W/m}^2)(2.0 \times 10^{-4} \text{ m}^2) = 4.2 \text{ mW}. \quad \text{(Answer)}$$

Sample Problem 18-5

In 1976, the Who set a record for the loudest concert—the sound level 46 m in front of the speaker systems was $\beta_2 = 120 \text{ dB}$. What is the ratio of the intensity I_2 of the band at that spot to the intensity I_1 of a jackhammer operating at sound level $\beta_1 = 92 \text{ dB}$?

SOLUTION: The Key Idea here is that for both the Who and the jackhammer, the sound level β is related to the intensity by the definition of sound level in Eq. 18-29. For the Who, we have

$$\beta_2 = (10 \text{ dB}) \log \frac{I_2}{I_0},$$

and for the jackhammer, we have

$$\beta_1 = (10 \text{ dB}) \log \frac{I_1}{I_0}.$$

The difference in the sound levels is

$$\beta_2 - \beta_1 = (10 \text{ dB})\left(\log \frac{I_2}{I_0} - \log \frac{I_1}{I_0} \right). \quad (18\text{-}36)$$

Using the identity

$$\log \frac{a}{b} - \log \frac{c}{d} = \log \frac{ad}{bc},$$

we can rewrite Eq. 18-36 as

$$\beta_2 - \beta_1 = (10 \text{ dB}) \log \frac{I_2}{I_1}. \quad (18\text{-}37)$$

Rearranging and substituting the known sound levels now yield

$$\log \frac{I_2}{I_1} = \frac{\beta_2 - \beta_1}{10 \text{ dB}} = \frac{120 \text{ dB} - 92 \text{ dB}}{10 \text{ dB}} = 2.8.$$

Taking the antilog of the far left and far right sides of this equation (the antilog key on your calculator is probably marked as 10^x), we find

$$\frac{I_2}{I_1} = \log^{-1} 2.8 = 630. \quad \text{(Answer)}$$

Thus, the Who was *very* loud.

Temporary exposure to sound intensities as great as those of a jackhammer and the 1976 Who concert results in a temporary

Fig. 18-12 Sample Problem 18-5. Peter Townshend of the Who, playing in front of a speaker system. He suffered a permanent reduction in his hearing ability due to his exposure to high-intensity sound, not so much during on-stage performances as from wearing headphones in recording studios and at home.

reduction of hearing. Repeated or prolonged exposure can result in permanent reduction of hearing (Fig. 18-12). Loss of hearing is a clear risk for anyone continually listening to, say, heavy metal at high volume, especially on headphones.

Fig. 18-13 The air column within a fujara oscillates when that traditional Slovakian instrument is played.

18-6 Sources of Musical Sound

Musical sounds can be set up by oscillating strings (guitar, piano, violin), membranes (kettledrum, snare drum), air columns (flute, oboe, pipe organ, and the fujara of Fig. 18-13), wooden blocks or steel bars (marimba, xylophone), and many other oscillating bodies. Most instruments involve more than a single oscillating part. In the violin, for example, both the strings and the body of the instrument participate in producing the music.

Recall from Chapter 17 that standing waves can be set up on a stretched string that is fixed at both ends. They arise because waves traveling along the string are reflected back onto the string at each end. If the wavelength of the waves is suitably matched to the length of the string, the superposition of waves traveling in opposite directions produces a standing wave pattern (or oscillation mode). The wavelength required of the waves for such a match is one that corresponds to a *resonant frequency* of the string. The advantage of setting up standing waves is that the string then oscillates with a large, sustained amplitude, pushing back and forth against the surrounding air and thus generating a noticeable sound wave with the same frequency as the oscillations of the string. This production of sound is of obvious importance to, say, a guitarist.

We can set up standing waves of sound in an air-filled pipe in a similar way. As sound waves travel through the air in the pipe, they are reflected at each end and travel back through the pipe. (The reflection occurs even if an end is open, but the reflection is not as complete as when the end is closed.) If the wavelength of the sound waves is suitably matched to the length of the pipe, the superposition of waves traveling in opposite directions through the pipe sets up a standing wave pattern. The wavelength required of the sound waves for such a match is one that corresponds to a resonant frequency of the pipe. The advantage of such a standing wave is that the air in the pipe oscillates with a large, sustained amplitude, emitting at any open end a sound wave that has the same frequency as the oscillations in the pipe. This emission of sound is of obvious importance to, say, an organist.

Many other aspects of standing sound wave patterns are similar to those of string waves: The closed end of a pipe is like the fixed end of a string in that there must be a node (zero displacement) there, and the open end of a pipe is like the end of a string attached to a freely moving ring, as in Fig. 17-19b, in that there must be an antinode there. (Actually, the antinode for the open end of a pipe is located slightly beyond the end, but we shall not dwell on that detail here.)

The simplest standing wave pattern that can be set up in a pipe with two open ends is shown in Fig. 18-14a. There is an antinode across each open end, as required. There is also a node across the middle of the pipe. An easier way of representing this standing longitudinal sound wave is shown in Fig. 18-14b—by drawing it as a standing transverse string wave.

The standing wave pattern of Fig. 18-14a is called the *fundamental mode* or *first harmonic*. For it to be set up, the sound waves in a pipe of length L must have a wavelength given by $L = \lambda/2$, so that $\lambda = 2L$. Several more standing sound wave patterns for a pipe with two open ends are shown in Fig. 18-15a using string wave representations. The *second harmonic* requires sound waves of wavelength $\lambda = L$, the *third harmonic* requires wavelength $\lambda = 2L/3$, and so on.

More generally, the resonant frequencies for a pipe of length L with two open ends correspond to the wavelengths

$$\lambda = \frac{2L}{n}, \qquad \text{for } n = 1, 2, 3, \ldots, \tag{18-38}$$

where n is called the *harmonic number*. The resonant frequencies for a pipe with

Fig. 18-14 (a) The simplest standing wave pattern of displacement for (longitudinal) sound waves in a pipe with both ends open has an antinode (A) across each end and a node (N) across the middle. (The longitudinal displacements represented by the double arrows are greatly exaggerated.) (b) The corresponding standing wave pattern for (transverse) string waves.

$\lambda = 2L/2 = L$ ($n = 2$)

$\lambda = 2L/3$ ($n = 3$)

$\lambda = 2L/4 = L/2$ ($n = 4$)

(a)

$\lambda = 4L$ ($n = 1$)

$\lambda = 4L/3$ ($n = 3$)

$\lambda = 4L/5$ ($n = 5$)

$\lambda = 4L/7$ ($n = 7$)

(b)

Fig. 18-15 Standing wave patterns for string waves superimposed on pipes to represent standing sound wave patterns in the pipes. (a) With *both* ends of the pipe open, any harmonic can be set up in the pipe. (b) With only *one* end open, only odd harmonics can be set up.

two open ends are then given by

$$f = \frac{v}{\lambda} = \frac{nv}{2L}, \qquad \text{for } n = 1, 2, 3, \ldots \qquad \text{(pipe, two open ends),} \qquad (18\text{-}39)$$

where v is the speed of sound.

Figure 18-15b shows (using string wave representations) some of the standing sound wave patterns that can be set up in a pipe with only one open end. As required, across the open end there is an antinode and across the closed end there is a node. The simplest pattern requires sound waves having a wavelength given by $L = \lambda/4$, so that $\lambda = 4L$. The next simplest pattern requires a wavelength given by $L = 3\lambda/4$, so that $\lambda = 4L/3$, and so on.

More generally, the resonant frequencies for a pipe of length L with only one open end correspond to the wavelengths

$$\lambda = \frac{4L}{n}, \qquad \text{for } n = 1, 3, 5, \ldots, \qquad (18\text{-}40)$$

in which the harmonic number n *must be an odd number*. The resonant frequencies are then given by

$$f = \frac{v}{\lambda} = \frac{nv}{4L}, \qquad \text{for } n = 1, 3, 5, \ldots \qquad \text{(pipe, one open end).} \qquad (18\text{-}41)$$

Note again that only odd harmonics can exist in a pipe with one open end. For example, the second harmonic, with $n = 2$, cannot be set up in such a pipe. Note also that for such a pipe the adjective in a phrase such as "the third harmonic" still refers to the harmonic number n (and not to, say, the third possible harmonic).

The length of a musical instrument reflects the range of frequencies over which the instrument is designed to function, and smaller length implies higher frequencies. Figure 18-16, for example, shows the saxophone and violin families, with their frequency ranges suggested by the piano keyboard. Note that, for every instrument, there is overlap with its higher- and lower-frequency neighbors.

In any oscillating system that gives rise to a musical sound, whether it is a violin string or the air in an organ pipe, the fundamental and one or more of the higher

Fig. 18-16 The saxophone and violin families, showing the relations between instrument length and frequency range. The frequency range of each instrument is indicated by a horizontal bar along a frequency scale suggested by the keyboard at the bottom; the frequency increases toward the right.

Bass saxophone
Baritone saxophone
Tenor saxophone
Alto saxophone
Soprano saxophone

Violin
Viola
Cello
Bass

Time

Fig. 18-17 The wave forms produced by (*a*) a flute, (*b*) an oboe, and (*c*) a saxophone when they all play the same note, with the same first harmonic frequency.

harmonics are usually generated simultaneously. Thus, you hear them together—that is, superimposed as a net wave. When different instruments are played at the same note, they produce the same fundamental frequency but different intensities for the higher harmonics. For example, the fourth harmonic of middle C might be relatively loud on one instrument and relatively quiet or even missing on another. Thus, because different instruments produce different net waves, they sound different to you even when they are played at the same note. That would be the case for the three net waves shown in Fig. 18-17, which were produced at the same note by different instruments.

✓CHECKPOINT 4:   Pipe *A*, with length *L*, and pipe *B*, with length 2*L*, both have two open ends. Which harmonic of pipe *B* has the same frequency as the fundamental of pipe *A*?

Sample Problem 18-6

Weak background noises from a room set up the fundamental standing wave in a cardboard tube of length $L = 67.0$ cm with two open ends. Assume that the speed of sound in the air within the tube is 343 m/s.

(a) What frequency do you hear from the tube?

SOLUTION: The **Key Idea** here is that, with both pipe ends open, we have a symmetric situation in which the standing wave has an antinode at each end of the tube. The standing wave pattern (in string wave style) is that of Fig. 18-14*b*. The frequency is given by Eq. 18-39 with $n = 1$ for the fundamental mode:

$$f = \frac{nv}{2L} = \frac{(1)(343 \text{ m/s})}{2(0.670 \text{ m})} = 256 \text{ Hz.} \qquad \text{(Answer)}$$

If the background noises set up any higher harmonics, such as the

second harmonic, you must also hear frequencies that are *integer* multiples of 256 Hz.

(b) If you jam your ear against one end of the tube, what fundamental frequency do you hear from the tube?

SOLUTION: The **Key Idea** now is that, with your ear effectively closing one end of the tube, we have an asymmetric situation—an antinode still exists at the open end but a node is now at the other (closed) end. The standing wave pattern is the top one in Fig. 18-15*b*. The frequency is given by Eq. 18-41 with $n = 1$ for the fundamental mode:

$$f = \frac{nv}{4L} = \frac{(1)(343 \text{ m/s})}{4(0.670 \text{ m})} = 128 \text{ Hz.} \qquad \text{(Answer)}$$

If the background noises set up any higher harmonics, they will be *odd* multiples of 128 Hz. That means that the frequency of 256 Hz (which is an even multiple) cannot now occur.

18-7 Beats

If we listen, a few minutes apart, to two sounds whose frequencies are, say, 552 and 564 Hz, most of us cannot tell one from the other. However, if the sounds reach our ears simultaneously, what we hear is a sound whose frequency is 558 Hz, the *average* of the two combining frequencies. We also hear a striking variation in the intensity of this sound—it increases and decreases in slow, wavering **beats** that repeat at a frequency of 12 Hz, the *difference* between the two combining frequencies. Figure 18-18 shows this beat phenomenon.

Let the time-dependent variations of the displacements due to two sound waves at a particular location be

$$s_1 = s_m \cos \omega_1 t \quad \text{and} \quad s_2 = s_m \cos \omega_2 t, \qquad (18\text{-}42)$$

where $\omega_1 > \omega_2$. We have assumed, for simplicity, that the waves have the same

(a)

(b)

Time

(c)

Fig. 18-18 (a, b) The pressure variations Δp of two sound waves as they would be detected separately. The frequencies of the waves are nearly equal. (c) The resultant pressure variation if the two waves are detected simultaneously.

amplitude. According to the superposition principle, the resultant displacement is

$$s = s_1 + s_2 = s_m(\cos \omega_1 t + \cos \omega_2 t).$$

Using the trigonometric identity (see Appendix E)

$$\cos \alpha + \cos \beta = 2 \cos \tfrac{1}{2}(\alpha - \beta) \cos \tfrac{1}{2}(\alpha + \beta)$$

allows us to write the resultant displacement as

$$s = 2s_m \cos \tfrac{1}{2}(\omega_1 - \omega_2)t \cos \tfrac{1}{2}(\omega_1 + \omega_2)t. \tag{18-43}$$

If we write

$$\omega' = \tfrac{1}{2}(\omega_1 - \omega_2) \quad \text{and} \quad \omega = \tfrac{1}{2}(\omega_1 + \omega_2), \tag{18-44}$$

we can then write Eq. 18-43 as

$$s(t) = [2s_m \cos \omega' t] \cos \omega t. \tag{18-45}$$

We now assume that the angular frequencies ω_1 and ω_2 of the combining waves are almost equal, which means that $\omega \gg \omega'$ in Eq. 18-44. We can then regard Eq. 18-45 as a cosine function whose angular frequency is ω and whose amplitude (which is not constant but varies with angular frequency ω') is the quantity in the brackets.

A maximum amplitude will occur whenever $\cos \omega' t$ in Eq. 18-45 has the value $+1$ or -1, which happens twice in each repetition of the cosine function. Because $\cos \omega' t$ has angular frequency ω', the angular frequency ω_{beat} at which beats occur is $\omega_{\text{beat}} = 2\omega'$. Then, with the aid of Eq. 18-44, we can write

$$\omega_{\text{beat}} = 2\omega' = (2)(\tfrac{1}{2})(\omega_1 - \omega_2) = \omega_1 - \omega_2.$$

Because $\omega = 2\pi f$, we can recast this as

$$f_{\text{beat}} = f_1 - f_2 \quad \text{(beat frequency).} \tag{18-46}$$

Musicians use the beat phenomenon in tuning their instruments. If an instrument is sounded against a standard frequency (for example, the lead oboe's reference A) and tuned until the beat disappears, then the instrument is in tune with that standard. In musical Vienna, concert A (440 Hz) is available as a telephone service for the benefit of the city's many professional and amateur musicians.

Sample Problem 18-7

You wish to tune the note A_3 on a piano to its proper frequency of 220 Hz. You have available a tuning fork whose frequency is 440 Hz. How should you proceed?

SOLUTION: We need two **Key Ideas** here: (1) The two frequencies are too far apart to produce beats. (2) However, the piano string will oscillate not only in its fundamental mode (at 220 Hz when tuned) but also in its second harmonic mode (at 440 Hz when in tune). Thus, with the string somewhat out of tune, the frequency of its

second harmonic will beat against the 440 Hz of the tuning fork. To tune the string, you can listen for those beats and then either tighten or loosen the string to decrease the beat frequency until the beating disappears.

✓**CHECKPOINT 5:**   In this sample problem, you tighten the string and the beat frequency increases from 6 Hz. Should you continue to tighten the string or should you loosen the string to put the string in tune?

18-8   The Doppler Effect

A police car is parked by the side of the highway, sounding its 1000 Hz siren. If you are also parked by the highway, you will hear that same frequency. However, if there is relative motion between you and the police car, either toward or away from each other, you will hear a different frequency. For example, if you are driving *toward* the police car at 120 km/h (about 75 mi/h), you will hear a *higher* frequency (1096 Hz, an *increase* of 96 Hz). If you are driving *away from* the police car at that same speed, you will hear a *lower* frequency (904 Hz, a *decrease* of 96 Hz).

These motion-related frequency changes are examples of the **Doppler effect.** The effect was proposed (although not fully worked out) in 1842 by Austrian physicist Johann Christian Doppler. It was tested experimentally in 1845 by Buys Ballot in Holland, "using a locomotive drawing an open car with several trumpeters."

The Doppler effect holds not only for sound waves but also for electromagnetic waves, including microwaves, radio waves, and visible light. Here, however, we shall consider only sound waves, and we shall take as a reference frame the body of air through which these waves travel. This means that we shall measure the speeds of a source S of sound waves and a detector D of those waves *relative to that body of air.* (Unless otherwise stated, the body of air is stationary relative to the ground, so the speeds can also be measured relative to the ground.) We shall assume that S and D move either directly toward or directly away from each other, at speeds less than the speed of sound.

If either the detector or the source is moving, or both are moving, the emitted frequency f and the detected frequency f' are related by

$$f' = f\frac{v \pm v_D}{v \pm v_S} \qquad \text{(general Doppler effect)}, \qquad (18\text{-}47)$$

where v is the speed of sound through the air, v_D is the detector's speed relative to the air, and v_S is the source's speed relative to the air. The choice of plus or minus signs is set by this rule:

> When the motion of detector or source is toward the other, the sign on its speed must give an upward shift in frequency. When the motion of detector or source is away from the other, the sign on its speed must give a downward shift in frequency.

In short, *toward* means *shift up,* and *away* means *shift down.*

Here are some examples of the rule. If the detector moves toward the source, use the plus sign in the numerator of Eq. 18-47 to get a shift up in the frequency. If it moves away, use the minus sign in the numerator to get a shift down. If it is stationary, substitute 0 for v_D. If the source moves toward the detector, use the minus sign in the denominator of Eq. 18-47 to get a shift up in the frequency. If it moves away, use the plus sign in the denominator to get a shift down. If the source is stationary, substitute 0 for v_S.

Next, we derive equations for the Doppler effect for the following two specific situations and then derive Eq. 18-47 for the general situation.

1. When the detector moves relative to the air and the source is stationary relative to the air, the motion changes the frequency at which the detector intercepts wavefronts and thus the detected frequency of the sound wave.

2. When the source moves relative to the air and the detector is stationary relative to the air, the motion changes the wavelength of the sound wave and thus the detected frequency (recall that frequency is related to wavelength).

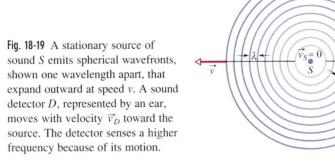

Fig. 18-19 A stationary source of sound S emits spherical wavefronts, shown one wavelength apart, that expand outward at speed v. A sound detector D, represented by an ear, moves with velocity $\vec{v}_D$ toward the source. The detector senses a higher frequency because of its motion.

Detector Moving; Source Stationary

In Fig. 18-19, a detector D (represented by an ear) is moving at speed v_D toward a stationary source S that emits spherical wavefronts, of wavelength λ and frequency f, moving at the speed v of sound in air. The wavefronts are drawn one wavelength apart. The frequency detected by detector D is the rate at which D intercepts wavefronts (or individual wavelengths). If D were stationary, that rate would be f, but since D is moving into the wavefronts, the rate of interception is greater, and thus the detected frequency f' is greater than f.

Let us for the moment consider the situation in which D is stationary (Fig. 18-20). In time t, the wavefronts move to the right a distance vt. The number of wavelengths in that distance vt is the number of wavelengths intercepted by D in time t, and that number is vt/λ. The rate at which D intercepts wavelengths, which is the frequency f detected by D, is

$$f = \frac{vt/\lambda}{t} = \frac{v}{\lambda}. \tag{18-48}$$

In this situation, with D stationary, there is no Doppler effect—the frequency detected by D is the frequency emitted by S.

Now let us again consider the situation in which D moves opposite the wavefronts (Fig. 18-21). In time t, the wavefronts move to the right a distance vt as previously, but now D moves to the left a distance $v_D t$. Thus, in this time t, the distance moved by the wavefronts relative to D is $vt + v_D t$. The number of wavelengths in this relative distance $vt + v_D t$ is the number of wavelengths intercepted by D in time t, and is $(vt + v_D t)/\lambda$. The *rate* at which D intercepts wavelengths in this situation is the frequency f', given by

$$f' = \frac{(vt + v_D t)/\lambda}{t} = \frac{v + v_D}{\lambda}. \tag{18-49}$$

From Eq. 18-48, we have $\lambda = v/f$. Then Eq. 18-49 becomes

$$f' = \frac{v + v_D}{v/f} = f\frac{v + v_D}{v}. \tag{18-50}$$

Note that in Eq. 18-50, f' must be greater than f unless $v_D = 0$ (the detector is stationary).

Similarly, we can find the frequency detected by D if D moves away from the source. In this situation, the wavefronts move a distance $vt - v_D t$ relative to D in time t, and f' is given by

$$f' = f\frac{v - v_D}{v}. \tag{18-51}$$

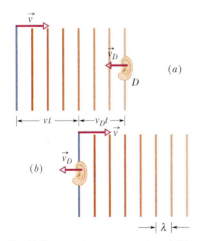

Fig. 18-20 Wavefronts of Fig. 18-19, assumed planar, (*a*) reach and (*b*) pass a stationary detector D; they move a distance vt to the right in time t.

Fig. 18-21 Wavefronts (*a*) reach and (*b*) pass detector D, which moves opposite the wavefronts. In time t, the wavefronts move a distance vt to the right and D moves a distance $v_D t$ to the left.

In Eq. 18-51, f' must be less than f unless $v_D = 0$.

We can summarize Eqs. 18-50 and 18-51 with

$$f' = f\frac{v \pm v_D}{v} \qquad \text{(detector moving; source stationary).} \qquad (18\text{-}52)$$

Source Moving; Detector Stationary

Let detector D be stationary with respect to the body of air, and let source S move toward D at speed v_S (Fig. 18-22). The motion of S changes the wavelength of the sound waves it emits, and thus the frequency detected by D.

To see this change, let $T\ (= 1/f)$ be the time between the emission of any pair of successive wavefronts W_1 and W_2. During T, wavefront W_1 moves a distance vT and the source moves a distance $v_S T$. At the end of T, wavefront W_2 is emitted. In the direction in which S moves, the distance between W_1 and W_2, which is the wavelength λ' of the waves moving in that direction, is $vT - v_S T$. If D detects those waves, it detects frequency f' given by

$$f' = \frac{v}{\lambda'} = \frac{v}{vT - v_S T} = \frac{v}{v/f - v_S/f}$$

$$= f\frac{v}{v - v_S}. \qquad (18\text{-}53)$$

Note that f' must be greater than f unless $v_S = 0$.

In the direction opposite that taken by S, the wavelength λ' of the waves is $vT + v_S T$. If D detects those waves, it detects frequency f' given by

$$f' = f\frac{v}{v + v_S}. \qquad (18\text{-}54)$$

Now f' must be less than f unless $v_S = 0$.

We can summarize Eqs. 18-53 and 18-54 with

$$f' = f\frac{v}{v \pm v_S} \qquad \text{(source moving; detector stationary).} \qquad (18\text{-}55)$$

General Doppler Effect Equation

We can now derive the general Doppler effect equation by replacing f in Eq. 18-55 (the frequency of the source) with f' of Eq. 18-52 (the frequency associated with motion of the detector). The result is Eq. 18-47 for the general Doppler effect.

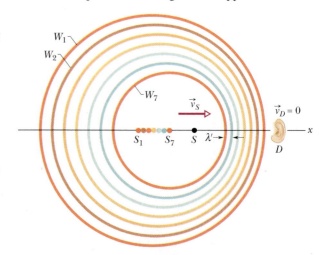

Fig. 18-22 A detector D is stationary, and a source S is moving toward it at speed v_S. Wavefront W_1 was emitted when the source was at S_1, wavefront W_7 when it was at S_7. At the moment depicted, the source is at S. The detector senses a higher frequency because the moving source, chasing its own wavefronts, emits a reduced wavelength λ' in the direction of its motion.

That general equation holds not only when both detector and source are moving but also in the two specific situations we just discussed. For the situation in which the detector is moving and the source is stationary, substitution of $v_S = 0$ into Eq. 18-47 gives us Eq. 18-52, which we previously found. For the situation in which the source is moving and the detector is stationary, substitution of $v_D = 0$ into Eq. 18-47 gives us Eq. 18-55, which we previously found. Thus, Eq. 18-47 is the equation to remember.

Bat Navigation

Bats navigate and search out prey by emitting, and then detecting reflections of, ultrasonic waves. These are sound waves with frequencies greater than can be heard by a human. For example, a horseshoe bat emits ultrasonic waves at 83 kHz, well above the 20 kHz limit of human hearing.

After the sound is emitted through the bat's nostrils, it might reflect (echo) from a moth, and then return to the bat's ears. The motions of the bat and the moth relative to the air cause the frequency heard by the bat to differ by a few kilohertz from the frequency it emitted. The bat automatically translates this difference into a relative speed between itself and the moth, so it can zero in on the moth.

Some moths evade capture by flying away from the direction in which they hear ultrasonic waves. That choice of flight path reduces the frequency difference between what the bat emits and what it hears, and then the bat may not notice the echo. Some moths avoid capture by clicking to produce their own ultrasonic waves, thus "jamming" the detection system and confusing the bat. (Surprisingly, moths and bats do all this without first studying physics.)

✔CHECKPOINT 6:   The figure indicates the directions of motion of a sound source and a detector for six situations in stationary air. For each situation, is the detected frequency greater than or less than the emitted frequency, or can't we tell without more information about the actual speeds?

	Source	Detector		Source	Detector
(a)	→	• 0 speed	(d)	←	←
(b)	←	• 0 speed	(e)	→	←
(c)	→	→	(f)	←	→

Sample Problem 18-8

A rocket moves at a speed of 242 m/s directly toward a stationary pole (through stationary air) while emitting sound waves at frequency $f = 1250$ Hz.

(a) What frequency f' is measured by a detector that is attached to the pole?

SOLUTION: We can find f' with Eq. 18-47 for the general Doppler effect. The **Key Idea** here is that, because the sound source (the rocket) moves through the air *toward* the stationary detector on the pole, we need to choose the sign on v_S that gives a *shift up* in the frequency of the sound. Thus, in Eq. 18-47 we use the minus sign in the denominator. We then substitute 0 for the detector speed v_D, 242 m/s for the source speed v_S, 343 m/s for the speed of sound v (from Table 18-1), and 1250 Hz for the emitted frequency f.

We find

$$f' = f\frac{v \pm v_D}{v \pm v_S} = (1250 \text{ Hz})\frac{343 \text{ m/s} \pm 0}{343 \text{ m/s} - 242 \text{ m/s}}$$

$$= 4245 \text{ Hz} \approx 4250 \text{ Hz}, \qquad \text{(Answer)}$$

which, indeed, is a greater frequency than the emitted frequency.

(b) Some of the sound reaching the pole reflects back to the rocket as an echo. What frequency f'' does a detector on the rocket detect for the echo?

SOLUTION: Two **Key Ideas** here are the following:

1. The pole is now the source of sound (because it is the source of the echo), and the rocket's detector is now the detector (because it detects the echo).

2. The frequency of the sound emitted by the source (the pole) is equal to f', the frequency of the sound the pole intercepts and reflects.

We can rewrite Eq. 18-47 in terms of the source frequency f' and the detected frequency f'' as

$$f'' = f' \frac{v \pm v_D}{v \pm v_S}. \qquad (18\text{-}56)$$

A third **Key Idea** here is that, because the detector (on the rocket) moves through the air *toward* the stationary source, we need to use the sign on v_D that gives a *shift up* in the frequency of the sound. Thus, we use the plus sign in the numerator of Eq. 18-56. Also, we substitute $v_D = 242$ m/s, $v_S = 0$, $v = 343$ m/s, and

$f' = 4245$ Hz. We find

$$f'' = (4245 \text{ Hz}) \frac{343 \text{ m/s} + 242 \text{ m/s}}{343 \text{ m/s} \pm 0}$$

$$= 7240 \text{ Hz}, \qquad \text{(Answer)}$$

which, indeed, is greater than the frequency of the sound reflected by the pole.

✔**CHECKPOINT 7:** If the air in this sample problem is moving toward the pole at speed 20 m/s, (a) what value for the source speed v_S should be used in the solution of part (a), and (b) what value for the detector speed v_D should be used in the solution of part (b)?

18-9 Supersonic Speeds; Shock Waves

If a source is moving toward a stationary detector at a speed equal to the speed of sound—that is, if $v_S = v$—Eqs. 18-47 and 18-55 predict that the detected frequency f' will be infinitely great. This means that the source is moving so fast that it keeps pace with its own spherical wavefronts, as Fig. 18-23a suggests. What happens when the speed of the source *exceeds* the speed of sound?

For such *supersonic* speeds, Eqs. 18-47 and 18-55 no longer apply. Figure 18-23b depicts the spherical wavefronts that originated at various positions of the source. The radius of any wavefront in this figure is vt, where v is the speed of sound and t is the time that has elapsed since the source emitted that wavefront. Note that all the wavefronts bunch along a V-shaped envelope in the two-dimensional drawing of Fig. 18-23b. The wavefronts actually extend in three dimensions, and the bunching actually forms a cone called the *Mach cone*. A *shock wave* is said to exist along the surface of this cone, because the bunching of wavefronts causes an abrupt rise and fall of air pressure as the surface passes through any point. From Fig. 18-23b, we see that the half-angle θ of the cone, called the *Mach cone angle*, is given by

$$\sin \theta = \frac{vt}{v_S t} = \frac{v}{v_S} \qquad \text{(Mach cone angle)}. \qquad (18\text{-}57)$$

The ratio v_S/v is called the *Mach number*. When you hear that a particular plane has flown at Mach 2.3, it means that its speed was 2.3 times the speed of sound in the air through which the plane was flying. The shock wave generated by a super-

Fig. 18-23 (a) A source of sound S moves at speed v_S equal to the speed of sound and thus as fast as the wavefronts it generates. (b) A source S moves at speed v_S faster than the speed of sound and thus faster than the wavefronts. When the source was at position S_1 it generated wavefront W_1, and at position S_6 it generated W_6. All the spherical wavefronts expand at the speed of sound v and bunch along the surface of a cone called the Mach cone, forming a shock wave. The surface of the cone has half-angle θ and is tangent to all the wavefronts.

(a)

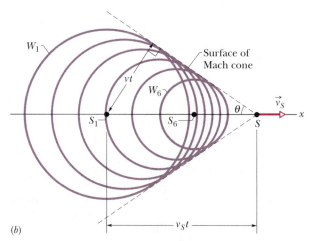

(b)

sonic aircraft (Fig. 18-24) or projectile produces a burst of sound, called a *sonic boom*, in which the air pressure first suddenly increases and then suddenly decreases below normal before returning to normal. Part of the sound that is heard when a rifle is fired is the sonic boom produced by the bullet. A sonic boom can also be heard from a long bullwhip when it is snapped quickly: Near the end of the whip's motion, its tip is moving faster than sound and produces a small sonic boom—the *crack* of the whip.

Fig. 18-24 Shock waves produced by the wings of a Navy FA 18 jet. They are visible because the sudden decrease in air pressure in the shock waves caused water molecules in the air to condense, forming a fog.

REVIEW & SUMMARY

Sound Waves Sound waves are longitudinal mechanical waves that can travel through solids, liquids, or gases. The speed v of a sound wave in a medium having **bulk modulus** B and density ρ is

$$v = \sqrt{\frac{B}{\rho}} \qquad \text{(speed of sound)}. \qquad (18\text{-}3)$$

In air at 20°C, the speed of sound is 343 m/s.

A sound wave causes a longitudinal displacement s of a mass element in a medium as given by

$$s = s_m \cos(kx - \omega t), \qquad (18\text{-}13)$$

where s_m is the **displacement amplitude** (maximum displacement) from equilibrium, $k = 2\pi/\lambda$, and $\omega = 2\pi f$, λ and f being the wavelength and frequency, respectively, of the sound wave. The sound wave also causes a pressure change Δp of the medium from the equilibrium pressure:

$$\Delta p = \Delta p_m \sin(kx - \omega t), \qquad (18\text{-}14)$$

where the **pressure amplitude** is

$$\Delta p_m = (v\rho\omega)s_m. \qquad (18\text{-}15)$$

Interference The interference of two sound waves with identical wavelengths passing through a common point depends on their phase difference ϕ there. If the sound waves were emitted in phase and are traveling in approximately the same direction, ϕ is given by

$$\phi = \frac{\Delta L}{\lambda} 2\pi, \qquad (18\text{-}21)$$

where ΔL is their **path length difference** (the difference in the distances traveled by the waves to reach the common point). Fully constructive interference occurs when ϕ is an integer multiple of 2π,

$$\phi = m(2\pi), \qquad \text{for } m = 0, 1, 2, \ldots, \qquad (18\text{-}22)$$

and, equivalently, when ΔL is related to wavelength λ by

$$\frac{\Delta L}{\lambda} = 0, 1, 2, \ldots. \qquad (18\text{-}23)$$

Fully destructive interference occurs when ϕ is an odd multiple of π,

$$\phi = (2m + 1)\pi, \qquad \text{for } m = 0, 1, 2, \ldots, \qquad (18\text{-}24)$$

and, equivalently, when ΔL is related to λ by

$$\frac{\Delta L}{\lambda} = 0.5, 1.5, 2.5, \ldots. \qquad (18\text{-}25)$$

Sound Intensity The **intensity** I of a sound wave at a surface is the average rate per unit area at which energy is transferred by the wave through or onto the surface:

$$I = \frac{P}{A}, \qquad (18\text{-}26)$$

where P is the time rate of energy transfer (power) of the sound wave and A is the area of the surface intercepting the sound. The intensity I is related to the displacement amplitude s_m of the sound wave by

$$I = \tfrac{1}{2}\rho v \omega^2 s_m^2. \qquad (18\text{-}27)$$

The intensity at a distance r from a point source that emits sound waves of power P_s is

$$I = \frac{P_s}{4\pi r^2}. \qquad (18\text{-}28)$$

Sound Level in Decibels The *sound level* β in *decibels* (dB) is defined as

$$\beta = (10 \text{ dB}) \log \frac{I}{I_0}, \qquad (18\text{-}29)$$

where I_0 (= 10^{-12} W/m²) is a reference intensity level to which all intensities are compared. For every factor-of-10 increase in intensity, 10 dB is added to the sound level.

Standing Wave Patterns in Pipes   Standing sound wave patterns can be set up in pipes. A pipe open at both ends will resonate at frequencies

$$f = \frac{v}{\lambda} = \frac{nv}{2L}, \qquad n = 1, 2, 3, \ldots, \qquad (18\text{-}39)$$

where v is the speed of sound in the air in the pipe. For a pipe closed at one end and open at the other, the resonant frequencies are

$$f = \frac{v}{\lambda} = \frac{nv}{4L}, \qquad n = 1, 3, 5, \ldots. \qquad (18\text{-}41)$$

Beats   *Beats* arise when two waves having slightly different frequencies, f_1 and f_2, are detected together. The beat frequency is

$$f_{\text{beat}} = f_1 - f_2. \qquad (18\text{-}46)$$

The Doppler Effect   The *Doppler effect* is a change in the observed frequency of a wave when the source or the detector moves relative to the transmitting medium (such as air). For sound the observed frequency f' is given in terms of the source frequency f by

$$f' = f\frac{v \pm v_D}{v \pm v_S} \qquad \text{(general Doppler effect),} \qquad (18\text{-}47)$$

where v_D is the speed of the detector relative to the medium, v_S is that of the source, and v is the speed of sound in the medium. The signs are chosen such that f' tends to be *greater* for motion (of detector or source) "toward" and *less* for motion "away."

Shock Wave   If the speed of a source relative to the medium exceeds the speed of sound in the medium, the Doppler equation no longer applies. In such a case, shock waves result. The half angle θ of the Mach cone is given by

$$\sin \theta = \frac{v}{v_S} \qquad \text{(Mach cone angle).} \qquad (18\text{-}57)$$

QUESTIONS

1. Figure 18-25 shows the paths taken by two pulses of sound that begin simultaneously and then race each other through equal distances in air. The only difference between the paths is that a region of hot (low density) air lies along path 2. Which pulse wins the race?

Fig. 18-25 Question 1.

2. A sound wave of wavelength λ and displacement amplitude s_m begins to travel down a passageway (a tube, the opening of the ear, etc.). When a small device in the passageway detects this wave, it issues a second sound wave (said to be *antisound*) that is able to cancel the first wave, so that nothing is heard at the far end of the passageway. For such cancellation, what must be (a) the direction of travel, (b) the wavelength, and (c) the displacement amplitude of the second wave? (d) What must be the phase difference between the two waves? (Such antisound devices are used to eliminate unwanted sound in a noisy environment.)

3. In Fig. 18-26, two point sources S_1 and S_2, which are in phase, emit identical sound waves of wavelength 2.0 m. In terms of wavelengths, what is the phase difference between the waves arriving at point P if (a) $L_1 = 38$ m and $L_2 = 34$ m, and (b) $L_1 = 39$ m and $L_2 = 36$ m? (c) Assuming that the source separation is much smaller than L_1 and L_2, what type of interference occurs at P in situations (a) and (b), respectively?

Fig. 18-26 Question 3.

4. In Fig. 18-27, sound waves of wavelength λ are emitted by a point source S and travel to a detector D directly along path 1 and via reflection from a panel along path 2. Initially, the panel is al-

Fig. 18-27 Question 4.

most along path 1 and the waves arriving at D along the two paths are almost exactly in phase. Then the panel is moved away from path 1 as shown until the waves arriving at D are exactly out of phase. What then is the path length difference ΔL of the waves along the two paths?

5. In Fig. 18-28, two point sources S_1 and S_2, which are in phase, emit identical sound waves of wavelength λ, and point P is at equal distances from them. Then S_2 is moved directly away from P by a distance equal to $\lambda/4$. Are the waves at P then exactly in phase, exactly out of phase, or do they have some intermediate phase relation if (a) S_1 is moved directly toward P by a distance equal to $\lambda/4$ and (b) S_1 is moved directly away from P by a distance equal to $3\lambda/4$?

Fig. 18-28 Question 5.

6. In Sample Problem 18-3 and Fig. 18-9a, the waves arriving at point P_1 on the perpendicular bisector are exactly in phase; that is, the waves from S_1 and S_2 always tend to move an element of air at P_1 in the same direction. Let the intersection of the perpendicular bisector and the line through S_1 and S_2 be point P_3. (a) Are the waves arriving at P_3 exactly in phase, exactly out of phase, or do they have some intermediate relation? (b) What is the answer if we increase the separation between the sources to 1.7λ?

7. A standing sound wave in a pipe has five nodes and five antinodes. (a) How many open ends does the pipe have? (b) What is the harmonic number n for this standing wave?

8. The sixth harmonic is set up in a pipe. (a) How many open ends does the pipe have (it has at least one)? (b) Is there a node, antinode, or some intermediate state at the midpoint?

9. (a) When an orchestra warms up, the players' warm breath increases the temperature of the air within the wind instruments (and thus decreases the density of that air). Do the resonant frequencies of those instruments increase or decrease? (b) When the slide of a slide trombone is pushed outward, do the resonant frequencies of the instrument increase or decrease?

10. For a particular tube, here are four of the six harmonic frequencies below 1000 Hz: 300, 600, 750, and 900 Hz. What two frequencies are missing from the list?

11. Pipe *A* has length *L* and one open end. Pipe *B* has length 2*L* and two open ends. Which harmonics of pipe *B* have a frequency that matches a resonant frequency of pipe *A*?

12. Figure 18-29 shows a stretched string of length *L* and pipes *a*, *b*, *c*, and *d* of lengths *L*, 2*L*, *L*/2, and *L*/2, respectively. The string's tension is adjusted until the speed of waves on the string equals the speed of sound waves in the air. The fundamental mode of oscillation is then set up on the string. In which pipe will the sound produced by the string cause resonance, and what oscillation mode will that sound set up?

Fig. 18-29 Question 12.

13. Sound waves of frequency *f* are reflected by a fluid moving through a narrow tube along an *x* axis (Fig. 18-30*a*). The tube's inside diameter varies with *x*. The change in frequency Δf of the sound, due to the Doppler effect, also varies with *x*, as shown in Fig. 18-30*b*. Rank the five indicated regions in terms of the tube's inside diameter, greatest first. (*Hint:* See Section 15-10.)

Fig. 18-30 Question 13.

14. A friend rides, in turn, the rims of three fast merry-go-rounds while holding a sound source that emits isotropically at a certain frequency. You stand far from each merry-go-round. The frequency you hear for each of your friend's three rides varies as the merry-go-round rotates. The variations in frequency for the three rides are given by the three curves in Fig. 18-31. Rank the curves according to (a) the linear speed *v* of the sound source, (b) the angular speeds ω of the merry-go-rounds, and (c) the radii *r* of the merry-go-rounds, greatest first.

Fig. 18-31 Question 14.

EXERCISES & PROBLEMS

Where needed in the problems, use

$$\text{speed of sound in air} = 343 \text{ m/s}$$

and $\qquad$ density of air = 1.21 kg/m³

unless otherwise specified.

SEC. 18-2 The Speed of Sound

1E. Devise a rule for finding your distance in kilometers from a lightning flash by counting the seconds from the time you see the flash until you hear the thunder. Assume that the sound travels to you along a straight line.

2E. You are at a large outdoor concert, seated 300 m from the speaker system. The concert is also being broadcast live via satellite (at the speed of light, 3.0×10^8 m/s). Consider a listener 5000 km away who receives the broadcast. Who hears the music first, you or the listener, and by what time difference?

3E. Two spectators at a soccer game in Montjuic Stadium see, and a moment later hear, the ball being kicked on the playing field. The time delay for one spectator is 0.23 s and for the other 0.12 s. Sight lines from the two spectators to the player kicking the ball meet at an angle of 90°. (a) How far is each spectator from the player? (b) How far are the spectators from each other? **ssm**

4E. A column of soldiers, marching at 120 paces per minute, keep in step with the beat of a drummer at the head of the column. It is observed that the soldiers in the rear end of the column are striding forward with the left foot when the drummer is advancing with the right. What is the approximate length of the column?

5P. Earthquakes generate sound waves inside Earth. Unlike a gas, Earth can experience both transverse (S) and longitudinal (P) sound waves. Typically, the speed of S waves is about 4.5 km/s, and that of P waves 8.0 km/s. A seismograph records P and S waves from an earthquake. The first P waves arrive 3.0 min before the first S waves (Fig. 18-32). Assuming the waves travel in a straight line, how far away does the earthquake occur? **ssm** **ilw**

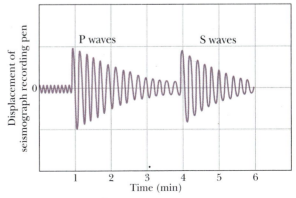

Fig. 18-32 Problem 5.

6P. The speed of sound in a certain metal is V. One end of a long pipe of that metal of length L is struck a hard blow. A listener at the other end hears two sounds, one from the wave that travels along the pipe and the other from the wave that travels through the air. (a) If v is the speed of sound in air, what time interval t elapses between the arrivals of the two sounds? (b) Suppose that $t = 1.00$ s and the metal is steel. Find the length L.

7P. A stone is dropped into a well. The sound of the splash is heard 3.00 s later. What is the depth of the well?   ssm

SEC. 18-3   Traveling Sound Waves

8E. The audible frequency range for normal hearing is from about 20 Hz to 20 kHz. What are the wavelengths of sound waves at these frequencies?

9E. Diagnostic ultrasound of frequency 4.50 MHz is used to examine tumors in soft tissue. (a) What is the wavelength in air of such a sound wave? (b) If the speed of sound in tissue is 1500 m/s, what is the wavelength of this wave in tissue?   ssm

10P. (a) A continuous sinusoidal longitudinal wave is sent along a very long coiled spring from an oscillating source attached to it. The frequency of the source is 25 Hz, and at any time the distance between successive points of maximum expansion in the spring is 24 cm. Find the wave speed. (b) If the maximum longitudinal displacement of a particle in the spring is 0.30 cm and the wave moves in the negative direction of an x axis, write the equation for the wave. Place $x = 0$ at the source and take the displacement there to be zero when $t = 0$.

11P. The pressure in a traveling sound wave is given by the equation

$$\Delta p = (1.50 \text{ Pa}) \sin \pi[(0.900 \text{ m}^{-1})x - (315 \text{ s}^{-1})t].$$

Find (a) the pressure amplitude, (b) the frequency, (c) the wavelength, and (d) the speed of the wave.

SEC. 18-4   Interference

12P. Two point sources of sound waves of identical wavelength λ and amplitude are separated by distance $D = 2.0\lambda$. The sources are in phase. (a) How many points of maximum signal (that is, maximum constructive interference) lie along a large circle around the sources? (b) How many points of minimum signal (destructive interference) lie around the circle?

13P. In Fig. 18-33, two loudspeakers, separated by a distance of 2.00 m, are in phase. Assume the amplitudes of the sound from the speakers are approximately the same at the position of a listener, who is 3.75 m directly in front of one of the speakers. (a) For what frequencies in the audible range (20 Hz to 20 kHz) does the listener hear a minimum signal? (b) For what frequencies is the signal a maximum?   ssm   www

Fig. 18-33 Problem 13.

14P. Two sound waves, from two different sources with the same frequency, 540 Hz, travel in the same direction at 330 m/s. The sources are in phase. What is the phase difference of the waves at a point that is 4.40 m from one source and 4.00 m from the other?

15P. Two loudspeakers are located 3.35 m apart on an outdoor stage. A listener is 18.3 m from one and 19.5 m from the other. During the sound check, a signal generator drives the two speakers in phase with the same amplitude and frequency. The transmitted frequency is swept through the audible range (20 Hz to 20 kHz). (a) What are the three lowest frequencies at which the listener will hear a minimum signal because of destructive interference? (b) What are the three lowest frequencies at which the listener will hear a maximum signal?   ilw

16P. In Fig. 18-34, sound with a 40.0 cm wavelength travels rightward from a source and through a tube that consists of a straight portion and a half-circle. Part of the sound wave travels through the half-circle and then rejoins the rest of the wave, which goes directly through the straight portion. This rejoining results in interference. What is the smallest radius r that results in an intensity minimum at the detector?

Fig. 18-34 Problem 16.

SEC. 18-5   Intensity and Sound Level

17E. A source emits sound waves isotropically. The intensity of the waves 2.50 m from the source is 1.91×10^{-4} W/m^2. Assuming that the energy of the waves is conserved, find the power of the source.   ssm

18E. A 1.0 W point source emits sound waves isotropically. Assuming that the energy of the waves is conserved, find the intensity (a) 1.0 m from the source and (b) 2.5 m from the source.

19E. A sound wave of frequency 300 Hz has an intensity of 1.00 μW/m^2. What is the amplitude of the air oscillations caused by this wave?   ssm

20E. Two sounds differ in sound level by 1.00 dB. What is the ratio of the greater intensity to the smaller intensity?

21E. A certain sound source is increased in sound level by 30 dB. By what multiple is (a) its intensity increased and (b) its pressure amplitude increased?   ssm

22E. The source of a sound wave has a power of 1.00 μW. If it is a point source, (a) what is the intensity 3.00 m away and (b) what is the sound level in decibels at that distance?

23E. (a) If two sound waves, one in air and one in (fresh) water, are equal in intensity, what is the ratio of the pressure amplitude of the wave in water to that of the wave in air? Assume the water and the air are at 20°C. (See Table 15-1.) (b) If the pressure amplitudes are equal instead, what is the ratio of the intensities of the waves?   ssm

24P. Assume that a noisy freight train on a straight track emits a cylindrical, expanding sound wave, and that the air absorbs no energy. How does the amplitude s_m of the wave depend on the perpendicular distance r from the source?

25P. (a) Show that the intensity I of a wave is the product of the wave's energy per unit volume u and its speed v. (b) Radio waves travel at a speed of 3.00×10^8 m/s. Find u for a radio wave 480 km from a 50 000 W source, assuming the wavefronts are spherical.   ssm

26P. Find the ratios (greater to smaller) of (a) the intensities,

(b) the pressure amplitudes, and (c) the particle displacement amplitudes for two sounds whose sound levels differ by 37 dB.

27P. A sound wave travels out uniformly in all directions from a point source. (a) Justify the following expression for the displacement s of the transmitting medium at any distance r from the source:

$$s = \frac{b}{r} \sin k(r - vt),$$

where b is a constant. Consider the speed, direction of propagation, periodicity, and intensity of the wave. (b) What is the dimension of the constant b? ssm www

28P. A point source emits 30.0 W of sound isotropically. A small microphone intercepts the sound in an area of 0.750 cm², 200 m from the source. Calculate (a) the sound intensity there and (b) the power intercepted by the microphone.

29P*. Figure 18-35 shows an air-filled, acoustic interferometer, used to demonstrate the interference of sound waves. Sound source S is an oscillating diaphragm; D is a sound detector, such as the ear or a microphone. Path SBD can be varied in length, but path SAD is fixed. At D, the sound wave coming along path SBD interferes with that coming along path SAD. In one demonstration, the sound intensity at D has a minimum value of 100 units at one position of the movable arm and continuously climbs to a maximum value of 900 units when that arm is shifted by 1.65 cm. Find (a) the frequency of the sound emitted by the source and (b) the ratio of the amplitude at D of the SAD wave to that of the SBD wave. (c) How can it happen that these waves have different amplitudes, considering that they originate at the same source? ssm

Fig. 18-35 Problem 29.

SEC. 18-6   Sources of Musical Sound

30E. A violin string 15.0 cm long and fixed at both ends oscillates in its $n = 1$ mode. The speed of waves on the string is 250 m/s, and the speed of sound in air is 348 m/s. What are (a) the frequency and (b) the wavelength of the emitted sound wave?

31E. Organ pipe A, with both ends open, has a fundamental frequency of 300 Hz. The third harmonic of organ pipe B, with one end open, has the same frequency as the second harmonic of pipe A. How long are (a) pipe A and (b) pipe B?

32E. The water level in a vertical glass tube 1.00 m long can be adjusted to any position in the tube. A tuning fork vibrating at 686 Hz is held just over the open top end of the tube, to set up a standing wave of sound in the air-filled top portion of the tube. (That air-filled top portion acts as a tube with one end closed and the other end open.) At what positions of the water level is there resonance?

33E. (a) Find the speed of waves on a violin string of mass 800 mg and length 22.0 cm if the fundamental frequency is 920 Hz. (b) What is the tension in the string? For the fundamental, what is the wavelength of (c) the waves on the string and (d) the sound waves emitted by the string? ssm ilw

34P. A certain violin string is 30 cm long between its fixed ends and has a mass of 2.0 g. The "open" string (no applied finger)

sounds an A note (440 Hz). (a) To play a C note (523 Hz), how far down the string must one place a finger? (b) What is the ratio of the wavelength of the string waves required for an A note to that required for a C note? (c) What is the ratio of the wavelength of the sound wave for an A note to that for a C note?

35P. In Fig. 18-36, S is a small loudspeaker driven by an audio oscillator and amplifier, adjustable in frequency from 1000 to 2000 Hz only. Tube D is a piece of cylindrical sheet-metal pipe 45.7 cm long and open at both ends. (a) If the speed of sound in air is 344 m/s at the existing temperature, at what frequencies will resonance occur in the pipe when the frequency emitted by the speaker is varied from 1000 Hz to 2000 Hz? (b) Sketch the standing wave (using the style of Fig. 18-14b) for each resonant frequency. ssm www

Fig. 18-36 Problem 35.

36P. A string on a cello has length L, for which the fundamental frequency is f. (a) By what length l must the string be shortened by fingering to change the fundamental frequency to rf? (b) What is l if $L = 0.80$ m and $r = 1.2$? (c) For $r = 1.2$, what is the ratio of the wavelength of the new sound wave emitted by the string to that of the wave emitted before fingering?

37P. A well with vertical sides and water at the bottom resonates at 7.00 Hz and at no lower frequency. (The air-filled portion of the well acts as a tube with one closed end and one open end.) The air in the well has a density of 1.10 kg/m³ and a bulk modulus of 1.33×10^5 Pa. How far down in the well is the water surface? ssm

38P. A tube 1.20 m long is closed at one end. A stretched wire is placed near the open end. The wire is 0.330 m long and has a mass of 9.60 g. It is fixed at both ends and oscillates in its fundamental mode. By resonance, it sets the air column in the tube into oscillation at that column's fundamental frequency. Find (a) that frequency and (b) the tension in the wire.

39P. The period of a pulsating variable star may be estimated by considering the star to be executing *radial* longitudinal pulsations in the fundamental standing wave mode; that is, the star's radius varies periodically with time, with a displacement antinode at the star's surface. (a) Would you expect the center of the star to be a displacement node or antinode? (b) By analogy with a pipe with one open end, show that the period of pulsation T is given by

$$T = \frac{4R}{v},$$

where R is the equilibrium radius of the star and v is the average sound speed in the material of the star. (c) Typical white dwarf stars are composed of material with a bulk modulus of 1.33×10^{22} Pa and a density of 10^{10} kg/m³. They have radii equal to 9.0×10^{-3} solar radius. What is the approximate pulsation period of a white dwarf? ssm

40P. Pipe A, which is 1.2 m long and open at both ends, oscillates at its third lowest harmonic frequency. It is filled with air for which the speed of sound is 343 m/s. Pipe B, which is closed at one end, oscillates at its second lowest harmonic frequency. The frequencies of pipes A and B happen to match. (a) If an x axis extends along the interior of pipe A, with $x = 0$ at one end, where along the axis are the displacement nodes? (b) How long is pipe B? (c) What is the lowest harmonic frequency of pipe A?

41P. A violin string 30.0 cm long with linear density 0.650 g/m is placed near a loudspeaker that is fed by an audio oscillator of variable frequency. It is found that the string is set into oscillation only at the frequencies 880 and 1320 Hz as the frequency of the oscillator is varied over the range 500–1500 Hz. What is the tension in the string? ssm

SEC. 18-7   Beats

42E. The A string of a violin is a little too tightly stretched. Four beats per second are heard when the string is sounded together with a tuning fork that is oscillating accurately at concert A (440 Hz). What is the period of the violin string oscillation?

43E. A tuning fork of unknown frequency makes three beats per second with a standard fork of frequency 384 Hz. The beat frequency decreases when a small piece of wax is put on a prong of the first fork. What is the frequency of this fork? ssm

44P. You have five tuning forks that oscillate at close but different frequencies. What are the (a) maximum and (b) minimum number of different beat frequencies you can produce by sounding the forks two at a time, depending on how the frequencies differ?

45P. Two identical piano wires have a fundamental frequency of 600 Hz when kept under the same tension. What fractional increase in the tension of one wire will lead to the occurrence of 6 beats/s when both wires oscillate simultaneously? ssm

SEC. 18-8   The Doppler Effect

46E. Trooper B is chasing speeder A along a straight stretch of road. Both are moving at a speed of 160 km/h. Trooper B, failing to catch up, sounds his siren again. Take the speed of sound in air to be 343 m/s and the frequency of the source to be 500 Hz. What is the Doppler shift in the frequency heard by speeder A?

47E. The 16 000 Hz whine of the turbines in the jet engines of an aircraft moving with speed 200 m/s is heard at what frequency by the pilot of a second craft trying to overtake the first at a speed of 250 m/s? ssm

48E. An ambulance with a siren emitting a whine at 1600 Hz overtakes and passes a cyclist pedaling a bike at 2.44 m/s. After being passed, the cyclist hears a frequency of 1590 Hz. How fast is the ambulance moving?

49P. A whistle of frequency 540 Hz moves in a circle of radius 60.0 cm at an angular speed of 15.0 rad/s. What are (a) the lowest and (b) the highest frequencies heard by a listener a long distance away, at rest with respect to the center of the circle? ilw

50P. A stationary motion detector sends sound waves of frequency 0.150 MHz toward a truck approaching at a speed of 45.0 m/s. What is the frequency of the waves reflected back to the detector?

51P. A French submarine and a U.S. submarine move toward each other during maneuvers in motionless water in the North Atlantic (Fig. 18-37). The French sub moves at 50.0 km/h, and the U.S. sub at 70.0 km/h. The French sub sends out a sonar signal (sound wave in water) at 1000 Hz. Sonar waves travel at 5470 km/h. (a) What is the signal's frequency as detected by the U.S. sub? (b) What

frequency is detected by the French sub in the signal reflected back to it by the U.S. sub?

French
50.0 km/h

U.S.
70.0 km/h

Fig. 18-37 Problem 51.

52P. A sound source A and a reflecting surface B move directly toward each other. Relative to the air, the speed of source A is 29.9 m/s, the speed of surface B is 65.8 m/s, and the speed of sound is 329 m/s. The source emits waves at frequency 1200 Hz as measured in the source frame. In the reflector frame, what are (a) the frequency and (b) the wavelength of the arriving sound waves? In the source frame, what are (c) the frequency and (d) the wavelength of the sound waves reflected back to the source?

53P. An acoustic burglar alarm consists of a source emitting waves of frequency 28.0 kHz. What is the beat frequency between the source waves and the waves reflected from an intruder walking at an average speed of 0.950 m/s directly away from the alarm? ilw

54P. A bat is flitting about in a cave, navigating via ultrasonic bleeps. Assume that the sound emission frequency of the bat is 39 000 Hz. During one fast swoop directly toward a flat wall surface, the bat is moving at 0.025 times the speed of sound in air. What frequency does the bat hear reflected off the wall?

55P. A girl is sitting near the open window of a train that is moving at a velocity of 10.00 m/s to the east. The girl's uncle stands near the tracks and watches the train move away. The locomotive whistle emits sound at frequency 500.0 Hz. The air is still. (a) What frequency does the uncle hear? (b) What frequency does the girl hear? A wind begins to blow from the east at 10.00 m/s. (c) What frequency does the uncle now hear? (d) What frequency does the girl now hear? ssm www

56P. A 2000 Hz siren and a civil defense official are both at rest with respect to the ground. What frequency does the official hear if the wind is blowing at 12 m/s (a) from source to official and (b) from official to source?

57P. Two trains are traveling toward each other at 30.5 m/s relative to the ground. One train is blowing a whistle at 500 Hz. (a) What frequency is heard on the other train in still air? (b) What frequency is heard on the other train if the wind is blowing at 30.5 m/s toward the whistle and away from the listener? (c) What frequency is heard if the wind direction is reversed?

SEC. 18-9   Supersonic Speeds; Shock Waves

58E. A bullet is fired with a speed of 685 m/s. Find the angle made by the shock cone with the line of motion of the bullet.

59P. A jet plane passes over you at a height of 5000 m and a speed of Mach 1.5. (a) Find the Mach cone angle. (b) How long after the jet passes directly overhead does the shock wave reach you? Use 331 m/s for the speed of sound. ssm

60P. A plane flies at 1.25 times the speed of sound. Its sonic boom reaches a man on the ground 1.00 min after the plane passes directly overhead. What is the altitude of the plane? Assume the speed of sound to be 330 m/s.

NEW PROBLEMS

Where needed in the problems, use

speed of sound in air $= 343$ m/s

and   density of air $= 1.21$ kg/m^3

unless otherwise specified.

N1. A sinusoidal sound wave moves at 343 m/s through air in the positive direction of an x axis. At one instant, air molecule A is at its maximum displacement in the negative direction of the axis while air molecule B is at its equilibrium position. The separation between those molecules is 15.0 cm, and the molecules between A and B have intermediate displacements in the negative direction of the axis. (a) What is the frequency of the sound wave?

In a similar arrangement, for a different sinusoidal sound wave, air molecule C is at its maximum displacement in the positive direction while molecule D is at its maximum displacement in the negative direction. The separation between the molecules is again 15.0 cm, and the molecules between C and D have intermediate displacements. (b) What is the frequency of the sound wave?

N2. A sound wave of the form $s = s_m \cos(kx - \omega t + \phi)$ travels at 343 m/s through air in a long horizontal tube. At one instant, air molecule A at $x = 2.000$ m is at its maximum positive displacement of 6.0 nm and air molecule B at $x = 2.070$ m is at a positive displacement of 2.0 nm. All the molecules between A and B are at intermediate displacements. What is the frequency of the wave?

N3. Figure 18N-1 shows the output from a pressure monitor mounted at a point along the path taken by a sound wave of a single frequency traveling at 343 m/s through air with a uniform density of 1.21 kg/m^3. (a) Write an equation for the *displacement* function of the wave, in the form of $s(x, t) = s_m \cos(kx - \omega t)$. The air is then cooled so that its density is 1.35 kg/m^3 and the speed of a sound wave through it is 320 m/s. The sound source again emits the sound wave at the same frequency and same pressure amplitude. (b) Write an equation for this sound wave.

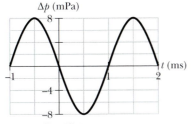

Fig. 18N-1 Problem N3.

N4. Suppose that the sound level of a conversation is initially at a polite 60 dB but then increases to an angry 70 dB and finally drops off to a soothing 50 dB, a variation of 20 dB (Fig. 18N-2). Estimating that the frequency of the sound is 500 Hz, determine the corresponding variation in (a) the sound intensity and (b) the sound wave amplitude.

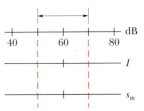

Fig. 18N-2 Problem N4.

N5. A stationary detector measures the frequency of a sound source that first moves at constant velocity directly toward it and then (after passing the detector) directly away from it. The emitted frequency is f. During the approach the detected frequency is f'_{app} and during the recession it is f'_{rec}. If $(f'_{app} - f'_{rec})/f = 0.500$, what is the speed of the source in terms of the speed of sound v?

N6. A detector initially moves at constant velocity directly toward a stationary sound source and then (after passing it) directly from it. The emitted frequency is f. During the approach the detected frequency is f'_{app} and during the recession it is f'_{rec}. If the frequencies are related by $(f'_{app} - f'_{rec})/f = 0.500$, what is the speed of the detector in terms of the speed of sound v?

N7. Two atmospheric sound sources A and B emit isotropically at constant power. The sound levels β of their emissions are plotted in Fig. 18N-3 versus the radial distance r from the sources. What are (a) the difference in their sound levels at $r = 10$ m and (b) the ratio of the larger power to the smaller power?

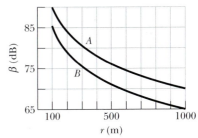

Fig. 18N-3 Problem N7.

N8. Figure 18N-4 shows four tubes with lengths 1.0 m or 2.0 m, with one or two open ends as drawn. The third harmonic is set up in each tube, and some of the sound that escapes from them is detected by detector D, which moves directly away from the tubes. In terms of the speed of sound v, what speed must the detector have such that the detected frequency of the sound from (a) tube 1, (b) tube 2, (c) tube 3, and (d) tube 4 is equal to the tube's fundamental frequency?

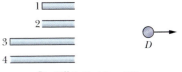

Fig. 18N-4 Problem N8.

N9. Two sound waves with an amplitude of 12 nm and a wavelength of 35 cm travel in the same direction through a long tube, with a phase difference of $\pi/3$ rad. What are (a) the amplitude and (b) the wavelength of the net sound wave produced by their interference? If, instead, the sound waves travel through the tube in opposite directions, what are (c) the amplitude and (d) the wavelength of the net wave?

N10. If the form of a sound wave traveling through air is

$$s(x, t) = (6.0 \text{ nm}) \cos(kx + (3000 \text{ rad/s})t + \phi),$$

how much time does any given air molecule along the path take to move between displacements $s = +2.0$ nm and $s = -2.0$ nm?

N11. Figure 18N-5 shows two isotropic point sources of sound, S_1 and S_2. The sources emit waves in phase at wavelength 0.50 m; they are separated by $D = 1.75$ m. If we move a sound detector along a large circle centered at the midpoint between the sources, at how many points do waves arrive at the detector (a) exactly in phase and (b) exactly out of phase?

Fig. 18N-5 Problems N11 and N12.

N12. In Fig. 18N-5, S_1 and S_2 are two isotropic point sources of sound. They emit waves in phase at wavelength 0.50 m; they are separated by $D = 1.60$ m. If we move a sound detector along a large circle centered at the midpoint between the sources, at how many points do waves arrive at the detector (a) exactly in phase and (b) exactly out of phase?

N13. A sound source sends a sinusoidal sound wave of angular frequency 3000 rad/s and amplitude 12.0 nm through a tube of air. The density of the air is 1.21 kg/m³, the speed of sound through the air is 343 m/s, and the internal radius of the tube is 2.00 cm. (a) What is the average rate at which energy (the sum of the kinetic and potential energies) is transported to the opposite end of the tube? (b) If, simultaneously, an identical wave travels along an adjacent, identical tube, what is the total average rate at which energy is transported to the opposite ends of the two tubes by the waves? If, instead, those two waves are sent along the *same* tube simultaneously, what is the total average rate at which they transport energy when their phase difference is (c) 0, (d) 0.40π rad, and (e) π rad?

N14. A point source that is stationary on an x axis emits a sinusoidal sound wave at a frequency of 686 Hz and speed 343 m/s. The wave travels radially outward from the source, causing air molecules to oscillate radially inward and outward. Let us define a wavefront as a line that connects points where the air molecules have the maximum, radially outward displacement. At any given instant, the wavefronts are concentric circles that are centered on the source. (a) Along the x axis, what is the separation between adjacent wavefronts? Next, the source moves along the x axis at a speed of 110 m/s. Along the x axis, what are the wavefront separations (b) in front of the source and (c) behind the source?

N15. The speed of sound in a certain metal is V. One end of a long pipe of that metal of length L is struck a hard blow. A listener at the other end hears two sounds, one from the wave that has traveled along the pipe and the other from the wave that has traveled through the air. (a) If v is the speed of sound in air, what time interval t elapses between the arrivals of the two sounds? (b) Suppose that $t = 1.00$ s and the metal is steel. Find length L.

N16. Figure 18N-6 shows four isotropic point sources of sound that are uniformly spaced on an x axis. The sources emit sound at the same wavelength λ and same amplitude s_m, and they emit in phase. A point P is shown on the x axis. Assume that as the sound waves travel to P, the decrease in their intensity (as given by Eq. 18-28) is negligible. What is the amplitude of the net wave at P if distance d in the figure is (a) $\lambda/4$, (b) $\lambda/2$, and (c) λ?

Fig. 18N-6 Problem N16.

N17. In Fig. 18N-7, a point source S of sound waves lies near a reflecting wall AB. A sound detector D intercepts sound ray R_1 traveling directly from S. It also intercepts sound ray R_2 that reflects from the wall such that the *angle of incidence* θ_i is equal to the *angle of reflection* θ_r. Assume that the reflection of sound by the wall causes a phase shift of 0.500λ. Find the lowest two frequencies at which there is maximum constructive interference of R_1 and R_2 at D.

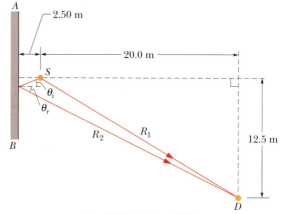

Fig. 18N-7 Problem N17.

N18. One of the harmonic frequencies of tube A with two open ends is 325 Hz. The next higher harmonic frequency is 390 Hz. (a) What harmonic frequency is next higher after the harmonic frequency 195 Hz? (b) What is the corresponding harmonic number?

One of the harmonic frequencies of tube B with only one open end is 1080 Hz. The next higher harmonic frequency is 1320 Hz. (c) What harmonic frequency is next higher after the harmonic frequency 600 Hz? (d) What is the corresponding harmonic number of the tube?

N19. A sound wave in a fluid medium is reflected at a barrier so that a standing wave is formed. The distance between nodes is 3.8 cm, and the speed of propagation is 1500 m/s. Find the frequency of the sound wave.

N20. Four sound waves are to be sent through the same tube of air, in the same direction:

$$s_1(x, t) = (9.00 \text{ nm}) \cos(2\pi x - 700\pi t)$$
$$s_2(x, t) = (9.00 \text{ nm}) \cos(2\pi x - 700\pi t + 0.7\pi)$$
$$s_3(x, t) = (9.00 \text{ nm}) \cos(2\pi x - 700\pi t + \pi)$$
$$s_4(x, t) = (9.00 \text{ nm}) \cos(2\pi x - 700\pi t + 1.7\pi).$$

What is the resultant wave? (*Hint:* Use a phasor diagram to simplify the problem.)

N21. An audio engineer has designed a loudspeaker that is spherical and emits sound isotropically. The speaker emits 10 W of acoustic power into a room with completely absorbent walls, floor, and ceiling (an *anechoic chamber*). (a) What is the intensity (W/m²) of the sound waves 3.0 m from the center of the source? (b) How does the amplitude of the waves at 4.0 m compare with that at 3.0 m from the center of the source?

N22. In Fig. 18N-8, sound waves *A* and *B*, both of wavelength λ, are initially in phase and traveling rightward, as indicated by the two rays. One of the waves is reflected from four surfaces but ends up traveling in its original direction. The other ends in that direction after reflecting from two surfaces. (a) What is the least value of the distance *L* in the figure that puts waves *A* and *B* exactly out of phase with each other after the reflections? (b) What is the next greater value of *L* that gives that phase relation?

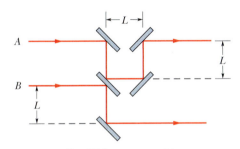

Fig. 18N-8 Problem N22.

N23. A salesperson claimed that a stereo system had a maximum audio power of 120 W. Testing the system with several speakers set up so as to simulate a point source, the consumer noted that she could get as close as 1.2 m with the volume full on before the sound hurt her ears. Was the salesperson truthful? Explain your answer with a calculation.

N24. A sound source moves along an *x* axis, between detectors *A* and *B*. The wavelength of the sound detected at *A* is 0.500 that of the sound detected at *B*. In terms of the speed of sound *v*, what is the speed of the source?

N25. Pipe *A* has only one open end; pipe *B* is four times as long and has two open ends. Which of the first 10 harmonic frequencies of pipe *B* equal one of the first 10 harmonic frequencies of pipe *A*?

N26. In Fig. 18N-9, sound of wavelength 0.85 m is emitted istropically by point source *S*. Sound ray 1 extends directly to detector *D*, at distance $L = 10$ m. Sound ray 2 extends to *D* via a reflection (effectively, a "bouncing") of the sound at a flat surface. That reflection occurs on a perpendicular bisector to the *SD* line, at distance *d* from the line. Assume that the reflection shifts the sound wave by 0.50λ. For what least values of *d* (other than zero) do the direct sound and the reflected sound arrive at *D* (a) exactly out of phase and (b) exactly in phase?

Fig. 18N-9 Problem N26.

N27. An experimenter wishes to measure the speed of sound in an aluminum rod 10 cm long by measuring the time it takes for a sound pulse to travel the length of the rod. If results good to four significant figures are desired, how precisely must the length of the rod be known and how closely must she be able to resolve time intervals?

N28. Figure 18N-10*a* shows two point sound sources S_1 and S_2, located on a line. The sources emit sound isotropically and at the same wavelength λ and same amplitude. Detection point P_1 is on a perpendicular bisector to the line between the sources; waves arrive there from the two sources with zero phase difference. Detection point P_2 is on the line through the sources; waves arrive there with a phase difference of 5.0 wavelengths. (a) In terms of wavelengths, what is the distance between the sources? (b) What type of interference occurs at point P_1?

The sources are now both moved directly away from point P_1 by a distance of $\lambda/2$, at an angle $\theta = 30°$ (Fig. 18N-10*b*). (c) In terms of wavelengths, what is the phase difference between the waves arriving at point P_3, which is on the new line through the sources? (d) What type of interference occurs at point P_3?

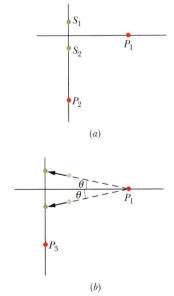

Fig. 18N-10 Problem N28.

N29. Figure 18N-11 shows a transmitter and receiver of waves contained in a single instrument. It is used to measure the speed u of a target object (idealized as a flat plate) that is moving directly toward the unit, by analyzing the waves reflected from the target. (a) Show that the frequency f_r of the reflected waves at the receiver is related to their source frequency f_s by

$$f_r = f_s \left(\frac{v + u}{v - u} \right),$$

where v is the speed of the waves. (b) In a great many practical situations, $u \ll v$. In this case, show that the equation above becomes

$$\frac{f_r - f_s}{f_s} \approx \frac{2u}{v}.$$

N30. The shock wave off the cockpit of the FA 18 in Fig. 18-24 appears to have an angle of about 60°. The airplane was traveling at about 1350 km/h when the photograph was taken. Approximately what was the speed of sound at the airplane's altitude?

Fig. 18N-11 Problem N29.

N31. A police car is chasing a speeding Porsche 911. Assume that the Porsche's maximum speed is 80 m/s (180 mi/h) and the police car's is 54 m/s (120 mi/h). At the moment both cars reach their maximum speed, what frequency will the Porsche driver hear if the frequency of the police car's siren is 440 Hz? The speed of sound in air is 340 m/s.

N32. In pipe A, the ratio of a particular harmonic frequency to the next lower harmonic frequency is 1.2. In pipe B, the ratio of a particular harmonic frequency to the next lower harmonic frequency is 1.4. How many open ends do (a) pipe A and (b) pipe B have?

19 Temperature, Heat, and the First Law of Thermodynamics

The giant hornet *Vespa mandarinia japonica* preys on Japanese bees. However, if one of the hornets attempts to invade a bee hive, several hundred of the bees quickly form a compact ball around the hornet to stop it.

After about 20 minutes the hornet is dead, although the bees do not sting, bite, crush, or suffocate it.

Why, then, does the hornet die?

The answer is in this chapter.

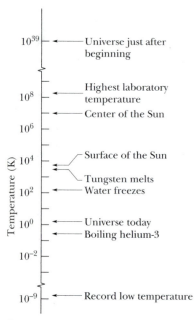

Fig. 19-1 Some temperatures on the Kelvin scale. Temperature $T = 0$ corresponds to $10^{-\infty}$ and cannot be plotted on this logarithmic scale.

19-1   Thermodynamics

In this and the next two chapters we focus on a new subject—**thermodynamics,** the study of the *thermal energy* (often called the *internal energy*) of systems. The central concept of thermodynamics is temperature. This word is so familiar that most of us—because of our built-in sense of hot and cold—tend to be overconfident in our understanding of it. Our "temperature sense" is in fact not always reliable. On a cold winter day, for example, an iron railing seems much colder to the touch than a wooden fence post, yet both are at the same temperature. This error in our perception comes about because iron removes energy from our fingers more quickly than wood does. Here, we shall develop the concept of temperature from its foundations, without relying in any way on our temperature sense.

Temperature is one of the seven SI base quantities. Physicists measure temperature on the **Kelvin scale,** which is marked in units called *kelvins.* Although the temperature of a body apparently has no upper limit, it does have a lower limit; this limiting low temperature is taken as the zero of the Kelvin temperature scale. Room temperature is about 290 kelvins, or 290 K as we write it, above this *absolute zero.* Figure 19-1 shows a wide range of temperatures, either measured or conjectured.

When the universe began, some 10 to 20 billion years ago, its temperature was about 10^{39} K. As the universe expanded it cooled, and it has now reached an average temperature of about 3 K. We on Earth are a little warmer than that because we happen to live near a star. Without our Sun, we too would be at 3 K (or, rather, we could not exist).

19-2   The Zeroth Law of Thermodynamics

The properties of many bodies change as we alter their temperature, perhaps by moving them from a refrigerator to a warm oven. To give a few examples: As their temperatures increase, the volume of a liquid increases, a metal rod grows a little longer, and the electrical resistance of a wire increases, as does the pressure exerted by a confined gas. We can use any one of these properties as the basis of an instrument that will help us to pin down the concept of temperature.

Figure 19-2 shows such an instrument. Any resourceful engineer could design and construct it, using any one of the properties listed above. The instrument is fitted with a digital readout display and has the following properties: If you heat it (say, with a Bunsen burner), the displayed number starts to increase; if you then put it into a refrigerator, the displayed number starts to decrease. The instrument is not calibrated in any way, and the numbers have (as yet) no physical meaning. The device is a *thermoscope* but not (as yet) a *thermometer.*

Suppose that, as in Fig. 19-3a, we put the thermoscope (which we shall call body T) into intimate contact with another body (body A). The entire system is confined within a thick-walled insulating box. The numbers displayed by the thermoscope roll by until, eventually, they come to rest (let us say the reading is "137.04") and no further change takes place. In fact, we suppose that every measurable property of body T and of body A has assumed a stable, unchanging value. Then we say that the two bodies are in *thermal equilibrium* with each other. Even though the displayed readings for body T have not been calibrated, we conclude that bodies T and A must be at the same (unknown) temperature.

Suppose that we next put body T in intimate contact with body B (Fig. 19-3b) and find that the two bodies come to thermal equilibrium *at the same reading of the thermoscope.* Then bodies T and B must be at the same (still unknown) temperature. If we now put bodies A and B into intimate contact (Fig. 19-3c), are they immediately in thermal equilibrium with each other? Experimentally, we find that they are.

Fig. 19-2 A thermoscope. The numbers increase when the device is heated and decrease when it is cooled. The thermally sensitive element could be—among many possibilities—a coil of wire whose electrical resistance is measured and displayed.

Fig. 19-3 (a) Body T (a thermoscope) and body A are in thermal equilibrium. (Body S is a thermally insulating screen.) (b) Body T and body B are also in thermal equilibrium, at the same reading of the thermoscope. (c) If (a) and (b) are true, the zeroth law of thermodynamics states that body A and body B are also in thermal equilibrium.

The experimental fact shown in Fig. 19-3 is summed up in the **zeroth law of thermodynamics:**

⮞ If bodies A and B are each in thermal equilibrium with a third body T, then they are in thermal equilibrium with each other.

In less formal language, the message of the zeroth law is: "Every body has a property called **temperature.** When two bodies are in thermal equilibrium, their temperatures are equal. And vice versa." We can now make our thermoscope (the third body T) into a thermometer, confident that its readings will have physical meaning. All we have to do is calibrate it.

We use the zeroth law constantly in the laboratory. If we want to know whether the liquids in two beakers are at the same temperature, we measure the temperature of each with a thermometer. We do not need to bring the two liquids into intimate contact and observe whether they are or are not in thermal equilibrium.

The zeroth law, which has been called a logical afterthought, came to light only in the 1930s, long after the first and second laws of thermodynamics had been discovered and numbered. Because the concept of temperature is fundamental to those two laws, the law that establishes temperature as a valid concept should have the lowest number—hence the zero.

19-3 Measuring Temperature

Here we first define and measure temperatures on the Kelvin scale. Then we calibrate a thermoscope so as to make it a thermometer.

The Triple Point of Water

To set up a temperature scale, we pick some reproducible thermal phenomenon and, quite arbitrarily, assign a certain Kelvin temperature to its environment; that is, we select a *standard fixed point* and give it a standard fixed-point *temperature.* We could, for example, select the freezing point or the boiling point of water but, for various technical reasons, we select instead the **triple point of water.**

Liquid water, solid ice, and water vapor (gaseous water) can coexist, in thermal equilibrium, at only one set of values of pressure and temperature. Figure 19-4 shows a triple-point cell, in which this so-called triple point of water can be achieved in the laboratory. By international agreement, the triple point of water has been assigned a value of 273.16 K as the standard fixed-point temperature for the calibration of thermometers; that is,

$$T_3 = 273.16 \text{ K} \qquad \text{(triple-point temperature)}, \qquad (19\text{-}1)$$

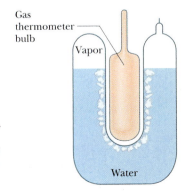

Fig. 19-4 A triple-point cell, in which solid ice, liquid water, and water vapor coexist in thermal equilibrium. By international agreement, the temperature of this mixture has been defined to be 273.16 K. The bulb of a constant-volume gas thermometer is shown inserted into the well of the cell.

Fig. 19-5 A constant-volume gas thermometer, its bulb immersed in a liquid whose temperature T is to be measured.

in which the subscript 3 means "triple point." This agreement also sets the size of the kelvin as 1/273.16 of the difference between absolute zero and the triple-point temperature of water.

Note that we do not use a degree mark in reporting Kelvin temperatures. It is 300 K (not 300°K), and it is read "300 kelvins" (not "300 degrees Kelvin"). The usual SI prefixes apply. Thus, 0.0035 K is 3.5 mK. No distinction in nomenclature is made between Kelvin temperatures and temperature differences, so we can write, "the boiling point of sulfur is 717.8 K" and "the temperature of this water bath was raised by 8.5 K."

The Constant-Volume Gas Thermometer

The standard thermometer, against which all other thermometers are calibrated, is based on the pressure of a gas in a fixed volume. Figure 19-5 shows such a **constant-volume gas thermometer**; it consists of a gas-filled bulb connected by a tube to a mercury manometer. By raising and lowering reservoir R, the mercury level on the left can always be brought to the zero of the scale to keep the gas volume constant (variations in the gas volume can affect temperature measurements).

The temperature of any body in thermal contact with the bulb (like the liquid in Fig. 19-5) is then defined to be

$$T = Cp, \qquad (19\text{-}2)$$

in which p is the pressure within the gas and C is a constant. From Eq. 15-10, the pressure p is

$$p = p_0 - \rho g h, \qquad (19\text{-}3)$$

in which p_0 is the atmospheric pressure, ρ is the density of the mercury in the manometer, and h is the measured difference between the mercury levels in the two arms of the tube.*

If we next put the bulb in a triple-point cell (Fig. 19-4), the temperature now being measured is

$$T_3 = Cp_3, \qquad (19\text{-}4)$$

in which p_3 is the gas pressure now. Eliminating C between Eqs. 19-2 and 19-4 gives us the temperature as

$$T = T_3\left(\frac{p}{p_3}\right) = (273.16 \text{ K})\left(\frac{p}{p_3}\right) \qquad \text{(provisional)}. \qquad (19\text{-}5)$$

We still have a problem with this thermometer. If we use it to measure, say, the boiling point of water, we find that different gases in the bulb give slightly different results. However, as we use smaller and smaller amounts of gas to fill the bulb, the readings converge nicely to a single temperature, no matter what gas we use. Figure 19-6 shows this convergence for three gases.

Thus the recipe for measuring a temperature with a gas thermometer is

$$T = (273.16 \text{ K})\left(\lim_{\text{gas}\to 0}\frac{p}{p_3}\right). \qquad (19\text{-}6)$$

The recipe instructs us to measure an unknown temperature T as follows: Fill the thermometer bulb with an arbitrary amount of *any* gas (for example, nitrogen) and measure p_3 (using a triple-point cell) and p, the gas pressure at the temperature being

---

*For pressure units, we shall use units introduced in Section 15-3. The SI unit for pressure is the newton per square meter, which is called the pascal (Pa). The pascal is related to other common pressure units by

$$1 \text{ atm} = 1.01 \times 10^5 \text{ Pa} = 760 \text{ torr} = 14.7 \text{ lb/in.}^2.$$

Fig. 19-6 Temperatures measured by a constant-volume gas thermometer, with its bulb immersed in boiling water. For temperature calculations using Eq. 19-5, pressure p_3 was measured at the triple point of water. Three different gases in the thermometer bulbs gave generally different results at different gas pressures, but as the amount of gas was decreased (decreasing p_3), all three curves converged to 373.125 K.

measured. (Keep the gas volume the same.) Calculate the ratio p/p_3. Then repeat both measurements with a smaller amount of gas in the bulb, and again calculate this ratio. Continue this way, using smaller and smaller amounts of gas, until you can extrapolate to the ratio p/p_3 that you would find if there were approximately no gas in the bulb. Calculate the temperature T by substituting that extrapolated ratio into Eq. 19-6. (The temperature is called the *ideal gas temperature.*)

19-4 The Celsius and Fahrenheit Scales

So far, we have discussed only the Kelvin scale, used in basic scientific work. In nearly all countries of the world, the Celsius scale (formerly called the centigrade scale) is the scale of choice for popular and commercial use and much scientific use. Celsius temperatures are measured in degrees, and the Celsius degree has the same size as the kelvin. However, the zero of the Celsius scale is shifted to a more convenient value than absolute zero. If T_C represents a Celsius temperature and T a Kelvin temperature, then

$$T_C = T - 273.15°. \tag{19-7}$$

In expressing temperatures on the Celsius scale, the degree symbol is commonly used. Thus, we write 20.00°C for a Celsius reading but 293.15 K for a Kelvin reading.

The Fahrenheit scale, used in the United States, employs a smaller degree than the Celsius scale and a different zero of temperature. You can easily verify both these differences by examining an ordinary room thermometer on which both scales are marked. The relation between the Celsius and Fahrenheit scales is

$$T_F = \tfrac{9}{5}T_C + 32°, \tag{19-8}$$

where T_F is Fahrenheit temperature. Transferring between these two scales can be done easily by remembering a few corresponding points, such as the freezing and boiling points of water (see Table 19-1). Figure 19-7 compares the Kelvin, Celsius, and Fahrenheit scales.

TABLE 19-1 Some Corresponding Temperatures

Temperature	°C	°F
Boiling point of water[a]	100	212
Normal body temperature	37.0	98.6
Accepted comfort level	20	68
Freezing point of water[a]	0	32
Zero of Fahrenheit scale	≈ −18	0
Scales coincide	−40	−40

[a]Strictly, the boiling point of water on the Celsius scale is 99.975°C, and the freezing point is 0.00°C. Thus, there is slightly less than 100 C° between those two points.

Fig. 19-7 The Kelvin, Celsius, and Fahrenheit temperature scales compared.

We use the letters C and F to distinguish measurements and degrees on the two scales. Thus,

$$0°C = 32°F$$

means that 0° on the Celsius scale measures the same temperature as 32° on the Fahrenheit scale, whereas

$$5\ C° = 9\ F°$$

means that a temperature difference of 5 Celsius degrees (note the degree symbol appears *after* C) is equivalent to a temperature difference of 9 Fahrenheit degrees.

Sample Problem 19-1

Suppose you come across old scientific notes that describe a temperature scale called Z on which the boiling point of water is 65.0°Z and the freezing point is −14.0°Z. To what temperature on the Fahrenheit scale would a temperature of $T = -98.0°Z$ correspond? Assume that the Z scale is linear; that is, the size of a Z degree is the same everywhere on the Z scale.

SOLUTION: One Key Idea here is to relate the given temperature T to *either* of the two known temperatures on the Z scale. Since $T = -98.0°Z$ is closer to the freezing point of −14.0°Z, we use that point for simplicity. Then we note that T is *below the freezing point* by the difference $-14.0°Z - (-98.0°Z) = 84.0\ Z°$ (Fig. 19-8). (Read this difference as "84.0 Z degrees.")

Another Key Idea is to set up a conversion factor between the Z and Fahrenheit scales to convert this difference. To do so, we use *both* known temperatures on the Z scale and the corresponding temperatures on the Fahrenheit scale. On the Z scale, the difference between the boiling and freezing points is $65.0°Z - (-14.0°Z) = 79.0\ Z°$. On the Fahrenheit scale, it is $212°F - 32.0°F = 180\ F°$. Thus, a temperature difference of 79.0 Z° is equivalent to a temperature difference of 180 F° (Fig. 19-8), and we can use the ratio (180 F°)/(79.0 Z°) as our conversion factor.

Now, since T is below the freezing point by 84.0 Z°, it must also be below the freezing point by

$$(84.0\ Z°)\ \frac{180\ F°}{79.0\ Z°} = 191\ F°.$$

Because the freezing point is at 32.0°F, this means that

$$T = 32.0°F - 191\ F° = -159°F. \qquad \text{(Answer)}$$

✔CHECKPOINT 1:   The figure here shows three temperature scales with the freezing and boiling points of water indicated. (a) Rank the degrees on these scales by size, greatest first. (b) Rank the following temperatures, highest first: 50°X, 50°W, and 50°Y.

Fig. 19-8 Sample Problem 19-1. An unknown temperature scale compared to the Fahrenheit temperature scale.

Tactic 1: *Temperature Changes*

Between the boiling and freezing points of water, there are (approximately) 100 kelvins and 100 Celsius degrees. Thus, a kelvin is the same size as a Celsius degree. From this or from Eq. 19-7, we then know that any temperature change is the same number whether expressed in kelvins or Celsius degrees. For example, a temperature change of 10 K is exactly equivalent to a temperature change of 10 C°.

Between the boiling and freezing points of water, there are 180 Fahrenheit degrees. Thus, 180 F° = 100 K, and a Fahrenheit degree must be 100/180, or 5/9, the size of a kelvin or Celsius degree. From this or from Eq. 19-8, we then know that any temperature change expressed in Fahrenheit degrees must be $\frac{9}{5}$ times that same temperature change expressed in either kelvins or Celsius degrees. For example, in Fahrenheit degrees, a temperature change of 10 K is (9/5)(10 K), or 18 F°.

You should take care not to confuse a *temperature* with a temperature *change* or *difference*. A temperature of 10 K is certainly not the same as one of 10°C or 18°F but, as above, a temperature *change* of 10 K is the same as one of 10 C° or 18 F°. This distinction is very important in an equation containing a temperature T instead of a temperature change or difference such as $T_2 - T_1$: A temperature T by itself should generally be in kelvins and not degrees Celsius or Fahrenheit. In short, beware the "bare T."

Fig. 19-9 Railroad tracks in Asbury Park, New Jersey, distorted because of thermal expansion on a very hot July day.

19-5 Thermal Expansion

You can often loosen a tight metal jar lid by holding it under a stream of hot water. Both the metal of the lid and the glass of the jar expand as the hot water adds energy to their atoms. (With the added energy, the atoms can move a bit farther from each other than usual, against the spring-like interatomic forces that hold every solid together.) However, because the atoms in the metal move farther apart than those in the glass, the lid expands more than the jar and thus is loosened.

Such **thermal expansion** is not always desirable, as Fig. 19-9 suggests. To preclude buckling, therefore, expansion slots are placed in bridges to accommodate roadway expansion on hot days. Dental materials used for fillings must be matched in their thermal expansion properties to those of tooth enamel (otherwise consuming hot coffee or cold ice cream would be quite painful). In aircraft manufacture, however, rivets and other fasteners are often cooled in dry ice before insertion and then allowed to expand to a tight fit.

Thermometers and thermostats may be based on the differences in expansion between the components of a *bimetal strip* (Fig. 19-10). Also, the familiar liquid-in-glass thermometers are based on the fact that liquids such as mercury and alcohol expand to a different (greater) extent than their glass containers.

Linear Expansion

If the temperature of a metal rod of length L is raised by an amount ΔT, its length is found to increase by an amount

$$\Delta L = L\alpha \, \Delta T, \tag{19-9}$$

in which α is a constant called the **coefficient of linear expansion.** The coefficient α has the unit "per degree" or "per kelvin" and depends on the material. Although

Brass

Steel

$T = T_0$

(a)

$T > T_0$

(b)

Fig. 19-10 (a) A bimetal strip, consisting of a strip of brass and a strip of steel welded together, at temperature T_0. (b) The strip bends as shown at temperatures above this reference temperature. Below the reference temperature the strip bends the other way. Many thermostats operate on this principle, making and breaking an electrical contact as the temperature rises and falls.

Fig. 19-11 The same steel ruler at two different temperatures. When it expands, the scale, the numbers, the thickness, and the diameters of the circle and circular hole are all increased by the same factor. (The expansion has been exaggerated for clarity.)

TABLE 19-2   Some Coefficients of Linear Expansion[a]

Substance	$\alpha\ (10^{-6}/C°)$	Substance	$\alpha\ (10^{-6}/C°)$
Ice (at 0°C)	51	Steel	11
Lead	29	Glass (ordinary)	9
Aluminum	23	Glass (Pyrex)	3.2
Brass	19	Diamond	1.2
Copper	17	Invar[b]	0.7
Concrete	12	Fused quartz	0.5

[a]Room temperature values except for the listing for ice.
[b]This alloy was designed to have a low coefficient of expansion. The word is a shortened form of "invariable."

α varies somewhat with temperature, for most practical purposes it can be taken as constant for a particular material. Table 19-2 shows some coefficients of linear expansion. Note that the unit C° there could be replaced with the unit K.

The thermal expansion of a solid is like (three-dimensional) photographic enlargement. Figure 19-11b shows the (exaggerated) expansion of a steel ruler after its temperature is increased from that of Fig. 19-11a. Equation 19-9 applies to every linear dimension of the ruler, including its edge, thickness, diagonals, and the diameters of the circle etched on it and the circular hole cut in it. If the disk cut from that hole originally fits snugly in the hole, it will continue to fit snugly if it undergoes the same temperature increase as the ruler.

Volume Expansion

If all dimensions of a solid expand with temperature, the volume of that solid must also expand. For liquids, volume expansion is the only meaningful expansion parameter. If the temperature of a solid or liquid whose volume is V is increased by an amount ΔT, the increase in volume is found to be

$$\Delta V = V\beta\,\Delta T, \qquad (19\text{-}10)$$

where β is the **coefficient of volume expansion** of the solid or liquid. The coefficients of volume expansion and linear expansion for a solid are related by

$$\beta = 3\alpha. \qquad (19\text{-}11)$$

The most common liquid, water, does not behave like other liquids. Above about 4°C, water expands as the temperature rises, as we would expect. Between 0 and about 4°C, however, water *contracts* with increasing temperature. Thus, at about 4°C, the density of water passes through a maximum. At all other temperatures, the density of water is less than this maximum value.

This behavior of water is the reason why lakes freeze from the top down rather than from the bottom up. As water on the surface is cooled from, say, 10°C toward the freezing point, it becomes denser ("heavier") than lower water and sinks to the bottom. Below 4°C, however, further cooling makes the water then on the surface *less* dense ("lighter") than the lower water, so it stays on the surface until it freezes. Thus the surface freezes while the lower water is still liquid. If lakes froze from the bottom up, the ice so formed would tend not to melt completely during the summer, because it would be insulated by the water above. After a few years, many bodies of open water in the temperate zones of Earth would be frozen solid all year round—and aquatic life as we know it could not exist.

CHECKPOINT 2: The figure here shows four rectangular metal plates, with sides of L, $2L$, or $3L$. They are all made of the same material, and their temperature is to be increased by the same amount. Rank the plates according to the expected increase in (a) their vertical heights and (b) their areas, greatest first.

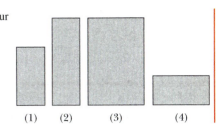

(1)   (2)   (3)   (4)

Sample Problem 19-2

On a hot day in Las Vegas, an oil trucker loaded 37,000 L of diesel fuel. He encountered cold weather on the way to Payson, Utah, where the temperature was 23.0 K lower than in Las Vegas, and where he delivered his entire load. How many liters did he deliver? The coefficient of volume expansion for diesel fuel is $9.50 \times 10^{-4}/C°$, and the coefficient of linear expansion for his steel truck tank is $11 \times 10^{-6}/C°$.

SOLUTION: The **Key Idea** here is that the volume of the diesel fuel depends directly on the temperature. Thus, because the temperature decreased, the volume of the fuel did also. From Eq. 19-10, the volume change is

$$\Delta V = V\beta \, \Delta T$$
$$= (37,000 \text{ L})(9.50 \times 10^{-4}/C°)(-23.0 \text{ K}) = -808 \text{ L}.$$

Thus, the amount delivered was

$$V_{del} = V + \Delta V = 37,000 \text{ L} - 808 \text{ L}$$
$$= 36,190 \text{ L}. \qquad \text{(Answer)}$$

Note that the thermal expansion of the steel tank has nothing to do with the problem. Question: Who paid for the "missing" diesel fuel?

19-6 Temperature and Heat

If you take a can of cola from the refrigerator and leave it on the kitchen table, its temperature will rise—rapidly at first but then more slowly—until the temperature of the cola equals that of the room (the two are then in thermal equilibrium). In the same way, the temperature of a cup of hot coffee, left sitting on the table, will fall until it also reaches room temperature.

In generalizing this situation, we describe the cola or the coffee as a *system* (with temperature T_S) and the relevant parts of the kitchen as the *environment* (with temperature T_E) of that system. Our observation is that if T_S is not equal to T_E, then T_S will change (T_E may also change some) until the two temperatures are equal and thus thermal equilibrium is reached.

Such a change in temperature is due to the transfer of energy between the thermal energy of the system and the system's environment. (*Thermal energy* is an internal energy that consists of the kinetic and potential energies associated with the random motions of the atoms, molecules, and other microscopic bodies within an object.) The transferred energy is called **heat** and is symbolized Q. Heat is *positive* when energy is transferred to a system's thermal energy from its environment (we say that heat is absorbed). Heat is *negative* when energy is transferred from a system's thermal energy to its environment (we say that heat is released or lost).

This transfer of energy is shown in Fig. 19-12. In the situation of Fig. 19-12a, in which $T_S > T_E$, energy is transferred from the system to the environment, so Q is negative. In Fig. 19-12b, in which $T_S = T_E$, there is no such transfer, Q is zero, and heat is neither released nor absorbed. In Fig. 19-12c, in which $T_S < T_E$, the transfer is to the system from the environment, so Q is positive.

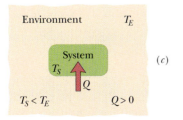

Fig. 19-12 If the temperature of a system exceeds that of its environment as in (a), heat Q is lost by the system to the environment until thermal equilibrium (b) is established. (c) If the temperature of the system is below that of the environment, heat is absorbed by the system until thermal equilibrium is established.

We are led then to this definition of heat:

> Heat is the energy that is transferred between a system and its environment because of a temperature difference that exists between them.

Recall that energy can also be transferred between a system and its environment as *work W* via a force acting on a system. Heat and work, unlike temperature, pressure, and volume, are not intrinsic properties of a system. They have meaning only as they describe the transfer of energy into or out of a system. Thus, it is proper to say: "During the last 3 min, 15 J of heat was transferred to the system from its environment" or "During the last minute, 12 J of work was done on the system by its environment." It is meaningless to say: "This system contains 450 J of heat" or "This system contains 385 J of work."

Before scientists realized that heat is transferred energy, heat was measured in terms of its ability to raise the temperature of water. Thus, the **calorie** (cal) was defined as the amount of heat that would raise the temperature of 1 g of water from 14.5°C to 15.5°C. In the British system, the corresponding unit of heat was the **British thermal unit** (Btu), defined as the amount of heat that would raise the temperature of 1 lb of water from 63°F to 64°F.

In 1948, the scientific community decided that since heat (like work) is transferred energy, the SI unit for heat should be the one we use for energy, namely, the **joule.** The calorie is now defined to be 4.1860 J (exactly), with no reference to the heating of water. (The "calorie" used in nutrition, sometimes called the Calorie (Cal), is really a kilocalorie.) The relations among the various heat units are

$$1 \text{ cal} = 3.969 \times 10^{-3} \text{ Btu} = 4.1860 \text{ J}. \tag{19-12}$$

19-7 The Absorption of Heat by Solids and Liquids

Heat Capacity

The **heat capacity** C of an object is the proportionality constant between the heat Q that the object absorbs or loses and the resulting temperature change ΔT of the object; that is,

$$Q = C \Delta T = C(T_f - T_i), \tag{19-13}$$

in which T_i and T_f are the initial and final temperatures of the object. Heat capacity C has the unit of energy per degree or energy per kelvin. The heat capacity C of, say, a marble slab used in a bun warmer might be 179 cal/C°, which we can also write as 179 cal/K or as 749 J/K.

The word "capacity" in this context is really misleading in that it suggests analogy with the capacity of a bucket to hold water. *That analogy is false,* and you should not think of the object as "containing" heat or being limited in its ability to absorb heat. Heat transfer can proceed without limit as long as the necessary temperature difference is maintained. The object may, of course, melt or vaporize during the process.

Specific Heat

Two objects made of the same material—say, marble—will have heat capacities proportional to their masses. It is therefore convenient to define a "heat capacity per

TABLE 19-3 Specific Heats of Some Substances at Room Temperature

Substance	Specific Heat		Molar Specific Heat
	cal / g·K	J / kg·K	J / mol·K
Elemental Solids			
Lead	0.0305	128	26.5
Tungsten	0.0321	134	24.8
Silver	0.0564	236	25.5
Copper	0.0923	386	24.5
Aluminum	0.215	900	24.4
Other Solids			
Brass	0.092	380	
Granite	0.19	790	
Glass	0.20	840	
Ice (−10°C)	0.530	2220	
Liquids			
Mercury	0.033	140	
Ethyl alcohol	0.58	2430	
Seawater	0.93	3900	
Water	1.00	4190	

unit mass" or **specific heat** c that refers not to an object but to a unit mass of the material of which the object is made. Equation 19-13 then becomes

$$Q = cm \, \Delta T = cm(T_f - T_i). \tag{19-14}$$

Through experiment we would find that although the heat capacity of a particular marble slab might be 179 cal/C° (or 749 J/K), the specific heat of marble itself (in that slab or in any other marble object) is 0.21 cal/g · C° (or 880 J/kg · K).

From the way the calorie and the British thermal unit were initially defined, the specific heat of water is

$$c = 1 \text{ cal/g} \cdot \text{C}° = 1 \text{ Btu/lb} \cdot \text{F}° = 4190 \text{ J/kg} \cdot \text{K}. \tag{19-15}$$

Table 19-3 shows the specific heats of some substances at room temperature. Note that the value for water is relatively high. The specific heat of any substance actually depends somewhat on temperature, but the values in Table 19-3 apply reasonably well in a range of temperatures near room temperature.

CHECKPOINT 3: A certain amount of heat Q will warm 1 g of material A by 3 C° and 1 g of material B by 4 C°. Which material has the greater specific heat?

Molar Specific Heat

In many instances the most convenient unit for specifying the amount of a substance is the mole (mol), where

$$1 \text{ mol} = 6.02 \times 10^{23} \text{ elementary units}$$

of *any* substance. Thus 1 mol of aluminum means 6.02×10^{23} atoms (the atom being the elementary unit), and 1 mol of aluminum oxide means 6.02×10^{23} molecules of the oxide (because the molecule is the elementary unit of a compound).

When quantities are expressed in moles, specific heats must also involve moles (rather than a mass unit); they are then called **molar specific heats.** Table 19-3 shows the values for some elemental solids (each consisting of a single element) at room temperature.

An Important Point

In determining and then using the specific heat of any substance, we need to know the conditions under which energy is transferred as heat. For solids and liquids, we usually assume that the sample is under constant pressure (usually atmospheric) during the transfer. It is also conceivable that the sample is held at constant volume while the heat is absorbed. This means that thermal expansion of the sample is prevented by applying external pressure. For solids and liquids, this is very hard to arrange experimentally but the effect can be calculated, and it turns out that the specific heats under constant pressure and constant volume for any solid or liquid differ usually by no more than a few percent. Gases, as you will see, have quite different values for their specific heats under constant-pressure conditions and under constant-volume conditions.

Heats of Transformation

When energy is absorbed as heat by a solid or liquid, the temperature of the sample does not necessarily rise. Instead, the sample may change from one *phase*, or *state*, to another. Matter can exist in three common states: In the *solid state*, the molecules of a sample are locked into a fairly rigid structure by their mutual attraction. In the

TABLE 19-4   Some Heats of Transformation

	Melting		Boiling	
Substance	Melting Point (K)	Heat of Fusion L_F (kJ/kg)	Boiling Point (K)	Heat of Vaporization L_V (kJ/kg)
Hydrogen	14.0	58.0	20.3	455
Oxygen	54.8	13.9	90.2	213
Mercury	234	11.4	630	296
Water	273	333	373	2256
Lead	601	23.2	2017	858
Silver	1235	105	2323	2336
Copper	1356	207	2868	4730

liquid state, the molecules have more energy and move about more. They may form brief clusters, but the sample does not have a rigid structure and can flow or settle into a container. In the *gas* or *vapor state,* the molecules have even more energy, are free of one another, and can fill up the full volume of a container.

To *melt* a solid means to change it from the solid state to the liquid state. The process requires energy because the molecules of the solid must be freed from their rigid structure. Melting an ice cube to form liquid water is a common example. To *freeze* a liquid to form a solid is the reverse of melting and requires that energy be removed from the liquid, so that the molecules can settle into a rigid structure.

To *vaporize* a liquid means to change it from the liquid state to the vapor or gas state. This process, like melting, requires energy because the molecules must be freed from their clusters. Boiling liquid water to transfer it to water vapor (or steam — a gas of individual water molecules) is a common example. *Condensing* a gas to form a liquid is the reverse of vaporizing; it requires that energy be removed from the gas, so that the molecules can cluster instead of flying away from one another.

The amount of energy per unit mass that must be transferred as heat when a sample completely undergoes a phase change is called the **heat of transformation** L. Thus, when a sample of mass m completely undergoes a phase change, the total energy transferred is

$$Q = Lm. \qquad (19\text{-}16)$$

When the phase change is from liquid to gas (then the sample must absorb heat) or from gas to liquid (then the sample must release heat), the heat of transformation is called the **heat of vaporization** L_V. For water at its normal boiling or condensation temperature,

$$L_V = 539 \text{ cal/g} = 40.7 \text{ kJ/mol} = 2256 \text{ kJ/kg}. \qquad (19\text{-}17)$$

When the phase change is from solid to liquid (then the sample must absorb heat) or from liquid to solid (then the sample must release heat), the heat of transformation is called the **heat of fusion** L_F. For water at its normal freezing or melting temperature,

$$L_F = 79.5 \text{ cal/g} = 6.01 \text{ kJ/mol} = 333 \text{ kJ/kg}. \qquad (19\text{-}18)$$

Table 19-4 shows the heats of transformation for some substances.

Sample Problem 19-3

(a) How much heat must be absorbed by ice of mass $m = 720$ g at $-10°C$ to take it to liquid state at $15°C$?

SOLUTION: The first Key Idea is that the heating process is accomplished in three steps.

Step 1. The Key Idea here is that the ice cannot melt at a temperature below the freezing point—so initially, any energy transferred to the ice as heat can only increase the temperature of the ice. The heat Q_1 needed to increase that temperature from the initial value $T_i = -10°C$ to a final value $T_f = 0°C$ (so that the ice can then melt) is given by Eq. 19-14 ($Q = cm \, \Delta T$). Using the specific heat of ice c_{ice} in Table 19-3 gives us

$$Q_1 = c_{ice}m(T_f - T_i)$$
$$= (2220 \text{ J/kg} \cdot \text{K})(0.720 \text{ kg})[0°C - (-10°C)]$$
$$= 15{,}984 \text{ J} \approx 15.98 \text{ kJ}.$$

Step 2. The next Key Idea is that the temperature cannot increase from 0°C until all the ice melts—so any energy transferred to the ice as heat now can only change ice to liquid water. The heat Q_2 needed to melt all the ice is given by Eq. 19-16 ($Q = Lm$). Here L is the heat of fusion L_F, with the value given in Eq. 19-18 and Table 19-4. We find

$$Q_2 = L_F m = (333 \text{ kJ/kg})(0.720 \text{ kg}) \approx 239.8 \text{ kJ}.$$

Step 3. Now we have liquid water at 0°C. The next Key Idea is that the energy transferred to the liquid water as heat now can only increase the temperature of the liquid water. The heat Q_3 needed to increase the temperature of the water from the initial value $T_i = 0°C$ to the final value $T_f = 15°C$ is given by Eq. 19-14 (with the specific heat of liquid water c_{liq}):

$$Q_3 = c_{liq}m(T_f - T_i)$$
$$= (4190 \text{ J/kg} \cdot \text{K})(0.720 \text{ kg})(15°C - 0°C)$$
$$= 45{,}252 \text{ J} \approx 45.25 \text{ kJ}.$$

The total required heat Q_{tot} is the sum of the amounts required in the three steps:

$$Q_{tot} = Q_1 + Q_2 + Q_3$$
$$= 15.98 \text{ kJ} + 239.8 \text{ kJ} + 45.25 \text{ kJ}$$
$$\approx 300 \text{ kJ}. \qquad \text{(Answer)}$$

Note that the heat required to melt the ice is much greater than the heat required to raise the temperature of either the ice or the liquid water.

(b) If we supply the ice with a total energy of only 210 kJ (as heat), what then are the final state and temperature of the water?

SOLUTION: From step 1, we know that 15.98 kJ is needed to raise the temperature of the ice to the melting point. The remaining heat Q_{rem} is then 210 kJ − 15.98 kJ, or about 194 kJ. From step 2, we can see that this amount of heat is insufficient to melt all the ice. Then this Key Idea becomes important: Because the melting of the ice is incomplete, we must end up with a mixture of ice and liquid; the temperature of the mixture must be the freezing point, 0°C.

We can find the mass m of ice that is melted by the available energy Q_{rem} by using Eq. 19-16 with L_F:

$$m = \frac{Q_{rem}}{L_F} = \frac{194 \text{ kJ}}{333 \text{ kJ/kg}} = 0.583 \text{ kg} \approx 580 \text{ g}.$$

Thus, the mass of the ice that remains is 720 g − 580 g, or 140 g, and we have

$$580 \text{ g water} \quad \text{and} \quad 140 \text{ g ice}, \quad \text{at } 0°C. \qquad \text{(Answer)}$$

Sample Problem 19-4

A copper slug whose mass m_c is 75 g is heated in a laboratory oven to a temperature T of 312°C. The slug is then dropped into a glass beaker containing a mass $m_w = 220$ g of water. The heat capacity C_b of the beaker is 45 cal/K. The initial temperature T_i of the water and the beaker is 12°C. Assuming that the slug, beaker, and water are an isolated system and the water does not vaporize, find the final temperature T_f of the system at thermal equilibrium.

SOLUTION: One Key Idea here is that, with the system isolated, only internal transfers of energy can occur. There are three such transfers, all as heat. The slug loses energy, the water gains energy, and the beaker gains energy. Another Key Idea is that, because these transfers do not involve a phase change, the energy transfers can only change the temperatures. To relate the transfers to the temperature changes, we can use Eqs. 19-13 and 19-14 to write

$$\text{for the water:} \quad Q_w = c_w m_w(T_f - T_i); \qquad (19\text{-}19)$$
$$\text{for the beaker:} \quad Q_b = C_b(T_f - T_i); \qquad (19\text{-}20)$$
$$\text{for the copper:} \quad Q_c = c_c m_c(T_f - T). \qquad (19\text{-}21)$$

A third Key Idea is that, with the system isolated, the total energy of the system cannot change. This means that the sum of these three energy transfers is zero:

$$Q_w + Q_b + Q_c = 0. \qquad (19\text{-}22)$$

Substituting Eqs. 19-19 through 19-21 into Eq. 19-22 yields

$$c_w m_w(T_f - T_i) + C_b(T_f - T_i) + c_c m_c(T_f - T) = 0. \qquad (19\text{-}23)$$

Temperatures are contained in Eq. 19-23 only as differences. Thus, because the differences on the Celsius and Kelvin scales are identical, we can use either of those scales in this equation. Solving it for T_f, we obtain

$$T_f = \frac{c_c m_c T + C_b T_i + c_w m_w T_i}{c_w m_w + C_b + c_c m_c}.$$

Using Celsius temperatures and taking values for c_c and c_w from Table 19-3, we find the numerator to be

$$(0.0923 \text{ cal/g} \cdot \text{K})(75 \text{ g})(312°C) + (45 \text{ cal/K})(12°C)$$
$$+ (1.00 \text{ cal/g} \cdot \text{K})(220 \text{ g})(12°C) = 5339.8 \text{ cal},$$

and the denominator to be

$$(1.00 \text{ cal/g} \cdot \text{K})(220 \text{ g}) + 45 \text{ cal/K}$$
$$+ (0.0923 \text{ cal/g} \cdot \text{K})(75 \text{ g}) = 271.9 \text{ cal/C°}.$$

We then have

$$T_f = \frac{5339.8 \text{ cal}}{271.9 \text{ cal/C°}} = 19.6°C \approx 20°C. \qquad \text{(Answer)}$$

From the given data you can show that

$$Q_w \approx 1670 \text{ cal}, \qquad Q_b \approx 342 \text{ cal}, \qquad Q_c \approx -2020 \text{ cal}.$$

Apart from rounding errors, the algebraic sum of these three heat transfers is indeed zero, as Eq. 19-22 requires.

19-8 A Closer Look at Heat and Work

Here we look in some detail at how energy can be transferred as heat and work between a system and its environment. Let us take as our system a gas confined to a cylinder with a movable piston, as in Fig. 19-13. The upward force on the piston due to the pressure of the confined gas is equal to the weight of lead shot loaded onto the top of the piston. The walls of the cylinder are made of insulating material that does not allow any transfer of energy as heat. The bottom of the cylinder rests on a reservoir for thermal energy, a *thermal reservoir* (perhaps a hot plate) whose temperature T you can control by turning a knob.

The system (the gas) starts from an *initial state i*, described by a pressure p_i, a volume V_i, and a temperature T_i. You want to change the system to a *final state f*, described by a pressure p_f, a volume V_f, and a temperature T_f. The procedure by which you change the system from its initial state to its final state is called a *thermodynamic process*. During such a process, energy may be transferred into the system from the thermal reservoir (positive heat) or vice versa (negative heat). Also, work can be done by the system to raise the loaded piston (positive work) or lower it (negative work). We assume that all such changes occur slowly, with the result that the system is always in (approximate) thermal equilibrium (that is, every part of the system is always in thermal equilibrium with every other part).

Suppose that you remove a few lead shot from the piston of Fig. 19-13, allowing the gas to push the piston and remaining shot upward through a differential displacement $d\vec{s}$ with an upward force $\vec{F}$. Since the displacement is tiny, we can assume that $\vec{F}$ is constant during the displacement. Then $\vec{F}$ has a magnitude that is equal to pA, where p is the pressure of the gas and A is the face area of the piston. The differential work dW done by the gas during the displacement is

$$dW = \vec{F} \cdot d\vec{s} = (pA)(ds) = p(A\ ds)$$
$$= p\ dV, \tag{19-24}$$

in which dV is the differential change in the volume of the gas owing to the movement of the piston. When you have removed enough shot to allow the gas to change its volume from V_i to V_f, the total work done by the gas is

$$W = \int dW = \int_{V_i}^{V_f} p\ dV. \tag{19-25}$$

During the change in volume, the pressure and temperature of the gas may also change. To evaluate the integral in Eq. 19-25 directly, we would need to know how pressure varies with volume for the actual process by which the system changes from state i to state f.

There are actually many ways to take the gas from state i to state f. One way is shown in Fig. 19-14a, which is a plot of the pressure of the gas versus its volume and which is called a *p-V diagram*. In Fig. 19-14a, the curve indicates that the pressure decreases as the volume increases. The integral Eq. 19-25 (and thus the work W done by the gas) is represented by the shaded area under the curve between points i and f. Regardless of what exactly we do to take the gas along the curve, that work is positive, owing to the fact that the gas increases its volume by forcing the piston upward.

Another way to get from state i to state f is shown in Fig. 19-14b. There the change takes place in two steps—the first from state i to state a, and the second from state a to state f.

Fig. 19-13 A gas is confined to a cylinder with a movable piston. Heat Q can be added to, or withdrawn from, the gas by regulating the temperature T of the adjustable thermal reservoir. Work W can be done by the gas by raising or lowering the piston.

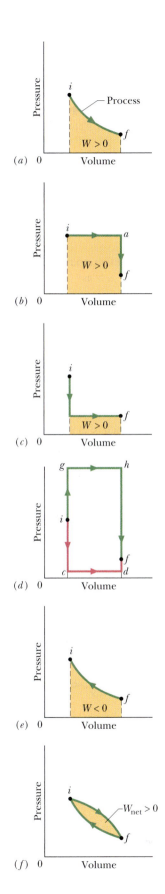

Fig. 19-14 (*a*) The shaded area represents the work *W* done by a system as it goes from an initial state *i* to a final state *f*. Work *W* is positive because the system's volume increases. (*b*) *W* is still positive, but now greater. (*c*) *W* is still positive, but now smaller. (*d*) *W* can be even smaller (path *icdf*) or larger (path *ighf*). (*e*) Here the system goes from state *f* to state *i*, as the gas is compressed to less volume by an external force. The work *W* done *by* the system is now negative. (*f*) The net work W_{net} done by the system during a complete cycle is represented by the shaded area.

Step *ia* of this process is carried out at constant pressure, which means that you leave undisturbed the lead shot that ride on top of the piston in Fig. 19-13. You cause the volume to increase (from V_i to V_f) by slowly turning up the temperature control knob, raising the temperature of the gas to some higher value T_a. (Increasing the temperature increases the force from the gas on the piston, moving it upward.) During this step, positive work is done by the expanding gas (to lift the loaded piston) and heat is absorbed by the system from the thermal reservoir (in response to the arbitrarily small temperature differences that you create as you turn up the temperature). This heat is positive because it is added to the system.

Step *af* of the process of Fig. 19-14*b* is carried out at constant volume, so you must wedge the piston, preventing it from moving. Then as you use the control knob to decrease the temperature, you find that the pressure drops from p_a to its final value p_f. During this step, heat is lost by the system to the thermal reservoir.

For the overall process *iaf*, the work *W*, which is positive and is carried out only during step *ia*, is represented by the shaded area under the curve. Energy is transferred as heat during both steps *ia* and *af*, with a net energy transfer *Q*.

Figure 19-14*c* shows a process in which the previous two steps are carried out in reverse order. The work *W* in this case is smaller than for Fig. 19-14*b*, as is the net heat absorbed. Figure 19-14*d* suggests that you can make the work done by the gas as small as you want (by following a path like *icdf*) or as large as you want (by following a path like *ighf*).

To sum up: A system can be taken from a given initial state to a given final state by an infinite number of processes. Heat may or may not be involved, and in general, the work *W* and the heat *Q* will have different values for different processes. We say that heat and work are *path-dependent* quantities.

Figure 19-14*e* shows an example in which negative work is done by a system as some external force compresses the system, reducing its volume. The absolute value of the work done is still equal to the area beneath the curve, but because the gas is *compressed*, the work done by the gas is negative.

Figure 19-14*f* shows a *thermodynamic cycle* in which the system is taken from some initial state *i* to some other state *f* and then back to *i*. The net work done by the system during the cycle is the sum of the *positive* work done during the expansion and the *negative* work done during the compression. In Fig. 19-14*f*, the net work is positive because the area under the expansion curve (*i* to *f*) is greater than the area under the compression curve (*f* to *i*).

✔**CHECKPOINT 4:**   The *p-V* diagram here shows six curved paths (connected by vertical paths) that can be followed by a gas. Which two of them should be part of a closed cycle if the net work done by the gas is to be at its maximum positive value?

19-9   The First Law of Thermodynamics

You have just seen that when a system changes from a given initial state to a given final state, both the work W and the heat Q depend on the nature of the process. Experimentally, however, we find a surprising thing. *The quantity $Q - W$ is the same for all processes.* It depends only on the initial and final states and does not depend at all on how the system gets from one to the other. All other combinations of Q and W, including Q alone, W alone, $Q + W$, and $Q - 2W$, are *path dependent;* only the quantity $Q - W$ is not.

The quantity $Q - W$ must represent a change in some intrinsic property of the system. We call this property the *internal energy E_{int}* and we write

$$\Delta E_{int} = E_{int,f} - E_{int,i} = Q - W \qquad \text{(first law).} \qquad (19\text{-}26)$$

Equation 19-26 is the **first law of thermodynamics.** If the thermodynamic system undergoes only a differential change, we can write the first law as*

$$dE_{int} = dQ - dW \qquad \text{(first law).} \qquad (19\text{-}27)$$

> The internal energy E_{int} of a system tends to increase if energy is added as heat Q and tends to decrease if energy is lost as work W done by the system.

In Chapter 8, we discussed the principle of energy conservation as it applies to isolated systems—that is, to systems in which no energy enters or leaves the system. The first law of thermodynamics is an extension of that principle to systems that are *not* isolated. In such cases, energy may be transferred into or out of the system as either work W or heat Q. In our statement of the first law of thermodynamics above, we assume that there are no changes in the kinetic energy or the potential energy of the system as a whole; that is, $\Delta K = \Delta U = 0$.

Before this chapter, the term *work* and the symbol W always meant the work done *on* a system. However, starting with Eq. 19-24 and continuing through the next two chapters about thermodynamics, we focus on the work done *by* a system, such as the gas in Fig. 19-13.

The work done *on* a system is always the negative of the work done *by* the system, so if we rewrite Eq. 19-26 in terms of the work W_{on} done *on* the system, we have $\Delta E_{int} = Q + W_{on}$. This tells us the following: The internal energy of a system tends to increase if heat is absorbed by the system or if positive work is done *on* the system. Conversely, the internal energy tends to decrease if heat is lost by the system or if negative work is done *on* the system.

✓CHECKPOINT 5:   The figure here shows four paths on a p-V diagram along which a gas can be taken from state i to state f. Rank the paths according to (a) the change ΔE_{int}, (b) the work W done by the gas, and (c) the magnitude of the energy transferred as heat Q, greatest first.

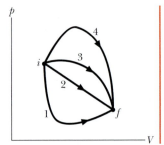

*Here dQ and dW, unlike dE_{int}, are not true differentials; that is, there are no such functions as $Q(p, V)$ and $W(p, V)$ that depend only on the state of the system. The quantities dQ and dW are called *inexact differentials* and are usually represented by the symbols $đQ$ and $đW$. For our purposes, we can treat them simply as infinitesimally small energy transfers.

19-10 Some Special Cases of the First Law of Thermodynamics

Here we look at four different thermodynamic processes, in each of which a certain restriction is imposed on the system. We then see what consequences follow when we apply the first law of thermodynamics to the process. The results are summarized in Table 19-5.

1. **Adiabatic processes.** An adiabatic process is one that occurs so rapidly or occurs in a system that is so well insulated that *no transfer of energy as heat* occurs between the system and its environment. Putting $Q = 0$ in the first law (Eq. 19-26) yields

$$\Delta E_{int} = -W \qquad \text{(adiabatic process)}. \qquad (19\text{-}28)$$

This tells us that if work is done *by* the system (that is, if W is positive), the internal energy of the system decreases by the amount of work. Conversely, if work is done *on* the system (that is, if W is negative), the internal energy of the system increases by that amount.

Figure 19-15 shows an idealized adiabatic process. Heat cannot enter or leave the system because of the insulation. Thus, the only way energy can be transferred between the system and its environment is by work. If we remove shot from the piston and allow the gas to expand, the work done by the system (the gas) is positive and the internal energy of the gas decreases. If, instead, we add shot and compress the gas, the work done by the system is negative and the internal energy of the gas increases.

2. **Constant-volume processes.** If the volume of a system (such as a gas) is held constant, that system can do no work. Putting $W = 0$ in the first law (Eq. 19-26) yields

$$\Delta E_{int} = Q \qquad \text{(constant-volume process)}. \qquad (19\text{-}29)$$

Thus, if heat is absorbed by a system (that is, if Q is positive), the internal energy of the system increases. Conversely, if heat is lost during the process (that is, if Q is negative), the internal energy of the system must decrease.

3. **Cyclical processes.** There are processes in which, after certain interchanges of heat and work, the system is restored to its initial state. In that case, no intrinsic property of the system—including its internal energy—can possibly change. Putting $\Delta E_{int} = 0$ in the first law (Eq. 19-26) yields

$$Q = W \qquad \text{(cyclical process)}. \qquad (19\text{-}30)$$

Thus, the net work done during the process must exactly equal the net amount of energy transferred as heat; the store of internal energy of the system remains unchanged. Cyclical processes form a closed loop on a p-V plot, as shown in Fig. 19-14f. We shall discuss such processes in some detail in Chapter 21.

Fig. 19-15 An adiabatic expansion can be carried out by slowly removing lead shot from the top of the piston. Adding lead shot reverses the process at any stage.

TABLE 19-5 **The First Law of Thermodynamics: Four Special Cases**

The Law: $\Delta E_{int} = Q - W$ (Eq. 19-26)		
Process	Restriction	Consequence
Adiabatic	$Q = 0$	$\Delta E_{int} = -W$
Constant volume	$W = 0$	$\Delta E_{int} = Q$
Closed cycle	$\Delta E_{int} = 0$	$Q = W$
Free expansion	$Q = W = 0$	$\Delta E_{int} = 0$

Fig. 19-16 The initial stage of a free-expansion process. After the stopcock is opened, the gas fills both chambers and eventually reaches an equilibrium state.

4. *Free expansions.* These are adiabatic processes in which no transfer of heat occurs between the system and its environment and no work is done on or by the system. Thus, $Q = W = 0$ and the first law requires that

$$\Delta E_{int} = 0 \qquad \text{(free expansion).} \qquad (19\text{-}31)$$

Figure 19-16 shows how such an expansion can be carried out. A gas, which is in thermal equilibrium within itself, is initially confined by a closed stopcock to one half of an insulated double chamber; the other half is evacuated. The stopcock is opened, and the gas expands freely to fill both halves of the chamber. No heat is transferred to or from the gas because of the insulation. No work is done by the gas because it rushes into a vacuum and thus does not meet any pressure.

A free expansion differs from all other processes we have considered because it cannot be done slowly and in a controlled way. As a result, at any given instant during the sudden expansion, the gas is not in thermal equilibrium and its pressure is not the same everywhere. Therefore, although we can plot the initial and final states on a p-V diagram, we cannot plot the expansion itself.

✓CHECKPOINT 6: For one complete cycle as shown in the p-V diagram here, are (a) ΔE_{int} for the gas and (b) the net energy transferred as heat Q positive, negative, or zero?

Sample Problem 19-5

Let 1.00 kg of liquid water at 100°C be converted to steam at 100°C by boiling at standard atmospheric pressure (which is 1.00 atm or 1.01×10^5 Pa) in the arrangement of Fig. 19-17. The volume of that water changes from an initial value of 1.00×10^{-3} m³ as a liquid to 1.671 m³ as steam.

(a) How much work is done by the system during this process?

SOLUTION: The Key Idea here is that the system must do positive work because the volume increases. In the general case we would calculate the work W done by integrating the pressure with respect to the volume (Eq. 19-25). However, here the pressure is constant at 1.01×10^5 Pa, so we can take p outside the integral. We then have

$$W = \int_{V_i}^{V_f} p \, dV = p \int_{V_i}^{V_f} dV = p(V_f - V_i)$$

$$= (1.01 \times 10^5 \text{ Pa})(1.671 \text{ m}^3 - 1.00 \times 10^{-3} \text{ m}^3)$$

$$= 1.69 \times 10^5 \text{ J} = 169 \text{ kJ.} \qquad \text{(Answer)}$$

(b) How much energy is transferred as heat during the process?

SOLUTION: The Key Idea here is that the heat causes only a phase change and not a change in temperature, so it is given fully by Eq. 19-16 ($Q = Lm$). Because the change is from liquid to gaseous phase, L is the heat of vaporization L_V, with the value given in Eq. 19-17 and Table 19-4. We find

$$Q = L_V m = (2256 \text{ kJ/kg})(1.00 \text{ kg})$$

$$= 2256 \text{ kJ} \approx 2260 \text{ kJ.} \qquad \text{(Answer)}$$

(c) What is the change in the system's internal energy during the process?

Fig. 19-17 Sample Problem 19-5. Water boiling at constant pressure. Energy is transferred from the thermal reservoir as heat until the liquid water has changed completely into steam. Work is done by the expanding gas as it lifts the loaded piston.

SOLUTION: The Key Idea here is that the change in the system's internal energy is related to the heat (here, this is energy transferred into the system) and the work (here, this is energy transferred out of the system) by the first law of thermodynamics (Eq. 19-26). Thus, we can write

$$\Delta E_{int} = Q - W = 2256 \text{ kJ} - 169 \text{ kJ}$$

$$\approx 2090 \text{ kJ} = 2.09 \text{ MJ.} \qquad \text{(Answer)}$$

This quantity is positive, indicating that the internal energy of the system has increased during the boiling process. This energy goes into separating the H_2O molecules, which strongly attract each other in the liquid state. We see that, when water is boiled, about 7.5% ($= 169$ kJ/2260 kJ) of the heat goes into the work of pushing back the atmosphere. The rest of the heat goes into the system's internal energy.

19-11 Heat Transfer Mechanisms

We have discussed the transfer of energy as heat between a system and its environment, but we have not yet described how that transfer takes place. There are three transfer mechanisms: conduction, convection, and radiation.

Conduction

If you leave the end of a metal poker in a fire for enough time, its handle will get hot. Energy is transferred from the fire to the handle by (thermal) **conduction** along the length of the poker. The vibration amplitudes of the atoms and electrons of the metal at the fire end of the poker become relatively large because of the high temperature of their environment. These increased vibrational amplitudes, and thus the associated energy, are passed along the poker, from atom to atom, during collisions between adjacent atoms. In this way, a region of rising temperature extends itself along the poker to the handle.

Consider a slab of face area A and thickness L, whose faces are maintained at temperatures T_H and T_C by a hot reservoir and a cold reservoir, as in Fig. 19-18. Let Q be the energy that is transferred as heat through the slab, from its hot face to its cold face, in time t. Experiment shows that the *conduction rate* P_{cond} (the amount of energy transferred per unit time) is

$$P_{cond} = \frac{Q}{t} = kA \frac{T_H - T_C}{L}, \qquad (19\text{-}32)$$

in which k, called the *thermal conductivity*, is a constant that depends on the material of which the slab is made. A material that readily transfers energy by conduction is a *good thermal conductor* and has a high value of k. Table 19-6 gives the thermal conductivities of some common metals, gases, and building materials.

Thermal Resistance to Conduction (R-Value)

If you are interested in insulating your house or in keeping cola cans cold on a picnic, you are more concerned with poor heat conductors than with good ones. For this reason, the concept of *thermal resistance R* has been introduced into engineering practice. The R-value of a slab of thickness L is defined as

$$R = \frac{L}{k}. \qquad (19\text{-}33)$$

The lower the thermal conductivity of the material of which a slab is made, the higher the R-value of the slab, so something that has a high R-value is a *poor thermal conductor* and thus a *good thermal insulator*.

Note that R is a property attributed to a slab of a specified thickness, not to a material. The commonly used unit for R (which, in the United States at least, is almost never stated) is the square foot–Fahrenheit degree–hour per British thermal unit (ft$^2 \cdot$ F$^\circ \cdot$ h/Btu). (Now you know why the unit is rarely stated.)

Conduction Through a Composite Slab

Figure 19-19 shows a composite slab, consisting of two materials having different thicknesses L_1 and L_2 and different thermal conductivities k_1 and k_2. The temperatures of the outer surfaces of the slab are T_H and T_C. Each face of the slab has area A. Let us derive an expression for the conduction rate through the slab under the assumption that the transfer is a *steady-state* process; that is, the temperatures everywhere in the slab and the rate of energy transfer do not change with time.

$T_H > T_C$

Fig. 19-18 Thermal conduction. Energy is transferred as heat from a reservoir at temperature T_H to a cooler reservoir at temperature T_C through a conducting slab of thickness L and thermal conductivity k.

TABLE 19-6 Some Thermal Conductivities[a]

Substance	k (W/m $\cdot$ K)
Metals	
Stainless steel	14
Lead	35
Aluminum	235
Copper	401
Silver	428
Gases	
Air (dry)	0.026
Helium	0.15
Hydrogen	0.18
Building Materials	
Polyurethane form	0.024
Rock wool	0.043
Fiberglass	0.048
White pine	0.11
Window glass	1.0

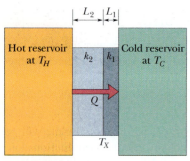

Fig. 19-19 Heat is transferred at a steady rate through a composite slab made up of two different materials with different thicknesses and different thermal conductivities. The steady-state temperature at the interface of the two materials is T_X.

In the steady state, the conduction rates through the two materials must be equal. This is the same as saying that the energy transferred through one material in a certain time must be equal to that transferred through the other material in the same time. If this were not true, temperatures in the slab would be changing and we would not have a steady-state situation. Letting T_X be the temperature of the interface between the two materials, we can now use Eq. 19-32 to write

$$P_{\text{cond}} = \frac{k_2 A(T_H - T_X)}{L_2} = \frac{k_1 A(T_X - T_C)}{L_1}. \tag{19-34}$$

Solving Eq. 19-34 for T_X yields, after a little algebra,

$$T_X = \frac{k_1 L_2 T_C + k_2 L_1 T_H}{k_1 L_2 + k_2 L_1}. \tag{19-35}$$

Substituting this expression for T_X into either equality of Eq. 19-34 yields

$$P_{\text{cond}} = \frac{A(T_H - T_C)}{L_1/k_1 + L_2/k_2}. \tag{19-36}$$

We can extend Eq. 19-36 to apply to any number n of materials making up a slab:

$$P_{\text{cond}} = \frac{A(T_H - T_C)}{\Sigma \, (L/k)}. \tag{19-37}$$

The summation sign in the denominator tells us to add the values of L/k for all the materials.

✔**CHECKPOINT 7:**   The figure shows the face and interface temperatures of a composite slab consisting of four materials, of identical thicknesses, through which the heat transfer is steady. Rank the materials according to their thermal conductivities, greatest first.

Convection

When you look at the flame of a candle or a match, you are watching thermal energy being transported upward by **convection.** Such energy transfer occurs when a fluid, such as air or water, comes in contact with an object whose temperature is higher than that of the fluid. The temperature of the part of the fluid that is in contact with the hot object increases, and (in most cases) that fluid expands and thus becomes less dense. Because this expanded fluid is now lighter than the surrounding cooler fluid, buoyant forces cause it to rise. Some of the surrounding cooler fluid then flows so as to take the place of the rising warmer fluid, and the process can then continue.

Convection is part of many natural processes. Atmospheric convection plays a fundamental role in determining global climate patterns and daily weather variations. Glider pilots and birds alike seek rising thermals (convection currents of warm air) that keep them aloft. Huge energy transfers take place within the oceans by the same process. Finally, energy is transported to the surface of the Sun from the nuclear furnace at its core by enormous cells of convection, in which hot gas rises to the surface along the cell core and cooler gas around the core descends below the surface.

Radiation

The third method by which an object and its environment can exchange energy as heat is via electromagnetic waves (visible light is one kind of electromagnetic wave).

Fig. 19-20 A false-color thermogram reveals the rate at which energy is radiated by houses along a street. The rates, from largest to smallest, are color coded as white, red, pink, blue, and black. You can tell where there is insulation in the walls, a heavy curtain over a window, and a higher air temperature at the ceiling on the second floor.

Energy transferred in this way is often called **thermal radiation** to distinguish it from electromagnetic *signals* (as in, say, television broadcasts) and from nuclear radiation (energy and particles emitted by nuclei). (To "radiate" generally means to emit.) When you stand in front of a big fire, you are warmed by absorbing thermal radiation from the fire; that is, your thermal energy increases as the fire's thermal energy decreases. No medium is required for heat transfer via radiation—the radiation can travel through vacuum from, say, the Sun to you.

The rate P_{rad} at which an object emits energy via electromagnetic radiation depends on the object's surface area A and the temperature T of that area in kelvins and is given by

$$P_{rad} = \sigma \varepsilon A T^4. \tag{19-38}$$

Here $\sigma = 5.6703 \times 10^{-8}$ W/m$^2 \cdot$ K^4 is called the *Stefan–Boltzmann constant* after Josef Stefan (who discovered Eq. 19-38 experimentally in 1879) and Ludwig Boltzmann (who derived it theoretically soon after). The symbol ε represents the *emissivity* of the object's surface, which has a value between 0 and 1, depending on the composition of the surface. A surface with the maximum emissivity of 1.0 is said to be a *blackbody radiator,* but such a surface is an ideal limit and does not occur in nature. Note again that the temperature in Eq. 19-38 must be in kelvins so that a temperature of absolute zero corresponds to no radiation. Note also that every object whose temperature is above 0 K—including you—emits thermal radiation. (See Fig. 19-20.)

The rate P_{abs} at which an object absorbs energy via thermal radiation from its environment, which we take to be at uniform temperature T_{env} (in kelvins), is

$$P_{abs} = \sigma \varepsilon A T_{env}^4. \tag{19-39}$$

The emissivity ε in Eq. 19-39 is the same as that in Eq. 19-38. An idealized blackbody radiator, with $\varepsilon = 1$, will absorb all the radiated energy it intercepts (rather than sending a portion back away from itself through reflection or scattering).

Because an object will radiate energy to the environment while it absorbs energy from the environment, the object's net rate P_{net} of energy exchange due to thermal radiation is

$$P_{net} = P_{abs} - P_{rad} = \sigma \varepsilon A (T_{env}^4 - T^4). \tag{19-40}$$

P_{net} is positive if net energy is being absorbed via radiation, and negative if it is being lost via radiation.

Sample Problem 19-6

Figure 19-21 shows the cross section of a wall made of white pine of thickness L_a and brick of thickness L_d ($= 2.0L_a$), sandwiching two layers of unknown material with identical thicknesses and thermal conductivities. The thermal conductivity of the pine is k_a and that of the brick is k_d ($= 5.0k_a$). The face area A of the wall is unknown. Thermal conduction through the wall has reached the steady state; the only known interface temperatures are $T_1 = 25°C$, $T_2 = 20°C$, and $T_5 = -10°C$. What is interface temperature T_4?

SOLUTION: One Key Idea here is that temperature T_4 helps determine the rate P_d at which energy is conducted through the brick, as given by Eq. 19-32. However, we lack enough data to solve Eq. 19-32 for T_4. A second Key Idea is that because the conduction is steady, the conduction rate P_d through the brick must equal the conduction rate P_a through the pine. From Eq. 19-32 and Fig. 19-21, we can write

$$P_a = k_aA\frac{T_1 - T_2}{L_a} \quad \text{and} \quad P_d = k_dA\frac{T_4 - T_5}{L_d}.$$

Setting $P_a = P_d$ and solving for T_4 yield

$$T_4 = \frac{k_aL_d}{k_dL_a}(T_1 - T_2) + T_5.$$

(a)   (b)   (c)   (d)

Fig. 19-21 Sample Problem 19-6. A wall of four layers through which there is steady-state heat transfer.

Letting $L_d = 2.0L_a$ and $k_d = 5.0k_a$, and inserting the known temperatures, we find

$$T_4 = \frac{k_a(2.0L_a)}{(5.0k_a)L_a}(25°C - 20°C) + (-10°C)$$
$$= -8.0°C. \quad \text{(Answer)}$$

Sample Problem 19-7

When hundreds of Japanese bees form a compact ball around a giant hornet that attempts to invade their hive, they can quickly raise their body temperature from the normal 35°C to 47°C or 48°C. That higher temperature is lethal to the hornet but not to the bees (Fig. 19-22). Assume the following: 500 bees form a ball of radius $R = 2.0$ cm for a time $t = 20$ min, the primary loss of energy by the ball is by thermal radiation, the ball's surface has emissivity $\varepsilon = 0.80$, and the ball has a uniform temperature. On average, how much additional energy must each bee produce during the 20 min to maintain 47°C?

SOLUTION: The Key Idea here is that, because the surface temperature of the bee ball increases after the ball forms, the rate at which energy is radiated by the ball also increases. Thus, the bees lose an additional amount of energy to thermal radiation. We can relate the surface temperature to the rate of radiation (energy per unit time)

Fig. 19-22 Sample Problem 19-7. The bees were unharmed by their increased body temperature, which the hornet could not withstand.

with Eq. 19-38 ($P_{rad} = \sigma\varepsilon AT^4$), in which A is the ball's surface area and T is the ball's surface temperature in kelvins. This rate is an energy per unit time; that is,

$$P_{rad} = \frac{E}{t}.$$

Thus, the amount of energy E radiated in time t is $E = P_{rad}t$.

At the normal temperature $T_1 = 35°C$, the radiation rate would be P_{r1} and the amount of energy radiated in time t would be $E_1 = P_{r1}t$. At the increased temperature $T_2 = 47°C$, the (greater) radiation rate is P_{r2} and the (greater) amount of energy radiated in time t is $E_2 = P_{r2}t$. Thus, in maintaining the ball at T_2 for time t, the bees must (together) provide an additional energy of $\Delta E = E_2 - E_1$.

We can now write

$$\Delta E = E_2 - E_1 = P_{r2}t - P_{r1}t$$
$$= (\sigma\varepsilon AT_2^4)t - (\sigma\varepsilon AT_1^4)t = \sigma\varepsilon At(T_2^4 - T_1^4). \quad (19\text{-}41)$$

The temperatures here *must* be in kelvins; thus, we write them as

$$T_2 = 47°C + 273\ C° = 320\ K$$

and

$$T_1 = 35°C + 273\ C° = 308\ K.$$

The surface area A of the ball is

$$A = 4\pi R^2 = (4\pi)(0.020\ \text{m})^2 = 5.027 \times 10^{-3}\ \text{m}^2,$$

and the time t is 20 min = 1200 s. Substituting these and other known values into Eq. 19-41, we find

$$\Delta E = (5.6703 \times 10^{-8}\ \text{W/m}^2 \cdot \text{K}^4)(0.80)(5.027 \times 10^{-3}\ \text{m}^2)$$
$$\times (1200\ \text{s})[(320\ \text{K})^4 - (308\ \text{K})^4] = 406.8\ \text{J}.$$

Thus, with 500 bees in the ball, each bee must produce an additional energy of

$$\frac{\Delta E}{500} = \frac{406.8\ \text{J}}{500} = 0.81\ \text{J}. \quad \text{(Answer)}$$

REVIEW & SUMMARY

Temperature; Thermometers Temperature is an SI base quantity related to our sense of hot and cold. It is measured with a thermometer, which contains a working substance with a measurable property, such as length or pressure, that changes in a regular way as the substance becomes hotter or colder.

Zeroth Law of Thermodynamics When a thermometer and some other object are placed in contact with each other, they eventually reach thermal equilibrium. The reading of the thermometer is then taken to be the temperature of the other object. The process provides consistent and useful temperature measurements because of the **zeroth law of thermodynamics**: If bodies A and B are each in thermal equilibrium with a third body C (the thermometer), then A and B are in thermal equilibrium with each other.

The Kelvin Temperature Scale In the SI system, temperature is measured on the **Kelvin scale,** which is based on the *triple point* of water (273.16 K). Other temperatures are then defined by use of a *constant-volume gas thermometer,* in which a sample of gas is maintained at constant volume so its pressure is proportional to its temperature. We define the *temperature T* as measured with a gas thermometer to be

$$T = (273.16 \text{ K}) \left(\lim_{\text{gas} \to 0} \frac{p}{p_3} \right). \tag{19-6}$$

Here T is measured in kelvins, and p_3 and p are the pressures of the gas at 273.16 K and the measured temperature, respectively.

Celsius and Fahrenheit Scales The Celsius temperature scale is defined by

$$T_C = T - 273.15°, \tag{19-7}$$

with T in kelvins. The Fahrenheit temperature scale is defined by

$$T_F = \tfrac{9}{5} T_C + 32°. \tag{19-8}$$

Thermal Expansion All objects change size with changes in temperature. For a temperature change ΔT, a change ΔL in any linear dimension L is given by

$$\Delta L = L\alpha \, \Delta T, \tag{19-9}$$

in which α is the **coefficient of linear expansion.** The change ΔV in the volume V of a solid or liquid is

$$\Delta V = V\beta \, \Delta T. \tag{19-10}$$

Here $\beta = 3\alpha$ is the material's **coefficient of volume expansion.**

Heat Heat Q is energy that is transferred between a system and its environment because of a temperature difference between them. It can be measured in **joules** (J), **calories** (cal), **kilocalories** (Cal or kcal), or **British thermal units** (Btu), with

$$1 \text{ cal} = 3.969 \times 10^{-3} \text{ Btu} = 4.1860 \text{ J}. \tag{19-12}$$

Heat Capacity and Specific Heat If heat Q is absorbed by an object, the object's temperature change $T_f - T_i$ is related to Q by

$$Q = C(T_f - T_i), \tag{19-13}$$

in which C is the **heat capacity** of the object. If the object has mass m, then

$$Q = cm(T_f - T_i), \tag{19-14}$$

where c is the **specific heat** of the material making up the object. The **molar specific heat** of a material is the heat capacity per mole, or per 6.02×10^{23} elementary units of the material.

Heat of Transformation Heat absorbed by a material may change the material's physical state or phase—for example, from solid to liquid or from liquid to gas. The amount of energy required per unit mass to change the phase (but not the temperature) of a particular material is its **heat of transformation** L. Thus,

$$Q = Lm. \tag{19-16}$$

The **heat of vaporization** L_V is the amount of energy per unit mass that must be added to vaporize a liquid or that must be removed to condense a gas. The **heat of fusion** L_F is the amount of energy per unit mass that must be added to melt a solid or that must be removed to freeze a liquid.

Work Associated with Volume Change A gas may exchange energy with its surroundings through work. The amount of work W done *by* a gas as it expands or contracts from an initial volume V_i to a final volume V_f is given by

$$W = \int dW = \int_{V_i}^{V_f} p \, dV. \tag{19-25}$$

The integration is necessary because the pressure p may vary during the volume change.

First Law of Thermodynamics The principle of conservation of energy for a thermodynamic process is expressed in the **first law of thermodynamics**, which may assume either of the forms

$$\Delta E_{\text{int}} = E_{\text{int},f} - E_{\text{int},i} = Q - W \quad \text{(first law)} \tag{19-26}$$

or

$$dE_{\text{int}} = dQ - dW \quad \text{(first law)}. \tag{19-27}$$

E_{int} represents the internal energy of the material, which depends only on its state (temperature, pressure, and volume). Q represents the energy exchanged as heat between the system and its surroundings; Q is positive if the system absorbs heat, and negative if the system loses heat. W is the work done *by* the system; W is positive if the system expands against some external force exerted by the surroundings, and negative if the system contracts because of some external force. Both Q and W are path dependent; ΔE_{int} is path *independent.*

Applications of the First Law The first law of thermodynamics finds application in several special cases:

$$
\begin{aligned}
\textit{adiabatic processes:} & \quad Q = 0, \quad \Delta E_{\text{int}} = -W \\
\textit{constant-volume processes:} & \quad W = 0, \quad \Delta E_{\text{int}} = Q \\
\textit{cyclical processes:} & \quad \Delta E_{\text{int}} = 0, \quad Q = W \\
\textit{free expansions:} & \quad Q = W = \Delta E_{\text{int}} = 0
\end{aligned}
$$

Conduction, Convection, and Radiation   The rate P_{cond} at which energy is *conducted* through a slab whose faces are maintained at temperatures T_H and T_C is

$$P_{cond} = \frac{Q}{t} = kA\frac{T_H - T_C}{L}, \qquad (19\text{-}32)$$

in which A and L are the face area and length of the slab, and k is the thermal conductivity of the material.

   Convection occurs when temperature differences cause an energy transfer by motion within a fluid. *Radiation* is an energy transfer via the emission of electromagnetic energy. The rate P_{rad} at which an object emits energy via thermal radiation is

$$P_{rad} = \sigma\varepsilon AT^4, \qquad (19\text{-}38)$$

where $\sigma\ (= 5.6703 \times 10^{-8}\ \text{W/m}^2\cdot\text{K}^4)$ is the Stefan–Boltzmann constant, ε is the emissivity of the object's surface, A is its surface area, and T is its surface temperature (in kelvins). The rate P_{abs} at which an object absorbs energy via thermal radiation from its environment, which is at the uniform temperature T_{env} (in kelvins), is

$$P_{abs} = \sigma\varepsilon AT_{env}^4. \qquad (19\text{-}39)$$

QUESTIONS

1. Figure 19-23 shows three linear temperature scales with the freezing and boiling points of water indicated. Rank changes of 25 R°, 25 S°, and 25 U° according to the corresponding changes in temperature, greatest first.

Fig. 19-23 Question 1.

2. The table gives the initial length L, change in temperature ΔT, and change in length ΔL of four rods. Rank the rods according to their coefficients of thermal expansion, greatest first.

Rod	L (m)	ΔT (C°)	ΔL (m)
a	2	10	4×10^{-4}
b	1	20	4×10^{-4}
c	2	10	8×10^{-4}
d	4	5	4×10^{-4}

3. In a thermally isolated container, material A of mass m is placed against material B, also of mass m but at a higher temperature. When thermal equilibrium is reached, the temperature changes ΔT_A and ΔT_B of A and B are recorded. Then the experiment is repeated, using A with other materials, all of the same mass m. The results are given in the table. Rank the four materials according to their specific heats, greatest first.

Experiment	Temperature Changes	
1	$\Delta T_A = +50$ C°	$\Delta T_B = -50$ C°
2	$\Delta T_A = +10$ C°	$\Delta T_C = -20$ C°
3	$\Delta T_A = +2$ C°	$\Delta T_D = -40$ C°

4. Materials A, B, and C are solids that are at their melting temperatures. Material A requires 200 J to melt 4 kg, material B requires 300 J to melt 5 kg, and material C requires 300 J to melt 6 kg. Rank the materials according to their heats of fusion, greatest first.

5. Figure 19-24 shows two closed cycles on p-V diagrams for a gas. The three parts of cycle 1 are of the same length and shape as those of cycle 2. For each cycle, should the cycle be traversed clockwise or counterclockwise if (a) the net work W done by the gas is to be positive and (b) the net energy transferred by the gas as heat Q is to be positive?

6. For which cycle in Fig. 19-24, traversed clockwise, is (a) W greater and (b) Q greater?

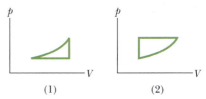

Fig. 19-24 Questions 5 and 6.

7. Figure 19-25 shows a composite slab of three different materials, a, b, and c, with identical thicknesses and with thermal conductivities $k_b > k_a > k_c$. The transfer of energy through them as heat is nonzero and steady. Rank the materials according to the temperature difference ΔT across them, greatest first.

Fig. 19-25 Question 7.

8. During an icicle's growth, its outer surface is covered with a thin sheath of liquid water that slowly seeps downward to form drops, one at a time, that hang at the icicle's tip (Fig. 19-26). Each drop straddles a thin tube of liquid water that extends up into the icicle, toward (but not all the way to) its *root* (its top). As the water at the top of the tube gradually freezes, energy is released. Is that energy conducted radially outward through the ice, downward through the water to the hanging

Fig. 19-26 Question 8.

drop, or upward toward the root? (Assume that the air temperature is below 0°C.)

9. The following solid objects, made of the same material, are maintained at a temperature of 300 K in an environment whose temperature is 350 K: a cube of edge length r, a sphere of radius r, and a hemisphere of radius r. Rank the objects according to the net rate at which thermal radiation is exchanged with the environment, greatest first.

10. Three different materials of identical masses are placed, in turn, in a special freezer that can extract energy from a material at a certain constant rate. During the cooling process, each material begins in the liquid state and ends in the solid state; Fig. 19-27 shows graphs of the temperature T versus time t for the three materials. (a) For material 1, is the specific heat for the liquid state greater than or less than that for the solid state? Rank the materials according to (b) their freezing-point temperatures, (c) their specific heats in the liquid state, (d) their specific heats in the solid state, and (e) their heats of fusion, all greatest first.

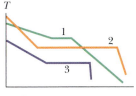
Fig. 19-27 Question 10.

11. A sample A of liquid water and a sample B of ice, of identical masses, are placed in a thermally isolated (insulated) container and allowed to come to thermal equilibrium. Figure 19-28a is a sketch of the temperature T of the samples versus time t. (a) Is the equilibrium temperature above, below, or at the freezing point of water? (b) In reaching equilibrium, does the liquid partly freeze or fully

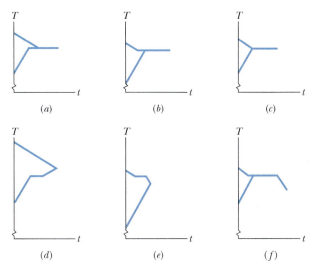
Fig. 19-28 Questions 11 and 12.

freeze, or does it undergo no freezing? (c) Does the ice partly melt or fully melt, or does it undergo no melting?

12. Question 11 continued: Figure 19-28 gives additional sketches of T versus t, of which one or more are impossible to produce. (a) Which sketch is impossible and why? (b) In the possible ones, is the equilibrium temperature above, below, or at the freezing point of water? (c) As the possible situations reach equilibrium, does the liquid partly freeze or fully freeze, or does it undergo no freezing? Does the ice partly melt or fully melt, or undergo no melting?

EXERCISES & PROBLEMS

SEC. 19-3 Measuring Temperature

1E. Two constant-volume gas thermometers are assembled, one with nitrogen and the other hydrogen. Both contain enough gas so that $p_3 = 80$ kPa. What is the difference between the pressures in the two thermometers if both bulbs are inserted into boiling water? (*Hint:* See Fig. 19-6.) Which gas is at higher pressure? ssm

2E. Suppose the temperature of a gas at the boiling point of water is 373.15 K. What then is the limiting value of the ratio of the pressure of the gas at that boiling point to its pressure at the triple point of water? (Assume the volume of the gas is the same at both temperatures.)

3E. A particular gas thermometer is constructed of two gas-containing bulbs, each of which is put into a water bath, as shown in Fig. 19-29. The pressure difference between the two bulbs is

Fig. 19-29 Exercise 3.

measured by a mercury manometer as shown. Appropriate reservoirs, not shown in the diagram, maintain constant gas volume in the two bulbs. There is no difference in pressure when both baths are at the triple point of water. The pressure difference is 120 torr when one bath is at the triple point and the other is at the boiling point of water. It is 90.0 torr when one bath is at the triple point and the other is at an unknown temperature to be measured. What is the unknown temperature? ssm

SEC. 19-4 The Celsius and Fahrenheit Scales

4E. At what temperature is the Fahrenheit scale reading equal to (a) twice that of the Celsius and (b) half that of the Celsius?

5E. At what temperature do the following pairs of scales read the same, if ever: (a) Fahrenheit and Celsius (verify the listing in Table 19-1), (b) Fahrenheit and Kelvin, and (c) Celsius and Kelvin? ssm

6E. (a) In 1964, the temperature in the Siberian village of Oymyakon reached −71°C. What temperature is this on the Fahrenheit scale? (b) The highest officially recorded temperature in the continental United States was 134°F in Death Valley, California. What is this temperature on the Celsius scale?

7P. It is an everyday observation that hot and cold objects cool down or warm up to the temperature of their surroundings. If the temperature difference ΔT between an object and its surroundings

$(\Delta T = T_{obj} - T_{sur})$ is not too great, the rate of cooling or warming of the object is proportional, approximately, to this temperature difference; that is,

$$\frac{d\,\Delta T}{dt} = -A(\Delta T),$$

where A is a constant. (The minus sign appears because ΔT decreases with time if ΔT is positive and increases if ΔT is negative.) This is known as *Newton's law of cooling*. (a) On what factors does A depend? What are its dimensions? (b) If at some instant $t = 0$ the temperature difference is ΔT_0, show that it is

$$\Delta T = \Delta T_0 e^{-At}$$

at a later time t. ssm

8P. The heater of a house breaks down one day when the outside temperature is 7.0°C. As a result, the inside temperature drops from 22°C to 18°C in 1.0 h. The owner fixes the heater and adds insulation to the house. Now she finds that, on a similar day, the house takes twice as long to drop from 22°C to 18°C when the heater is not operating. What is the ratio of the new value of constant A in Newton's law of cooling (see Problem 7) to the previous value?

9P. Suppose that on a linear temperature scale X, water boils at $-53.5°$X and freezes at $-170°$X. What is a temperature of 340 K on the X scale? ilw

SEC. 19-5   Thermal Expansion

10E. An aluminum flagpole is 33 m high. By how much does its length increase as the temperature increases by 15 C°?

11E. The Pyrex glass mirror in the telescope at the Mt. Palomar Observatory has a diameter of 200 in. The temperature ranges from $-10°$C to 50°C on Mt. Palomar. In micrometers, what is the maximum change in the diameter of the mirror, assuming that the glass can freely expand and contract? ssm

12E. An aluminum-alloy rod has a length of 10.000 cm at 20.000°C and a length of 10.015 cm at the boiling point of water. (a) What is the length of the rod at the freezing point of water? (b) What is the temperature if the length of the rod is 10.009 cm?

13E. A circular hole in an aluminum plate is 2.725 cm in diameter at 0.000°C. What is its diameter when the temperature of the plate is raised to 100.0°C? ilw

14E. What is the volume of a lead ball at 30°C if the ball's volume at 60°C is 50 cm³?

15E. Find the change in volume of an aluminum sphere with an initial radius of 10 cm when the sphere is heated from 0.0°C to 100°C. ssm

16E. The area A of a rectangular plate is ab. Its coefficient of linear expansion is α. After a temperature rise ΔT, side a is longer by Δa and side b is longer by Δb (Fig. 19-30). Show that if the small quantity $(\Delta a\,\Delta b)/ab$ is neglected, then $\Delta A = 2\alpha A\,\Delta T$.

Fig. 19-30 Exercise 16.

17E. An aluminum cup of 100 cm³ capacity is completely filled with glycerin at 22°C. How much glycerin, if any, will spill out of the cup if the temperature of both the cup and glycerin is increased to 28°C? (The coefficient of volume expansion of glycerin is $5.1 \times 10^{-4}/$C°.) ssm

18P. At 20°C, a rod is exactly 20.05 cm long on a steel ruler. Both the rod and the ruler are placed in an oven at 270°C, where the rod now measures 20.11 cm on the same ruler. What is the coefficient of thermal expansion for the material of which the rod is made?

19P. A steel rod is 3.000 cm in diameter at 25°C. A brass ring has an interior diameter of 2.992 cm at 25°C. At what common temperature will the ring just slide onto the rod? ssm ilw www

20P. When the temperature of a metal cylinder is raised from 0.0°C to 100°C, its length increases by 0.23%. (a) Find the percent change in density. (b) What is the metal? Use Table 19-2.

21P. Show that when the temperature of a liquid in a barometer changes by ΔT and the pressure is constant, the liquid's height h changes by $\Delta h = \beta h\,\Delta T$, where β is the coefficient of volume expansion. Neglect the expansion of the glass tube. ssm

22P. When the temperature of a copper coin is raised by 100 C°, its diameter increases by 0.18%. To two significant figures, give the percent increase in (a) the area of a face, (b) the thickness, (c) the volume, and (d) the mass of the coin. (e) Calculate the coefficient of linear expansion of the coin.

23P. A pendulum clock with a pendulum made of brass is designed to keep accurate time at 20°C. If the clock operates at 0.0°C, what is the magnitude of its error, in seconds per hour, and does the clock run fast or slow? ilw

24P. In a certain experiment, a small radioactive source must move at selected, extremely slow speeds. This motion is accomplished by fastening the source to one end of an aluminum rod and heating the central section of the rod in a controlled way. If the effective heated section of the rod in Fig. 19-31 is 2.00 cm, at what constant rate must the temperature of the rod be changed if the source is to move at a constant speed of 100 nm/s?

Fig. 19-31 Problem 24.

25P. As a result of a temperature rise of 32 C°, a bar with a crack at its center buckles upward (Fig. 19-32). If the fixed distance L_0 is 3.77 m and the coefficient of linear expansion of the bar is $25 \times 10^{-6}/$C°, find the rise x of the center. ssm ilw

Fig. 19-32 Problem 25.

SEC. 19-7   The Absorption of Heat by Solids and Liquids

26E. A certain diet doctor encourages people to diet by drinking ice water. His theory is that the body must burn off enough fat to raise the temperature of the water from 0.00°C to the body temperature of 37.0°C. How many liters of ice water would have to be consumed to burn off 454 g (about 1 lb) of fat, assuming that this much fat burning requires 3500 Cal be transferred to the ice water? Why is it not advisable to follow this diet? (One liter = 10^3 cm³. The density of water is 1.00 g/cm³.)

27E. A certain substance has a mass per mole of 50 g/mol. When 314 J is added as heat to a 30.0 g sample, the sample's temperature rises from 25.0°C to 45.0°C. What are (a) the specific heat and (b) the molar specific heat of this substance? (c) How many moles are present? ssm

28E. How much water remains unfrozen after 50.2 kJ is transferred as heat from 260 g of liquid water initially at its freezing point?

29E. Calculate the minimum amount of energy, in joules, required to completely melt 130 g of silver initially at 15.0°C. ssm

30E. A room is lighted by four 100 W incandescent lightbulbs. (The power of 100 W is the rate at which a bulb converts electrical energy to heat and the energy of visible light.) Assuming that 90% of the energy is converted to heat, how much heat does the room receive in 1.00 h?

31E. An energetic athlete can use up all the energy from a diet of 4000 Cal/day. If he were to use up this energy at a steady rate, how would his rate of energy use compare with the power of a 100 W bulb? (The power of 100 W is the rate at which the bulb converts electrical energy to heat and the energy of visible light.) ilw

32E. How many grams of butter, which has a usable energy content of 6.0 Cal/g (= 6000 cal/g), would be equivalent to the change in gravitational potential energy of a 73.0 kg man who ascends from sea level to the top of Mt. Everest, at elevation 8.84 km? Assume that the average value of g is 9.80 m/s^2.

33E. A power of 0.400 hp is required for 2.00 min to drill a hole in a 1.60-lb copper block. (a) If the full power is the rate at which thermal energy is generated, how much is generated in Btu? (b) What is the rise in temperature of the copper if the copper absorbs 75.0% of this energy? (Use the energy conversion 1 ft · lb = 1.285 × 10^{-3} Btu.) ssm

34E. One way to keep the contents of a garage from becoming too cold on a night when a severe subfreezing temperature is forecast is to put a tub of water in the garage. If the mass of the water is 125 kg and its initial temperature is 20°C, (a) how much energy must the water transfer to its surroundings in order to freeze completely and (b) what is the lowest possible temperature of the water and its surroundings until that happens?

35E. A small electric immersion heater is used to heat 100 g of water for a cup of instant coffee. The heater is labeled "200 watts," which means that it converts electrical energy to thermal energy at this rate. Calculate the time required to bring all this water from 23°C to 100°C, ignoring any heat losses. ssm

36P. A 150 g copper bowl contains 220 g of water, both at 20.0°C. A very hot 300 g copper cylinder is dropped into the water, causing the water to boil, with 5.00 g being converted to steam. The final temperature of the system is 100°C. Neglect energy transfers with the environment. (a) How much energy (in calories) is transferred to the water as heat? (b) How much to the bowl? (c) What is the original temperature of the cylinder?

37P. A chef, on finding his stove out of order, decides to boil the water for his wife's coffee by shaking it in a thermos flask. Suppose that he uses tap water at 15°C and that the water falls 30 cm each shake, the chef making 30 shakes each minute. Neglecting any loss of thermal energy by the flask, how long must he shake the flask until the water reaches 100°C? ssm www

38P. Nonmetric version: How long does a 2.0 × 10^5 Btu/h water heater take to raise the temperature of 40 gal of water from 70°F

to 100°F? Metric version: How long does a 59 kW water heater take to raise the temperature of 150 L of water from 21°C to 38°C?

39P. Ethyl alcohol has a boiling point of 78°C, a freezing point of −114°C, a heat of vaporization of 879 kJ/kg, a heat of fusion of 109 kJ/kg, and a specific heat of 2.43 kJ/kg · K. How much energy must be removed from 0.510 kg of ethyl alcohol that is initially a gas at 78°C so that it becomes a solid at −114°C?

40P. A 1500 kg Buick moving at 90 km/h brakes to a stop, at uniform deceleration and without skidding, over a distance of 80 m. At what average rate is mechanical energy transferred to thermal energy in the brake system?

41P. The specific heat of a substance varies with temperature according to $c = 0.20 + 0.14T + 0.023T^2$, with T in °C and c in cal/g · K. Find the energy required to raise the temperature of 2.0 g of this substance from 5.0°C to 15°C.

42P. In a solar water heater, energy from the Sun is gathered by water that circulates through tubes in a rooftop collector. The solar radiation enters the collector through a transparent cover and warms the water in the tubes; this water is pumped into a holding tank. Assume that the efficiency of the overall system is 20% (that is, 80% of the incident solar energy is lost from the system). What collector area is necessary to raise the temperature of 200 L of water in the tank from 20°C to 40°C in 1.0 h when the intensity of incident sunlight is 700 W/m^2?

43P. What mass of steam at 100°C must be mixed with 150 g of ice at its melting point, in a thermally insulated container, to produce liquid water at 50°C? ilw

44P. A person makes a quantity of iced tea by mixing 500 g of hot tea (essentially water) with an equal mass of ice at its melting point. If the initial hot tea is at a temperature of (a) 90°C and (b) 70°C, what are the temperature and mass of the remaining ice when the tea and ice reach a common temperature? Neglect energy transfers with the environment.

45P. (a) Two 50 g ice cubes are dropped into 200 g of water in a thermally insulated container. If the water is initially at 25°C, and the ice comes directly from a freezer at −15°C, what is the final temperature of the drink when the drink reaches thermal equilibrium? (b) What is the final temperature if only one ice cube is used? ssm

46P. An insulated Thermos contains 130 cm^3 of hot coffee, at a temperature of 80.0°C. You put in a 12.0 g ice cube at its melting point to cool the coffee. By how many degrees has your coffee cooled once the ice has melted? Treat the coffee as though it were pure water and neglect energy transfers with the environment.

47P. A 20.0 g copper ring has a diameter of 2.54000 cm at its temperature of 0.000°C. An aluminum sphere has a diameter of 2.54508 cm at its temperature of 100.0°C. The sphere is placed on top of the ring (Fig. 19-33), and the two are allowed to come to thermal equilibrium, with no heat lost to the surroundings. The sphere just passes through the ring at the equilibrium temperature. What is the mass of the sphere? ssm

Fig. 19-33 Problem 47.

SEC. 19-10 Some Special Cases of the First Law of Thermodynamics

48E. Consider that 200 J of work is done on a system and 70.0 cal is extracted from the system as heat. In the sense of the first law of thermodynamics, what are the values (including algebraic signs) of (a) W, (b) Q, and (c) ΔE_{int}?

49E. A sample of gas expands from 1.0 m³ to 4.0 m³ while its pressure decreases from 40 Pa to 10 Pa. How much work is done by the gas if its pressure changes with volume via each of the three paths shown in the p-V diagram in Fig. 19-34? ssm www

50E. A thermodynamic system is taken from an initial state A to another state B and back again to A, via state C, as shown by path $ABCA$ in the p-V diagram of Fig. 19-35a. (a) Complete the table in Fig. 19-35b by filling in either $+$ or $-$ for the sign of each thermodynamic quantity associated with each step of the cycle. (b) Calculate the numerical value of the work done by the system for the complete cycle $ABCA$.

51E. Gas within a closed chamber undergoes the cycle shown in the p-V diagram of Fig. 19-36. Calculate the net energy added to the system as heat during one complete cycle. ssm ilw

52E. Gas within a chamber passes through the cycle shown in Fig. 19-37. Determine the energy transferred by the system as heat during process CA if the energy added as heat Q_{AB} during process AB is 20.0 J, no energy is transferred as heat during process BC, and the net work done during the cycle is 15.0 J.

53P. When a system is taken from state i to state f along path iaf in Fig. 19-38, $Q = 50$ cal and $W = 20$ cal. Along path ibf, $Q = 36$ cal. (a) What is W along path ibf? (b) If $W = -13$ cal for the return path fi, what is Q for this path? (c) Take $E_{int,i} = 10$ cal. What is $E_{int,f}$? (d) If $E_{int,b} = 22$ cal, what

Fig. 19-34 Exercise 49.

(a)

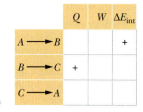

(b)

Fig. 19-35 Exercise 50.

Fig. 19-36 Exercise 51.

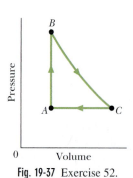

Fig. 19-37 Exercise 52.

are the values of Q for path ib and path bf? ssm

SEC. 19-11 Heat Transfer Mechanisms

54E. The average rate at which energy is conducted outward through the ground surface in North America is 54.0 mW/m², and the average thermal conductivity of the near-surface rocks is 2.50 W/m · K. Assuming a surface temperature of 10.0°C, find the temperature at a depth of 35.0 km (near the base of the crust). Ignore the heat generated by the presence of radioactive elements.

55E. The ceiling of a single-family dwelling in a cold climate should have an R-value of 30. To give such insulation, how thick would a layer of (a) polyurethane foam and (b) silver have to be?

56E. (a) Calculate the rate at which body heat is conducted through the clothing of a skier in a steady-state process, given the following data: the body surface area is 1.8 m² and the clothing is 1.0 cm thick; the skin surface temperature is 33°C and the outer surface of the clothing is at 1.0°C; the thermal conductivity of the clothing is 0.040 W/m · K. (b) How would the answer to (a) change if, after a fall, the skier's clothes became soaked with water of thermal conductivity 0.60 W/m · K?

57E. Consider the slab shown in Fig. 19-18. Suppose that $L = 25.0$ cm, $A = 90.0$ cm², and the material is copper. If $T_H = 125°C$, $T_C = 10.0°C$, and a steady state is reached, find the conduction rate through the slab. ssm

58E. If you were to walk briefly in space without a spacesuit while far from the Sun (as an astronaut does in the movie 2001), you would feel the cold of space—while you radiated energy, you would absorb almost none from your environment. (a) At what rate would you lose energy? (b) How much energy would you lose in 30 s? Assume that your emissivity is 0.90, and estimate other data needed in the calculations.

59E. A cylindrical copper rod of length 1.2 m and cross-sectional area 4.8 cm² is insulated to prevent heat loss through its surface. The ends are maintained at a temperature difference of 100 C° by having one end in a water–ice mixture and the other in boiling water and steam. (a) Find the rate at which energy is conducted along the rod. (b) Find the rate at which ice melts at the cold end. ilw

60E. Four square pieces of insulation of two different materials, all with the same thickness and area A, are available to cover an opening of area $2A$. This can be done in either of the two ways shown in Fig. 19-39. Which arrangement, (a) or (b), gives the lower energy flow if $k_2 \neq k_1$?

61P. Two identical rectangular rods of metal are welded end to

Fig. 19-38 Problem 53.

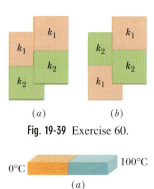

(a)   (b)

Fig. 19-39 Exercise 60.

(a)

(b)

Fig. 19-40 Problem 61.

end as shown in Fig. 19-40*a*, and 10 J is conducted (in a steady-state process) through the rods as heat in 2.0 min. How long would it take for 10 J to be conducted through the rods if they were welded together as shown in Fig. 19-40*b*?

62P. A sphere of radius 0.500 m, temperature 27.0°C, and emissivity 0.850 is located in an environment of temperature 77.0°C. At what rate does the sphere (a) emit and (b) absorb thermal radiation? (c) What is the sphere's net rate of energy exchange?

63P. (a) What is the rate of energy loss in watts per square meter through a glass window 3.0 mm thick if the outside temperature is −20°F and the inside temperature is +72°F? (b) A storm window having the same thickness of glass is installed parallel to the first window, with an air gap of 7.5 cm between the two windows. What now is the rate of energy loss if conduction is the only important energy-loss mechanism?   **ilw**

64P. Figure 19-41 shows (in cross section) a wall that consists of four layers. The thermal conductivities are $k_1 = 0.060$ W/m · K, $k_3 = 0.040$ W/m · K, and $k_4 = 0.12$ W/m · K (k_2 is not known). The layer thicknesses are $L_1 = 1.5$ cm, $L_3 = 2.8$ cm, and $L_4 = 3.5$ cm (L_2 is not known). Energy transfer through the wall is steady. What is the temperature of the interface indicated?

Fig. 19-41 Problem 64.

65P. A tank of water has been outdoors in cold weather, and a slab of ice 5.0 cm thick has formed on its surface (Fig. 19-42). The air above the ice is at −10°C. Calculate the rate of formation of ice (in centimeters per hour) on the ice slab. Take the thermal conductivity and density of ice to be 0.0040 cal/s · cm · C° and 0.92 g/cm³. Assume that energy is not transferred through the walls or bottom of the tank.   **ssm**
www

Fig. 19-42 Problem 65.

66P. Ice has formed on a shallow pond and a steady state has been reached, with the air above the ice at −5.0°C and the bottom of the pond at 4.0°C. If the total depth of *ice + water* is 1.4 m, how thick is the ice? (Assume that the thermal conductivities of ice and water are 0.40 and 0.12 cal/m · C° · s, respectively.)

Additional Problems

67. You can join the semi-secret "300 F" club at the Amundsen–Scott South Pole Station only when the outside temperature is below −70°C. On such a day, you first bask in a hot sauna and then run outside wearing only your shoes. (This is, of course, extremely

dangerous, but the rite is effectively a protest against the constant danger of the winter cold at the south pole.)

Assume that when you step out of the sauna, your skin temperature is 102°F and the walls, ceiling, and floor of the sauna room have a temperature of 30°C. Estimate your surface area, and take your skin emissivity to be 0.80. (a) What is the approximate net rate P_{net} at which you lose energy via thermal radiation exchanges with the room? Next, assume that when you are outside half your surface area exchanges thermal radiation with the sky at a temperature of −25°C and the other half exchanges thermal radiation with the snow and ground at a temperature of −80°C. What is the approximate net rate at which you lose energy via thermal radiation exchanges with (b) the sky and (c) the snow and ground?

68. Emperor penguins (Fig. 19-43), those large penguins that resemble stuffy English butlers, breed and hatch their young even during severe Antarctic winters. Once an egg is laid, the father balances the egg on his feet to prevent the egg from freezing. He must do this for the full incubation period of 105 to 115 days, during which he cannot eat because his food is in the water. He can survive this long without food only if he can reduce his consumption of internal energy significantly. If he is alone, he consumes that energy too quickly to stay warm, and eventually abandons the egg in order to eat. To protect themselves and each other from the cold so as to reduce the consumption of internal energy, penguin fathers huddle closely together, in groups of perhaps several thousand. In addition to providing other benefits, the huddling reduces the rate at which the penguins thermally radiate energy to their surroundings.

Assume that a penguin father is a circular cylinder with top surface area *a*, height *h*, surface temperature *T*, and emissivity *ε*. (a) Find an expression for the rate P_i at which an individual father would radiate energy to the environment from his top surface and his side surface were he alone with his egg.

If *N* identical fathers were well apart from one another, the total rate of energy loss via radiation would be NP_i. Suppose, instead, that they huddle closely to form a *huddled cylinder* with top surface area *Na* and height *h*. (b) Find an expression for the rate P_h at which energy is radiated by the top surface and the side surface of the huddled cylinder.

(c) Assuming $a = 0.34$ m² and $h = 1.1$ m and using the expressions you obtained for P_i and P_h, graph the ratio P_h/NP_i versus *N*. Of course, the penguins know nothing about algebra or graphing, but their instinctive huddling reduces this ratio so that more of their eggs survive to the hatching stage. From the graphs (as you will see, you probably need more than one version), approximate how many penguins must huddle so that P_h/NP_i is reduced to (d) 0.5, (e) 0.4, (f) 0.3, (g) 0.2, and (h) 0.15. (i) For the assumed data, what is the lower limiting value for P_h/NP_i?

Fig. 19-43 Problem 68.

NEW PROBLEMS

N1. Suppose that on a temperature scale X, water boils at $-53.5°$X and freezes at $-170°$X. What would a temperature of 340 K be on the X scale (assuming that the X scale is linear)?

N2. Samples A and B are at different initial temperatures when they are placed in a thermally isolated container and allowed to come to thermal equilibrium. Figure 19N-1a gives their temperatures T versus time t. Sample A has a mass of 5.0 kg; sample B has a mass of 1.5 kg. Figure 19N-1b is a general plot about the material of sample B. It shows the temperature change ΔT that the material undergoes when energy is transferred to it as heat Q. The change ΔT is plotted versus the energy Q per unit mass of the material. What is the specific heat of sample A?

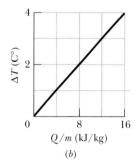

Fig. 19N-1 Problem N2.

N3. Density is mass divided by volume. If the volume V is temperature dependent, so is the density ρ. Show that a small change in density $\Delta\rho$ with a change in temperature ΔT is given by

$$\Delta\rho = -\beta\rho\,\Delta T,$$

where β is the coefficient of volume expansion. Explain the sign.

N4. A 0.300 kg sample is placed in a cooling apparatus that removes energy as heat at a constant rate of 2.81 W. Figure 19N-2 gives the temperature T of the sample versus time t. What is the specific heat of the sample?

Fig. 19N-2 Problem N4.

N5. It is possible to melt ice by rubbing one block of it against another. How much work, in joules, would you have to do to get 1.00 g of ice to melt?

N6. A 0.400 kg sample is placed in a cooling apparatus that removes energy as heat at a constant rate. Figure 19N-3 gives the temperature T of the sample versus time t; the sample freezes during the energy removal. The specific heat of the sample in its

Fig. 19N-3 Problem N6.

initial liquid phase is 3000 J/kg · K. What are (a) the sample's heat of fusion and (b) its specific heat in the frozen phase?

N7. If the heat necessary to warm a water sample of mass m from $68°$F to $78°$F were somehow converted to translational kinetic energy of that sample, what would be the speed of the sample?

N8. In a series of experiments, block B is to be placed in a thermally isolated container with block A, which has the same mass as block B. In each experiment, block B is initially at a certain temperature T_B, but temperature T_A of block A is changed from experiment to experiment. Let T_f represent the final temperature of the two blocks when they reach thermal equilibrium in any of the experiments. Figure 19N-4 gives temperature T_f versus the initial temperature T_A for a range of possible values of T_A. What are (a) temperature T_B and (b) the ratio c_B/c_A of the specific heats of the blocks?

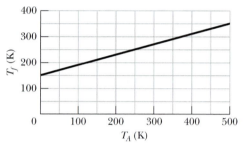

Fig. 19N-4 Problem N8.

N9. A pickup truck whose mass is 2200 kg is speeding along the highway at 105 km/h. (a) If you could use all this kinetic energy to vaporize water already at $100°$C, how much water could you vaporize? (b) If you had to buy this amount of energy from your local utility company at 0.12\$US/kW · h, how much would it cost you? Guess at the answers before you figure them out; you may be surprised.

N10. Figure 19N-5 shows the cross section of a wall made of three layers. The thicknesses of the layers are L_1, $L_2 = 0.70L_1$, and $L_3 = 0.35L_1$. The thermal conductivities are k_1, $k_2 = 0.90k_1$, $k_3 = 0.80k_1$. The temperatures at the left and right sides

Fig. 19N-5 Problem N10.

of the wall are $30°$C and $-15°$C. Thermal conduction through the wall has reached the steady state. (a) What is the temperature difference ΔT_2 across layer 2 (between the left and right sides of the layer)? If k_2 were, instead, equal to $1.1k_1$, (b) would the rate at which energy is conducted through the wall be greater than, less than, or the same as previously, and (c) what would be the value of ΔT_2?

N11. By means of a heating coil, energy is transferred at a constant rate to a substance in a thermally insulated container, beginning at time $t = 0$. The temperature of the substance is measured as a function of time. (a) Show how we can deduce from this information the way in which the heat capacity of the substance depends on the temperature. (b) Suppose that in a certain temperature range the temperature T is proportional to t^3, where t is the time. How does the heat capacity depend on T in this range?

N12. A 0.530 kg sample of liquid water and a sample of ice are placed in a thermally isolated container. The container also contains a device that transfers energy as heat from the liquid water to the ice at a constant rate P, until thermal equilibrium is reached. The temperatures T of the liquid water and the ice are given in Fig. 19N-6 as functions of time t. (a) What is rate P? (b) What is the initial mass of the ice in the container? (c) When thermal equilibrium is reached, what is the mass of the ice produced in this process?

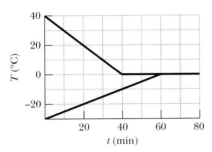

Fig. 19N-6 Problem N12.

N13. A clock pendulum made of Invar (see Table 19-2) has a period of 0.50 s and is accurate at 20°C. If the clock is used in a climate where the temperature averages 30°C, what correction (approximately) is necessary at the end of 30 days to the time given by the clock?

N14. Equation 19-9 ($\Delta L = \alpha L \, \Delta T$, with L interpreted as the initial length in any given process) only approximates a change in length due to a change in temperature. Let's check the approximation. Suppose that 20.000 00 kJ is transferred as heat to a 0.400 000 kg copper rod with an initial length $L = 3.000\ 000$ m. Assume that $\alpha = 17 \times 10^{-6}/C°$ is the exact coefficient of linear expansion for copper and 386 J/kg·K is the exact specific heat for copper. According to Eq. 19-9, what are (a) the increase in the rod's length and (b) the new length? Next, the same amount of energy is transferred as heat *from* the rod (so that the rod returns to its initial state). According to Eq. 19-9, what are (c) the decrease in the rod's length and (d) the new length? (e) What is the difference between the initial length and the answer to (d)?

N15. (a) Show that if the lengths of two rods of different solids are inversely proportional to their respective coefficients of linear expansion at the same initial temperature, the difference in length between them will be constant at all temperatures. What should be the lengths of (b) a steel and (c) a brass rod at 0.00°C so that at all temperatures their difference in length is 0.30 m?

N16. A ball of radius 2.00 cm, temperature 280 K, and emissivity 0.800 is suspended in an environment of temperature 300 K. What is the net rate of transfer of energy via radiation between the ball and the environment?

N17. A solid aluminum cube 20 cm on an edge floats face-down on mercury. How much higher on the cube's side will the mercury level be when the temperature rises from 270 to 320 K? (The coefficient of volume expansion of mercury is $1.8 \times 10^{-4}/$K.)

N18. The temperature of a 0.700 kg cube of ice is decreased to $-150°C$. Then energy is gradually transferred to it as heat while it is otherwise thermally isolated from its environment. The total transfer is 0.6993 MJ. Assume the value of c_{ice} given in Table 19-3 is valid for temperatures from $-150°C$ to 0°C. What is the final temperature of the water?

N19. Three equal-length straight rods, of aluminum, Invar, and steel, all at 20.0°C, form an equilateral triangle with hinge pins at the vertices. At what temperature will the angle opposite the Invar rod be 59.95°? See Appendix E for needed trigonometric formulas and Table 19-2 for needed data.

N20. Figure 19N-7 shows a closed cycle for a gas (the figure is not drawn to scale). The change in the internal energy of the gas as it moves from a to c along the path abc is -200 J. As it moves from c to d, 180 J must be transferred to it as heat. An additional transfer of 80 J as heat is needed as it moves from d to a. How much work is done on the gas as it moves from c to d?

Fig. 19N-7 Problem N20.

N21. A 15.0 kg sample of ice is initially at a temperature of $-20.0°C$. Then 7.0×10^6 J is added as heat to the sample, which is otherwise isolated. What then is the sample's temperature?

N22. The p-V diagram in Fig. 19N-8 shows two paths along which a sample of gas can be taken from state a to state b. Path 1 requires that an energy equal to $5.0p_1V_1$ be transferred to the gas as heat. Path 2 requires that an energy equal to $5.5p_1V_1$ be transferred to the gas as heat. What is pressure p_2 in terms of p_1?

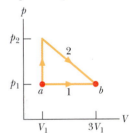

Fig. 19N-8 Problem N22.

N23. The temperature of a Pyrex disk is changed from 10.0°C to 60.0°C. Its initial radius is 8.00 cm; its initial thickness is 0.500 cm. Take these data as being exact. What is the change in the volume of the disk? (See Table 19-2.)

N24. A sample of gas is taken through the cycle $abca$ shown in the p-V diagram of Fig. 19N-9. The net work done is $+1.2$ J. Along path ab, the change in the internal energy is $+3.0$ J and the magnitude of the work done is 5.0 J. Along path ca, the energy transferred to the gas as heat is $+2.5$ J. How much energy is transferred as heat along (a) path ab and (b) path bc?

Fig. 19N-9 Problem N24.

N25. A 2.50 kg lump of aluminum is heated to 92.0°C and then dropped into 8.00 kg of water at 5.00°C. Assuming that the lump–water system is thermally isolated, what is the system's final (equilibrium) temperature?

20 The Kinetic Theory of Gases

When a container of cold champagne, soda pop, or any other carbonated drink is opened, a slight fog forms around the opening and some of the liquid sprays outward. (In the photograph, the fog is the white cloud that surrounds the stopper, and the spray has formed streaks within the cloud.)

What causes the fog?

The answer is in this chapter.

20-1 A New Way to Look at Gases

Classical thermodynamics—the subject of the last chapter—has nothing to say about atoms. Its laws are concerned only with such macroscopic variables as pressure, volume, and temperature. However, we know that a gas is made up of moving atoms or molecules (groups of atoms bound together). The pressure exerted by a gas must surely be related to the collisions of its molecules with the walls of its container. The ability of a gas to fill the volume of its container must surely be due to the freedom of motion of its molecules, and the temperature and internal energy of a gas must surely be related to the kinetic energy of these molecules. Thus, we can learn something about gases by approaching the subject from this direction. We call this molecular approach the **kinetic theory of gases.** It is the subject of this chapter.

20-2 Avogadro's Number

When our thinking is slanted toward molecules, it makes sense to measure the sizes of our samples in moles. If we do so, we can be certain that we are comparing samples that contain the same number of atoms or molecules. The *mole* is one of the seven SI base units and is defined as follows:

▶ One mole is the number of atoms in a 12 g sample of carbon-12.

The obvious question now is: "How many atoms or molecules are there in a mole?" The answer is determined experimentally and, as you saw in Chapter 19, is

$$N_A = 6.02 \times 10^{23} \text{ mol}^{-1} \qquad \text{(Avogadro's number)}, \qquad (20\text{-}1)$$

where mol^{-1} represents the inverse mole or "per mole," and mol is the abbreviation for mole. The number N_A is called **Avogadro's number** after Italian scientist Amadeo Avogadro (1776–1856), who suggested that all gases contain the same number of atoms or molecules when they occupy the same volume under the same conditions of temperature and pressure.

The number of moles n contained in a sample of any substance is equal to the ratio of the number of molecules N in the sample to the number of molecules N_A in 1 mole:

$$n = \frac{N}{N_A}. \qquad (20\text{-}2)$$

(*Caution:* The three symbols in this equation can easily be confused with one another, so you should sort them with their meanings now, before you end in "N-confusion.") We can find the number of moles in a sample from the mass M_{sam} of the sample and either the *molar mass M* (the mass of 1 mol) or the *molecular mass m* (the mass of one molecule):

$$n = \frac{M_{sam}}{M} = \frac{M_{sam}}{mN_A}. \qquad (20\text{-}3)$$

In Eq. 20-3, we used the fact that the mass M of 1 mol is the product of the mass m of one molecule and the number of molecules N_A in 1 mol:

$$M = mN_A. \qquad (20\text{-}4)$$

Tactic 1: *Avogadro's Number of What?*

In Eq. 20-1, Avogadro's number is expressed in terms of mol^{-1}, which is the inverse mole, or 1/mol. We could instead explicitly state the elementary unit involved in a given situation. For example, we might write $N_A = 6.02 \times 10^{23}$ atoms/mole if the elementary unit is an atom. If, instead, the elementary unit is a molecule, then we might write $N_A = 6.02 \times 10^{23}$ molecules/mole.

20-3 Ideal Gases

Our goal in this chapter is to explain the macroscopic properties of a gas—such as its pressure and its temperature—in terms of the behavior of the molecules that make it up. However, there is an immediate problem: which gas? Should it be hydrogen or oxygen, or methane, or perhaps uranium hexafluoride? They are all different. Experimenters have found, though, that if we confine 1 mole samples of various gases in boxes of identical volume and hold the gases at the same temperature, then their measured pressures are nearly—though not exactly—the same. If we repeat the measurements at lower gas densities, then these small differences in the measured pressures tend to disappear. Further experiments show that, at low enough densities, all real gases tend to obey the relation

$$pV = nRT \qquad \text{(ideal gas law),} \qquad (20\text{-}5)$$

in which p is the absolute (not gauge) pressure, n is the number of moles of gas present, and T is the temperature in kelvins. The symbol R is a constant called the **gas constant** that has the same value for all gases—namely,

$$R = 8.31 \text{ J/mol} \cdot \text{K.} \qquad (20\text{-}6)$$

Equation 20-5 is called the **ideal gas law.** Provided the gas density is low, this law holds for any single gas or for any mixture of different gases. (For a mixture, n is the total number of moles in the mixture.)

We can rewrite Eq. 20-5 in an alternative form, in terms of a constant called the **Boltzmann constant** k, which is defined as

$$k = \frac{R}{N_A} = \frac{8.31 \text{ J/mol} \cdot \text{K}}{6.02 \times 10^{23} \text{ mol}^{-1}} = 1.38 \times 10^{-23} \text{ J/K.} \qquad (20\text{-}7)$$

This allows us to write $R = kN_A$. Then, with Eq. 20-2 ($n = N/N_A$), we see that

$$nR = Nk. \qquad (20\text{-}8)$$

Substituting this into Eq. 20-5 gives a second expression for the ideal gas law:

$$pV = NkT \qquad \text{(ideal gas law).} \qquad (20\text{-}9)$$

(*Caution:* Note the difference between the two expressions for the ideal gas law— Eq. 20-5 involves the number of moles n and Eq. 20-9 involves the number of molecules N.)

You may well ask, "What is an *ideal gas* and what is so 'ideal' about one?" The answer lies in the simplicity of the law (Eqs. 20-5 and 20-9) that governs its macroscopic properties. Using this law—as you will see—we can deduce many properties of the ideal gas in a simple way. Although there is no such thing in nature as a truly ideal gas, *all real* gases approach the ideal state at low enough densities— that is, under conditions in which their molecules are far enough apart that they do not interact with one another. Thus, the ideal gas concept allows us to gain useful insights into the limiting behavior of real gases.

Work Done by an Ideal Gas at Constant Temperature

Suppose we put an ideal gas in a piston–cylinder arrangement like those in Chapter 19. Suppose also that we allow the gas to expand from an initial volume V_i to a final volume V_f while we keep the temperature T of the gas constant. Such a process, at *constant temperature*, is called an **isothermal expansion** (and the reverse is called an **isothermal compression**).

On a *p-V* diagram, an *isotherm* is a curve that connects points that have the same temperature. Thus, it is a graph of pressure versus volume for a gas whose temperature T is held constant. For n moles of an ideal gas, it is a graph of the equation

$$p = nRT \frac{1}{V} = (\text{a constant}) \frac{1}{V}. \tag{20-10}$$

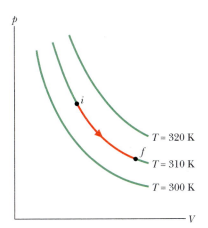

Fig. 20-1 Three isotherms on a *p-V* diagram. The path shown along the middle isotherm represents an isothermal expansion of a gas from an initial state *i* to a final state *f*. The path from *f* to *i* along the isotherm would represent the reverse process, that is, an isothermal compression.

Figure 20-1 shows three isotherms, each corresponding to a different (constant) value of T. (Note that the values of T for the isotherms increase upward to the right.) Superimposed on the middle isotherm is the path followed by a gas during an isothermal expansion from state *i* to state *f* at a constant temperature of 310 K.

To find the work done by an ideal gas during an isothermal expansion, we start with Eq. 19-25,

$$W = \int_{V_i}^{V_f} p \, dV. \tag{20-11}$$

This is a general expression for the work done during any change in volume of any gas. For an ideal gas, we can use Eq. 20-5 to substitute for p, obtaining

$$W = \int_{V_i}^{V_f} \frac{nRT}{V} \, dV. \tag{20-12}$$

Because we are considering an isothermal expansion, T is constant and we can move it in front of the integral sign to write

$$W = nRT \int_{V_i}^{V_f} \frac{dV}{V} = nRT \left[\ln V \right]_{V_i}^{V_f}. \tag{20-13}$$

By evaluating the expression in brackets at the limits and then using the relationship $\ln a - \ln b = \ln (a/b)$, we find that

$$W = nRT \ln \frac{V_f}{V_i} \qquad \text{(ideal gas, isothermal process).} \tag{20-14}$$

Recall that the symbol ln specifies a *natural* logarithm, which has base e.

For an expansion, V_f is greater than V_i, so the ratio V_f/V_i in Eq. 20-14 is greater than unity. The natural logarithm of a quantity greater than unity is positive, and so the work W done by an ideal gas during an isothermal expansion is positive, as we expect. For a compression, V_f is less than V_i, so the ratio of volumes in Eq. 20-14 is less than unity. The natural logarithm in that equation—hence the work W—is negative, again as we expect.

Work Done at Constant Volume and at Constant Pressure

Equation 20-14 does not give the work W done by an ideal gas during *every* thermodynamic process. Instead, it gives the work only for a process in which the temperature is held constant. If the temperature varies, then the symbol T in Eq. 20-12 cannot be moved in front of the integral symbol as in Eq. 20-13, and thus we do not end up with Eq. 20-14.

However, we can go back to Eq. 20-11 to find the work W done by an ideal gas (or any other gas) during two more processes—a constant-volume process and a constant-pressure process. If the volume of the gas is constant, then Eq. 20-11 yields

$$W = 0 \qquad \text{(constant-volume process)}. \qquad (20\text{-}15)$$

If, instead, the volume changes while the pressure p of the gas is held constant, then Eq. 20-11 becomes

$$W = p(V_f - V_i) = p\,\Delta V \qquad \text{(constant-pressure process)}. \qquad (20\text{-}16)$$

✓**CHECKPOINT 1:**   An ideal gas has an initial pressure of 3 pressure units and an initial volume of 4 volume units. The table gives the final pressure and volume of the gas (in those same units) in five processes. Which processes start and end on the same isotherm?

	a	b	c	d	e
p	12	6	5	4	1
V	1	2	7	3	12

Sample Problem 20-1

A cylinder contains 12 L of oxygen at 20°C and 15 atm. The temperature is raised to 35°C, and the volume is reduced to 8.5 L. What is the final pressure of the gas in atmospheres? Assume that the gas is ideal.

SOLUTION: The **Key Idea** here is that, because the gas is ideal, its pressure, volume, temperature, and number of moles are related by the ideal gas law, both in the initial state i and in the final state f (after the changes). Thus, from Eq. 20-5 we can write

$$p_i V_i = nRT_i \quad \text{and} \quad p_f V_f = nRT_f.$$

Dividing the second equation by the first equation and solving for p_f yields

$$p_f = \frac{p_i T_f V_i}{T_i V_f}. \qquad (20\text{-}17)$$

Note here that if we converted the given initial and final volumes from liters to the proper units of cubic meters, the multiplying conversion factors would cancel out of Eq. 20-17. The same would be true for conversion factors that convert the pressures from atmospheres to the proper pascals. However, to convert the given temperatures to kelvins requires the addition of an amount that would not cancel and thus must be included. Hence, we must write

$$T_i = (273 + 20)\ \text{K} = 293\ \text{K}$$

and

$$T_f = (273 + 35)\ \text{K} = 308\ \text{K}.$$

Inserting the given data into Eq. 20-17 then yields

$$p_f = \frac{(15\ \text{atm})(308\ \text{K})(12\ \text{L})}{(293\ \text{K})(8.5\ \text{L})} = 22\ \text{atm}. \qquad \text{(Answer)}$$

Sample Problem 20-2

One mole of oxygen (assume it to be an ideal gas) expands at a constant temperature T of 310 K from an initial volume V_i of 12 L to a final volume V_f of 19 L. How much work is done by the gas during the expansion?

SOLUTION: The **Key Idea** is this: Generally we find the work by integrating the gas pressure with respect to the gas volume, using Eq. 20-11. However, because the gas here is ideal and the expansion is isothermal, that integration leads to Eq. 20-14. Therefore, we can write

$$W = nRT \ln \frac{V_f}{V_i}$$

$$= (1\ \text{mol})(8.31\ \text{J/mol} \cdot \text{K})(310\ \text{K}) \ln \frac{19\ \text{L}}{12\ \text{L}}$$

$$= 1180\ \text{J}. \qquad \text{(Answer)}$$

The expansion is graphed in the p-V diagram of Fig. 20-2. The work done by the gas during the expansion is represented by the area beneath the curve if.

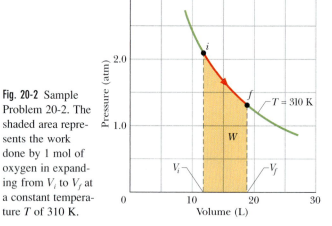

Fig. 20-2 Sample Problem 20-2. The shaded area represents the work done by 1 mol of oxygen in expanding from V_i to V_f at a constant temperature T of 310 K.

You can show that if the expansion is now reversed, with the gas undergoing an isothermal compression from 19 L to 12 L, the work done by the gas will be -1180 J. Thus, an external force would have to do 1180 J of work on the gas to compress it.

20-4 Pressure, Temperature, and RMS Speed

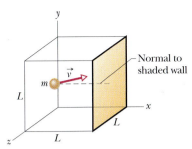

Fig. 20-3 A cubical box of edge L, containing n moles of an ideal gas. A molecule of mass m and velocity $\vec{v}$ is about to collide with the shaded wall of area L^2. A normal to that wall is shown.

Here is our first kinetic theory problem. Let n moles of an ideal gas be confined in a cubical box of volume V, as in Fig. 20-3. The walls of the box are held at temperature T. What is the connection between the pressure p exerted by the gas on the walls and the speeds of the molecules?

The molecules of gas in the box are moving in all directions and with various speeds, bumping into each other and bouncing from the walls of the box like balls in a racquetball court. We ignore (for the time being) collisions of the molecules with one another and consider only elastic collisions with the walls.

Figure 20-3 shows a typical gas molecule, of mass m and velocity $\vec{v}$, that is about to collide with the shaded wall. Because we assume that any collision of a molecule with a wall is elastic, when this molecule collides with the shaded wall, the only component of its velocity that is changed is the x component, and that component is reversed. This means that the only change in the particle's momentum is along the x axis, and that change is

$$\Delta p_x = (-mv_x) - (mv_x) = -2mv_x.$$

Hence, the momentum Δp_x delivered to the wall by the molecule during the collision is $+2mv_x$. (Because in this book the symbol p represents both momentum and pressure, we must be careful to note that here p represents momentum and is a vector quantity.)

The molecule of Fig. 20-3 will hit the shaded wall repeatedly. The time Δt between collisions is the time the molecule takes to travel to the opposite wall and back again (a distance of $2L$) at speed v_x. Thus, Δt is equal to $2L/v_x$. (Note that this result holds even if the molecule bounces off any of the other walls along the way, because those walls are parallel to x and so cannot change v_x.) Therefore, the average rate at which momentum is delivered to the shaded wall by this single molecule is

$$\frac{\Delta p_x}{\Delta t} = \frac{2mv_x}{2L/v_x} = \frac{mv_x^2}{L}.$$

From Newton's second law ($\vec{F} = d\vec{p}/dt$), the rate at which momentum is delivered to the wall is the force acting on that wall. To find the total force, we must add up the contributions of all the molecules that strike the wall, allowing for the possibility that they all have different speeds. Dividing the magnitude of the total force F_x by the area of the wall ($= L^2$) then gives the pressure p on that wall, where now and in the rest of this discussion, p represents pressure. Thus, using the expression for $\Delta p_x/\Delta t$, we can write this pressure as

$$p = \frac{F_x}{L^2} = \frac{mv_{x1}^2/L + mv_{x2}^2/L + \cdots + mv_{xN}^2/L}{L^2}$$

$$= \left(\frac{m}{L^3}\right)(v_{x1}^2 + v_{x2}^2 + \cdots + v_{xN}^2), \tag{20-18}$$

where N is the number of molecules in the box.

Since $N = nN_A$, there are nN_A terms in the second parentheses of Eq. 20-18. We can replace that quantity by $nN_A(v_x^2)_{avg}$, where $(v_x^2)_{avg}$ is the average value of the square of the x components of all the molecular speeds. Equation 20-18 then becomes

$$p = \frac{nmN_A}{L^3}\,(v_x^2)_{avg}.$$

However, mN_A is the molar mass M of the gas (that is, the mass of 1 mol of the gas). Also, L^3 is the volume of the box, so

$$p = \frac{nM(v_x^2)_{avg}}{V}.$$ (20-19)

For any molecule, $v^2 = v_x^2 + v_y^2 + v_z^2$. Because there are many molecules and because they are all moving in random directions, the average values of the squares of their velocity components are equal, so that $v_x^2 = \frac{1}{3}v^2$. Thus, Eq. 20-19 becomes

$$p = \frac{nM(v^2)_{avg}}{3V}.$$ (20-20)

The square root of $(v^2)_{avg}$ is a kind of average speed, called the **root-mean-square speed** of the molecules and symbolized by v_{rms}. Its name describes it rather well: You *square* each speed, you find the *mean* (that is, the average) of all these squared speeds, and then you take the square *root* of that mean. With $\sqrt{(v^2)_{avg}} = v_{rms}$, we can then write Eq. 20-20 as

$$p = \frac{nMv_{rms}^2}{3V}.$$ (20-21)

Equation 20-21 is very much in the spirit of kinetic theory. It tells us how the pressure of the gas (a purely macroscopic quantity) depends on the speed of the molecules (a purely microscopic quantity).

We can turn Eq. 20-21 around and use it to calculate v_{rms}. Combining Eq. 20-21 with the ideal gas law ($pV = nRT$) leads to

$$v_{rms} = \sqrt{\frac{3RT}{M}}.$$ (20-22)

Table 20-1 shows some rms speeds calculated from Eq. 20-22. The speeds are surprisingly high. For hydrogen molecules at room temperature (300 K), the rms speed is 1920 m/s or 4300 mi/h—faster than a speeding bullet! On the surface of the Sun, where the temperature is 2×10^6 K, the rms speed of hydrogen molecules would be 82 times greater than at room temperature were it not for the fact that at such high speeds, the molecules cannot survive collisions among themselves. Remember too that the rms speed is only a kind of average speed; many molecules move much faster than this, and some much slower.

TABLE 20-1   Some Molecular Speeds at Room Temperature ($T = 300$ K)[a]

Gas	Molar Mass (10^{-3} kg/mol)	v_{rms} (m/s)
Hydrogen (H_2)	2.02	1920
Helium (He)	4.0	1370
Water vapor (H_2O)	18.0	645
Nitrogen (N_2)	28.0	517
Oxygen (O_2)	32.0	483
Carbon dioxide (CO_2)	44.0	412
Sulfur dioxide (SO_2)	64.1	342

[a]For convenience, we often set room temperature = 300 K even though (at 27°C or 81°F) that represents a fairly warm room.

The speed of sound in a gas is closely related to the rms speed of the molecules of that gas. In a sound wave, the disturbance is passed on from molecule to molecule by means of collisions. The wave cannot move any faster than the "average" speed of the molecules. In fact, the speed of sound must be somewhat less than this "average" molecular speed because not all molecules are moving in exactly the same direction as the wave. As examples, at room temperature, the rms speeds of hydrogen and nitrogen molecules are 1920 m/s and 517 m/s, respectively. The speeds of sound in these two gases at this temperature are 1350 m/s and 350 m/s, respectively.

A question often arises: If molecules move so fast, why does it take as long as a minute or so before you can smell perfume when someone opens a bottle across a room? The answer is that, as we shall discuss in Section 20-6, each perfume molecule moves away from the bottle only very slowly because its repeated collisions with other molecules prevent it from moving directly across the room to you.

Sample Problem 20-3

Here are five numbers: 5, 11, 32, 67, and 89.

(a) What is the average value n_{avg} of these numbers?

SOLUTION: We find this from

$$n_{avg} = \frac{5 + 11 + 32 + 67 + 89}{5} = 40.8. \quad \text{(Answer)}$$

(b) What is the rms value n_{rms} of these numbers?

SOLUTION: We find this from

$$n_{rms} = \sqrt{\frac{5^2 + 11^2 + 32^2 + 67^2 + 89^2}{5}} = 52.1. \quad \text{(Answer)}$$

The rms value is greater than the average value because the larger numbers—being squared—are relatively more important in forming the rms value. To test this, let us replace 89 in our set of five numbers by 300. The average value of the new set of five numbers (as you should show) is 2.0 times the previous average value. The rms value, however, is 2.7 times the previous rms value.

20-5 Translational Kinetic Energy

We again consider a single molecule of an ideal gas as it moves around in the box of Fig. 20-3, but we now assume that its speed changes when it collides with other molecules. Its translational kinetic energy at any instant is $\frac{1}{2}mv^2$. Its *average* translational kinetic energy over the time that we watch it is

$$K_{avg} = (\tfrac{1}{2}mv^2)_{avg} = \tfrac{1}{2}m(v^2)_{avg} = \tfrac{1}{2}mv_{rms}^2, \tag{20-23}$$

in which we make the assumption that the average speed of the molecule during our observation is the same as the average speed of all the molecules at any given time. (Provided the total energy of the gas is not changing and we observe our molecule for long enough, this assumption is appropriate.) Substituting for v_{rms} from Eq. 20-22 leads to

$$K_{avg} = (\tfrac{1}{2}m)\frac{3RT}{M}.$$

However, M/m, the molar mass divided by the mass of a molecule, is simply Avogadro's number. Thus,

$$K_{avg} = \frac{3RT}{2N_A}.$$

Using Eq. 20-7 ($k = R/N_A$), we can then write

$$K_{avg} = \tfrac{3}{2}kT. \tag{20-24}$$

Fig. 20-4 A molecule traveling through a gas, colliding with other gas molecules in its path. Although the other molecules are shown as stationary, they are also moving in a similar fashion.

This equation tells us something unexpected:

> At a given temperature T, all ideal gas molecules—no matter what their mass—have the same average translational kinetic energy, namely, $\frac{3}{2}kT$. When we measure the temperature of a gas, we are also measuring the average translational kinetic energy of its molecules.

✔**CHECKPOINT 2:**   A gas mixture consists of molecules of types 1, 2, and 3, with molecular masses $m_1 > m_2 > m_3$. Rank the three types according to (a) average kinetic energy and (b) rms speed, greatest first.

20-6 Mean Free Path

We continue to examine the motion of molecules in an ideal gas. Figure 20-4 shows the path of a typical molecule as it moves through the gas, changing both speed and direction abruptly as it collides elastically with other molecules. Between collisions, our typical molecule moves in a straight line at constant speed. Although the figure shows all the other molecules as stationary, they are moving similarly.

One useful parameter to describe this random motion is the **mean free path** λ of the molecules. As its name implies, λ is the average distance traversed by a molecule between collisions. We expect λ to vary inversely with N/V, the number of molecules per unit volume (or density of molecules). The larger N/V is, the more collisions there should be and the smaller the mean free path. We also expect λ to vary inversely with the size of the molecules, say, with their diameter d. (If the molecules were points, as we have assumed them to be, they would never collide and the mean free path would be infinite.) Thus, the larger the molecules are, the smaller the mean free path. We can even predict that λ should vary (inversely) as the *square* of the molecular diameter because the cross section of a molecule—not its diameter—determines its effective target area.

The expression for the mean free path does, in fact, turn out to be

$$\lambda = \frac{1}{\sqrt{2}\,\pi d^2\, N/V} \qquad \text{(mean free path).} \qquad (20\text{-}25)$$

Fig. 20-5 (a) A collision occurs when the centers of two molecules come within a distance d of each other, d being the molecular diameter. (b) An equivalent but more convenient representation is to think of the moving molecule as having a *radius* d and all other molecules as being points. The condition for a collision is unchanged.

To justify Eq. 20-25, we focus attention on a single molecule and assume—as Fig. 20-4 suggests—that our molecule is traveling with a constant speed v and that all the other molecules are at rest. Later, we shall relax this assumption.

We assume further that the molecules are spheres of diameter d. A collision will then take place if the centers of the molecules come within a distance d of each other, as in Fig. 20-5a. Another, more helpful way to look at the situation is to consider our single molecule to have a *radius* of d and all the other molecules to be *points*, as in Fig. 20-5b. This does not change our criterion for a collision.

As our single molecule zigzags through the gas, it sweeps out a short cylinder of cross-sectional area πd^2 between successive collisions. If we watch this molecule for a time interval Δt, it moves a distance $v\,\Delta t$, where v is its assumed speed. Thus, if we align all the short cylinders swept out in interval Δt, we form a composite cylinder (Fig. 20-6) of length $v\,\Delta t$ and volume $(\pi d^2)(v\,\Delta t)$. The number of collisions that occur in time Δt is then equal to the number of (point) molecules that lie within this cylinder.

Since N/V is the number of molecules per unit volume, the number of molecules in the cylinder is N/V times the volume of the cylinder, or $(N/V)(\pi d^2 v\,\Delta t)$. This is also the number of collisions in time Δt. The mean free path is the length of the

Fig. 20-6 In time Δt the moving molecule effectively sweeps out a cylinder of length $v\,\Delta t$ and radius d.

path (and of the cylinder) divided by this number:

$$\lambda = \frac{\text{length of path during } \Delta t}{\text{number of collisions in } \Delta t} \approx \frac{v\,\Delta t}{\pi d^2 v\,\Delta t\,N/V}$$

$$= \frac{1}{\pi d^2\,N/V}. \tag{20-26}$$

This equation is only approximate because it is based on the assumption that all the molecules except one are at rest. In fact, *all* the molecules are moving; when this is taken properly into account, Eq. 20-25 results. Note that it differs from the (approximate) Eq. 20-26 only by a factor of $1/\sqrt{2}$.

We can even get a glimpse of what is "approximate" about Eq. 20-26. The v in the numerator and that in the denominator are—strictly—not the same. The v in the numerator is v_{avg}, the mean speed of the molecule *relative to the container*. The v in the denominator is v_{rel}, the mean speed of our single molecule *relative to the other molecules*, which are moving. It is this latter average speed that determines the number of collisions. A detailed calculation, taking into account the actual speed distribution of the molecules, gives $v_{rel} = \sqrt{2}\,v_{avg}$ and thus the factor $\sqrt{2}$.

The mean free path of air molecules at sea level is about $0.1\ \mu m$. At an altitude of 100 km, the density of air has dropped to such an extent that the mean free path rises to about 16 cm. At 300 km, the mean free path is about 20 km. A problem faced by those who would study the physics and chemistry of the upper atmosphere in the laboratory is the unavailability of containers large enough to hold gas samples that simulate upper atmospheric conditions. Yet studies of the concentrations of Freon, carbon dioxide, and ozone in the upper atmosphere are of vital public concern.

Sample Problem 20-4

(a) What is the mean free path λ for oxygen molecules at temperature $T = 300$ K and pressure $p = 1.0$ atm? Assume that the molecular diameter is $d = 290$ pm and the gas is ideal.

SOLUTION: The Key Idea here is that each oxygen molecule moves among other *moving* oxygen molecules in a zigzag path due to the resulting collisions. Thus, we use Eq. 20-25 for the mean free path, for which we need the number of molecules per unit volume, N/V. Because we assume the gas is ideal, we can use the ideal gas law of Eq. 20-9 ($pV = NkT$) to write $N/V = p/kT$. Substituting this into Eq. 20-25, we find

$$\lambda = \frac{1}{\sqrt{2}\,\pi d^2\,N/V} = \frac{kT}{\sqrt{2}\,\pi d^2 p}$$

$$= \frac{(1.38 \times 10^{-23}\ \text{J/K})(300\ \text{K})}{\sqrt{2}\,\pi(2.9 \times 10^{-10}\ \text{m})^2(1.01 \times 10^5\ \text{Pa})}$$

$$= 1.1 \times 10^{-7}\ \text{m}. \qquad \text{(Answer)}$$

This is about 380 molecular diameters.

(b) Assume the average speed of the oxygen molecules is $v = 450$ m/s. What is the average time t between successive collisions for any given molecule? At what rate does the molecule collide; that is, what is the frequency f of its collisions?

SOLUTION: To find the time t between collisions, we use this Key Idea: Between collisions, the molecule travels, on average, the mean free path λ at speed v. Thus, the average time between collisions is

$$t = \frac{\text{distance}}{\text{speed}} = \frac{\lambda}{v} = \frac{1.1 \times 10^{-7}\ \text{m}}{450\ \text{m/s}}$$

$$= 2.44 \times 10^{-10}\ \text{s} \approx 0.24\ \text{ns}. \qquad \text{(Answer)}$$

This tells us that, on average, any given oxygen molecule has less than a nanosecond between collisions.

To find the frequency f of the collisions, we use this Key Idea: The average rate or frequency at which the collisions occur is the inverse of the time t between collisions. Thus,

$$f = \frac{1}{t} = \frac{1}{2.44 \times 10^{-10}\ \text{s}} = 4.1 \times 10^9\ \text{s}^{-1}. \qquad \text{(Answer)}$$

This tells us that, on average, any given oxygen molecule makes about 4 billion collisions per second.

✔**CHECKPOINT 3:** One mole of gas A, with molecular diameter $2d_0$ and average molecular speed v_0, is placed inside a certain container. One mole of gas B, with molecular diameter d_0 and average molecular speed $2v_0$ (the molecules of B are smaller but faster), is placed in an identical container. Which gas has the greater average collision rate within its container?

20-7 The Distribution of Molecular Speeds

The root-mean-square speed v_{rms} gives us a general idea of molecular speeds in a gas at a given temperature. We often want to know more. For example, what fraction of the molecules have speeds greater than the rms value? Greater than twice the rms value? To answer such questions, we need to know how the possible values of speed are distributed among the molecules. Figure 20-7a shows this distribution for oxygen molecules at room temperature ($T = 300$ K); Fig. 20-7b compares it with the distribution at $T = 80$ K.

In 1852, Scottish physicist James Clerk Maxwell first solved the problem of finding the speed distribution of gas molecules. His result, known as **Maxwell's speed distribution law,** is

$$P(v) = 4\pi\left(\frac{M}{2\pi RT}\right)^{3/2} v^2 e^{-Mv^2/2RT}. \qquad (20\text{-}27)$$

Here v is the molecular speed, T is the gas temperature, M is the molar mass of the gas, and R is the gas constant. It is this equation that is plotted in Fig. 20-7a,b. The quantity $P(v)$ in Eq. 20-27 and Fig. 20-7 is a *probability distribution function:* For any speed v, the product $P(v)\,dv$ (a dimensionless quantity) is the fraction of molecules whose speeds lie in the interval of width dv centered on speed v.

As Fig. 20-7a shows, this fraction is equal to the area of a strip with height $P(v)$ and width dv. The total area under the distribution curve corresponds to the fraction of the molecules whose speeds lie between zero and infinity. All molecules fall into this category, so the value of this total area is unity; that is,

$$\int_0^\infty P(v)\,dv = 1. \qquad (20\text{-}28)$$

The fraction (frac) of molecules with speeds in an interval of, say, v_1 to v_2 is then

$$\text{frac} = \int_{v_1}^{v_2} P(v)\,dv. \qquad (20\text{-}29)$$

Average, RMS, and Most Probable Speeds

In principle, we can find the **average speed** v_{avg} of the molecules in a gas with the following procedure: We *weight* each value of v in the distribution; that is, we

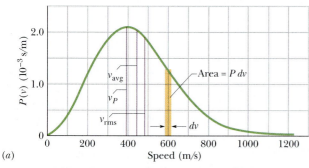

(a)

Fig. 20-7 (a) The Maxwell speed distribution for oxygen molecules at $T = 300$ K. The three characteristic speeds are marked. (b) The curves for 300 K and 80 K. Note that the molecules move more slowly at the lower temperature. Because these are probability distributions, the area under each curve has a numerical value of unity.

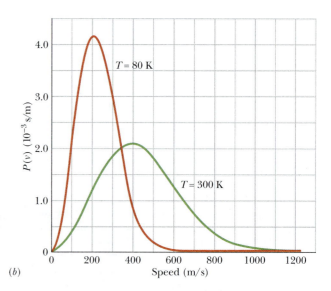

(b)

multiply it by the fraction $P(v) \, dv$ of molecules with speeds in a differential interval dv centered on v. Then we add up all these values of $v \, P(v) \, dv$. The result is v_{avg}. In practice, we do all this by evaluating

$$v_{avg} = \int_0^\infty v \, P(v) \, dv. \tag{20-30}$$

Substituting for $P(v)$ from Eq. 20-27 and using generic integral 20 from the list of integrals in Appendix E, we find

$$v_{avg} = \sqrt{\frac{8RT}{\pi M}} \qquad \text{(average speed)}. \tag{20-31}$$

Similarly, we can find the average of the square of the speeds $(v^2)_{avg}$ with

$$(v^2)_{avg} = \int_0^\infty v^2 \, P(v) \, dv. \tag{20-32}$$

Substituting for $P(v)$ from Eq. 20-27 and using generic integral 16 from the list of integrals in Appendix E, we find

$$(v^2)_{avg} = \frac{3RT}{M}. \tag{20-33}$$

The square root of $(v^2)_{avg}$ is the **root-mean-square speed** v_{rms}. Thus,

$$v_{rms} = \sqrt{\frac{3RT}{M}} \qquad \text{(rms speed)}, \tag{20-34}$$

which agrees with Eq. 20-22.

The **most probable speed** v_P is the speed at which $P(v)$ is maximum (see Fig. 20-7a). To calculate v_P, we set $dP/dv = 0$ (the slope of the curve in Fig. 20-7a is zero at the maximum of the curve) and then solve for v. Doing so, we find

$$v_P = \sqrt{\frac{2RT}{M}} \qquad \text{(most probable speed)}. \tag{20-35}$$

A molecule is more likely to have speed v_P than any other speed, but some molecules will have speeds that are many times v_P. These molecules lie in the *high-speed tail* of a distribution curve like that in Fig. 20-7a. We should be thankful for these few, higher speed molecules because they make possible both rain and sunshine (without which we could not exist). We next see why.

Rain: The speed distribution of water molecules in, say, a pond at summertime temperatures can be represented by a curve similar to that of Fig. 20-7a. Most of the molecules do not have nearly enough kinetic energy to escape from the water through its surface. However, small numbers of very fast molecules with speeds far out in the tail of the curve can do so. It is these water molecules that evaporate, making clouds and rain a possibility.

As the fast water molecules leave the surface, carrying energy with them, the temperature of the remaining water is maintained by heat transfer from the surroundings. Other fast molecules—produced in particularly favorable collisions—quickly take the place of those that have left, and the speed distribution is maintained.

Sunshine: Let the distribution curve of Fig. 20-7a now refer to protons in the core of the Sun. The Sun's energy is supplied by a nuclear fusion process that starts with the merging of two protons. However, protons repel each other because of their electrical charges, and protons of average speed do not have enough kinetic energy to overcome the repulsion and get close enough to merge. Very fast protons with speeds in the tail of the distribution curve can do so, however, and for that reason the Sun can shine.

Sample Problem 20-5

A container is filled with oxygen gas maintained at room temperature (300 K). What fraction of the molecules have speeds in the interval 599 to 601 m/s? The molar mass M of oxygen is 0.0320 kg/mol.

SOLUTION: The Key Ideas here are

1. The speeds of the molecules are distributed over a wide range of values, with the distribution $P(v)$ of Eq. 20-27.

2. The fraction of the molecules with speeds in a differential interval dv is $P(v)\,dv$.

3. For a larger interval, the fraction is found by integrating $P(v)$ over the interval.

4. However, the interval $\Delta v = 2$ m/s here is small compared to the speed $v = 600$ m/s on which it is centered.

Thus, we can avoid the integration by approximating the fraction as

$$\text{frac} = P(v)\,\Delta v = 4\pi \left(\frac{M}{2\pi RT}\right)^{3/2} v^2 e^{-Mv^2/2RT}\,\Delta v.$$

The function $P(v)$ is plotted in Fig. 20-7a. The total area between the curve and the horizontal axis represents the total fraction of molecules (unity). The area of the thin gold strip represents the fraction we seek.

To evaluate frac in parts, we can write

$$\text{frac} = (4\pi)(A)(v^2)(e^B)(\Delta v). \tag{20-36}$$

A and B are

$$A = \left(\frac{M}{2\pi RT}\right)^{3/2} = \left(\frac{0.0320 \text{ kg/mol}}{(2\pi)(8.31 \text{ J/mol}\cdot\text{K})(300 \text{ K})}\right)^{3/2}$$
$$= 2.92 \times 10^{-9} \text{ s}^3/\text{m}^3$$

and $B = -\dfrac{Mv^2}{2RT} = -\dfrac{(0.0320 \text{ kg/mol})(600 \text{ m/s})^2}{(2)(8.31 \text{ J/mol}\cdot\text{K})(300 \text{ K})} = -2.31.$

Substituting A and B into Eq. 20-36 yields

$$\text{frac} = (4\pi)(A)(v^2)(e^B)(\Delta v)$$
$$= (4\pi)(2.92 \times 10^{-9} \text{ s}^3/\text{m}^3)(600 \text{ m/s})^2(e^{-2.31})(2 \text{ m/s})$$
$$= 2.62 \times 10^{-3}. \tag{Answer}$$

Thus, at room temperature, 0.262% of the oxygen molecules will have speeds that lie in the narrow range between 599 and 601 m/s. If the gold strip of Fig. 20-7a were drawn to the scale of this problem, it would be a very thin strip indeed.

Sample Problem 20-6

The molar mass M of oxygen is 0.0320 kg/mol.

(a) What is the average speed v_{avg} of oxygen gas molecules at $T = 300$ K?

SOLUTION: The Key Idea here is that to find the average speed, we must weight speed v with the distribution function $P(v)$ of Eq. 20-27 and then integrate the resulting expression over the range of possible speeds (0 to ∞). That leads to Eq. 20-31, which gives us

$$v_{avg} = \sqrt{\frac{8RT}{\pi M}}$$
$$= \sqrt{\frac{8(8.31 \text{ J/mol}\cdot\text{K})(300 \text{ K})}{\pi(0.0320 \text{ kg/mol})}} = 445 \text{ m/s}. \tag{Answer}$$

This result is plotted in Fig. 20-7a.

(b) What is the root-mean-square speed v_{rms} at 300 K?

SOLUTION: The Key Idea here is that to find v_{rms}, we must first find $(v^2)_{avg}$ by weighting v^2 with the distribution function $P(v)$ of Eq. 20-27 and then integrating the expression over the range of possible speeds. Then we must take the square root of the result. That leads to Eq. 20-34, which gives us

$$v_{rms} = \sqrt{\frac{3RT}{M}}$$
$$= \sqrt{\frac{3(8.31 \text{ J/mol}\cdot\text{K})(300 \text{ K})}{0.0320 \text{ kg/mol}}} = 483 \text{ m/s}. \tag{Answer}$$

This result is plotted in Fig. 20-7a. It is greater than v_{avg} because the greater speed values influence the calculation more when we integrate the v^2 values than when we integrate the v values.

(c) What is the most probable speed v_P at 300 K?

SOLUTION: The Key Idea here is that v_P corresponds to the maximum of the distribution function $P(v)$, which we obtain by setting the derivative $dP/dv = 0$ and solving the result for v. That leads to Eq. 20-35, which gives us

$$v_P = \sqrt{\frac{2RT}{M}}$$
$$= \sqrt{\frac{2(8.31 \text{ J/mol}\cdot\text{K})(300 \text{ K})}{0.0320 \text{ kg/mol}}} = 395 \text{ m/s}. \tag{Answer}$$

This result is also plotted in Fig. 20-7a.

20-8 The Molar Specific Heats of an Ideal Gas

In this section, we want to derive from molecular considerations an expression for the internal energy E_{int} of an ideal gas. In other words, we want an expression for the energy associated with the random motions of the atoms or molecules in the gas. We shall then use that expression to derive the molar specific heats of an ideal gas.

Internal Energy E_{int}

Let us first assume that our ideal gas is a *monatomic gas* (which has individual atoms rather than molecules), such as helium, neon, or argon. Let us also assume that the internal energy E_{int} of our ideal gas is simply the sum of the translational kinetic energies of its atoms. (As explained by quantum theory, individual atoms do not have rotational kinetic energy.)

The average translational kinetic energy of a single atom depends only on the gas temperature and is given by Eq. 20-24 as $K_{avg} = \frac{3}{2}kT$. A sample of n moles of such a gas contains nN_A atoms. The internal energy E_{int} of the sample is then

$$E_{int} = (nN_A)K_{avg} = (nN_A)(\tfrac{3}{2}kT). \tag{20-37}$$

Using Eq. 20-7 ($k = R/N_A$), we can rewrite this as

$$E_{int} = \tfrac{3}{2}nRT \qquad \text{(monatomic ideal gas).} \tag{20-38}$$

Thus,

> The internal energy E_{int} of an ideal gas is a function of the gas temperature *only*; it does not depend on any other variable.

With Eq. 20-38 in hand, we are now able to derive an expression for the molar specific heat of an ideal gas. Actually, we shall derive two expressions. One is for the case in which the volume of the gas remains constant as energy is transferred to or from it as heat. The other is for the case in which the pressure of the gas remains constant as energy is transferred to or from it as heat. The symbols for these two molar specific heats are C_V and C_p, respectively. (By convention, the capital letter C is used in both cases, even though C_V and C_p represent types of specific heat and not heat capacities.)

Molar Specific Heat at Constant Volume

Figure 20-8a shows n moles of an ideal gas at pressure p and temperature T, confined to a cylinder of fixed volume V. This *initial state i* of the gas is marked on the p-V diagram of Fig. 20-8b. Suppose now that you add a small amount of energy to the gas as heat Q by slowly turning up the temperature of the thermal reservoir. The gas temperature rises a small amount to $T + \Delta T$, and its pressure rises to $p + \Delta p$, bringing the gas to *final state f*.

In such experiments, we would find that the heat Q is related to the temperature change ΔT by

$$Q = nC_V \Delta T \qquad \text{(constant volume),} \tag{20-39}$$

where C_V is a constant called the **molar specific heat at constant volume.** Substituting this expression for Q into the first law of thermodynamics as given by

Fig. 20-8 (*a*) The temperature of an ideal gas is raised from T to $T + \Delta T$ in a constant-volume process. Heat is added, but no work is done. (*b*) The process on a p-V diagram.

Eq. 19-26 ($\Delta E_{int} = Q - W$) yields

$$\Delta E_{int} = nC_V \Delta T - W. \tag{20-40}$$

With the volume held constant, the gas cannot expand and thus cannot do any work. Therefore, $W = 0$, and Eq. 20-40 gives us

$$C_V = \frac{\Delta E_{int}}{n \Delta T}. \tag{20-41}$$

From Eq. 20-38 we know that $E_{int} = \frac{3}{2}nRT$, so the change in internal energy must be

$$\Delta E_{int} = \frac{3}{2}nR \Delta T. \tag{20-42}$$

Substituting this result into Eq. 20-41 yields

$$C_V = \frac{3}{2}R = 12.5 \text{ J/mol} \cdot \text{K} \qquad \text{(monatomic gas)}. \tag{20-43}$$

As Table 20-2 shows, this prediction of the kinetic theory (for ideal gases) agrees very well with experiment for real monatomic gases, the case that we have assumed. The (predicted and) experimental values of C_V for *diatomic gases* (which have molecules with two atoms) and *polyatomic gases* (which have molecules with more than two atoms) are greater than those for monatomic gases for reasons that will be suggested in Section 20-9.

We can now generalize Eq. 20-38 for the internal energy of any ideal gas by substituting C_V for $\frac{3}{2}R$; we get

$$E_{int} = nC_V T \qquad \text{(any ideal gas)}. \tag{20-44}$$

This equation applies not only to an ideal monatomic gas but also to diatomic and polyatomic ideal gases, provided the appropriate value of C_V is used. Just as with Eq. 20-38, we see that the internal energy of a gas depends on the temperature of the gas but not on its pressure or density.

When an ideal gas that is confined to a container undergoes a temperature change ΔT, then from either Eq. 20-41 or Eq. 20-44 we can write the resulting change in its internal energy as

$$\Delta E_{int} = nC_V \Delta T \qquad \text{(ideal gas, any process)}. \tag{20-45}$$

TABLE 20-2   **Molar Specific Heats at Constant Volume**

Molecule		Example	C_V (J/mol $\cdot$ K)
Monatomic	Ideal		$\frac{3}{2}R = 12.5$
	Real	He	12.5
		Ar	12.6
Diatomic	Ideal		$\frac{5}{2}R = 20.8$
	Real	N_2	20.7
		O_2	20.8
Polyatomic	Ideal		$3R = 24.9$
	Real	NH_4	29.0
		CO_2	29.7

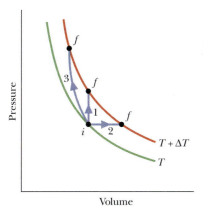

Fig. 20-9 Three paths for three different processes that take an ideal gas from an initial state i at temperature T to some final state f at temperature $T + \Delta T$. The change ΔE_{int} in the internal energy of the gas is the same for these three processes and for any others that result in the same change of temperature.

This equation tells us:

▶ A change in the internal energy E_{int} of a confined ideal gas depends on the change in the gas temperature only; it does *not* depend on what type of process produces the change in the temperature.

As examples, consider the three paths between the two isotherms in the p-V diagram of Fig. 20-9. Path 1 represents a constant-volume process. Path 2 represents a constant-pressure process (that we are about to examine). Path 3 represents a process in which no heat is exchanged with the system's environment (we discuss this in Section 20-11). Although the values of heat Q and work W associated with these three paths differ, as do p_f and V_f, the values of ΔE_{int} associated with the three paths are identical and are all given by Eq. 20-45, because they all involve the same temperature change ΔT. Therefore, no matter what path is actually taken between T and $T + \Delta T$, we can *always* use path 1 and Eq. 20-45 to compute ΔE_{int} easily.

Molar Specific Heat at Constant Pressure

We now assume that the temperature of the ideal gas is increased by the same small amount ΔT as previously, but that the necessary energy (heat Q) is added with the gas under constant pressure. An experiment for doing this is shown in Fig. 20-10a; the p-V diagram for the process is plotted in Fig. 20-10b. From such experiments we find that the heat Q is related to the temperature change ΔT by

$$Q = nC_p \, \Delta T \qquad \text{(constant pressure)}, \qquad (20\text{-}46)$$

where C_p is a constant called the **molar specific heat at constant pressure.** This C_p is *greater* than the molar specific heat at constant volume C_V, because energy must now be supplied not only to raise the temperature of the gas but also for the gas to do work—that is, to lift the weighted piston of Fig. 20-10a.

To relate molar specific heats C_p and C_V, we start with the first law of thermodynamics (Eq. 19-26):

$$\Delta E_{\text{int}} = Q - W. \qquad (20\text{-}47)$$

We next replace each term in Eq. 20-47. For ΔE_{int}, we substitute from Eq. 20-45. For Q, we substitute from Eq. 20-46. To replace W, we first note that since the pressure remains constant, Eq. 20-16 tells us that $W = p \, \Delta V$. Then we note that, using the ideal gas equation ($pV = nRT$), we can write

$$W = p \, \Delta V = nR \, \Delta T. \qquad (20\text{-}48)$$

Fig. 20-10 (a) The temperature of an ideal gas is raised from T to $T + \Delta T$ in a constant-pressure process. Heat is added and work is done in lifting the loaded piston. (b) The process on a p-V diagram. The work $p \, \Delta V$ is given by the shaded area.

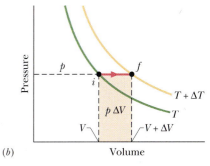

Making these substitutions in Eq. 20-47, and then dividing through by $n \, \Delta T$, we find

$$C_V = C_p - R$$

and then

$$C_p = C_V + R. \qquad (20\text{-}49)$$

This prediction of kinetic theory agrees well with experiment, not only for monatomic gases but for gases in general, as long as their density is low enough so that we may treat them as ideal.

✔**CHECKPOINT 4:**   The figure here shows five paths traversed by a gas on a $p\text{-}V$ diagram. Rank the paths according to the change in internal energy of the gas, greatest first.

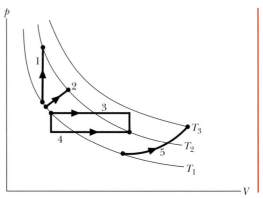

Sample Problem 20-7

A bubble of 5.00 mol of helium is submerged at a certain depth in liquid water when the water (and thus the helium) undergoes a temperature increase ΔT of 20.0 C° at constant pressure. As a result, the bubble expands. The helium is monatomic and ideal.

(a) How much energy is added as heat to the helium during the increase and expansion?

SOLUTION: One **Key Idea** here is that the heat Q is related to the temperature change ΔT by a molar specific heat of the gas. Because the pressure p is held constant during the addition of energy, we use the molar specific heat at constant pressure C_p and Eq. 20-46,

$$Q = nC_p \, \Delta T, \qquad (20\text{-}50)$$

to find Q. To evaluate C_p we go to Eq. 20-49, which tells us that for any ideal gas, $C_p = C_V + R$. Then from Eq. 20-43, we know that for any *monatomic* gas (like the helium here), $C_V = \frac{3}{2}R$. Thus, Eq. 20-50 gives us

$$Q = n(C_V + R) \, \Delta T = n(\tfrac{3}{2}R + R) \, \Delta T = n(\tfrac{5}{2}R) \, \Delta T$$
$$= (5.00 \text{ mol})(2.5)(8.31 \text{ J/mol} \cdot \text{K})(20.0 \text{ C°})$$
$$= 2077.5 \text{ J} \approx 2080 \text{ J}. \qquad \text{(Answer)}$$

(b) What is the change ΔE_{int} in the internal energy of the helium during the temperature increase?

SOLUTION: Because the bubble expands, this is not a constant-volume process. However, the helium is nonetheless confined (to the bubble). Thus, a **Key Idea** here is that the change ΔE_{int} is the same as *would occur* in a constant-volume process with the same temperature change ΔT. We can easily find the constant-volume change ΔE_{int} with Eq. 20-45:

$$\Delta E_{int} = nC_V \, \Delta T = n(\tfrac{3}{2}R) \, \Delta T$$
$$= (5.00 \text{ mol})(1.5)(8.31 \text{ J/mol} \cdot \text{K})(20.0 \text{ C°})$$
$$= 1246.5 \text{ J} \approx 1250 \text{ J}. \qquad \text{(Answer)}$$

(c) How much work W is done by the helium as it expands against the pressure of the surrounding water during the temperature increase?

SOLUTION: One **Key Idea** here is that the work done by *any* gas expanding against the pressure from its environment is given by Eq. 20-11, which tells us to integrate $p \, dV$. When the pressure is constant (as here), we can simplify that to $W = p \, \Delta V$. When the gas is *ideal* (as here), we can use the ideal gas law (Eq. 20-5) to write $p \, \Delta V = nR \, \Delta T$. We end up with

$$W = nR \, \Delta T$$
$$= (5.00 \text{ mol})(8.31 \text{ J/mol} \cdot \text{K})(20.0 \text{ C°})$$
$$= 831 \text{ J}. \qquad \text{(Answer)}$$

Because we happen to know Q and ΔE_{int}, we can work this problem another way: The **Key Idea** now is that we can account for the energy changes of the gas with the first law of thermodynamics, writing

$$W = Q - \Delta E_{int} = 2077.5 \text{ J} - 1246.5 \text{ J}$$
$$= 831 \text{ J}. \qquad \text{(Answer)}$$

Note that during the temperature increase, only a portion (1250 J) of the energy (2080 J) that is transferred to the helium as heat goes to increasing the internal energy of the helium and thus the temperature of the helium. The rest (831 J) is transferred out of the helium as work that the helium does during the expansion. If the water were frozen, it would not allow that expansion. Then the same temperature increase of 20.0 C° would require only 1250 J of heat, because no work would be done by the helium.

Fig. 20-11 Models of molecules as used in kinetic theory: (*a*) helium, a typical monatomic molecule; (*b*) oxygen, a typical diatomic molecule; and (*c*) methane, a typical polyatomic molecule. The spheres represent atoms, and the lines between them represent bonds. Two rotation axes are shown for the oxygen molecule.

20-9 Degrees of Freedom and Molar Specific Heats

As Table 20-2 shows, the prediction that $C_V = \frac{3}{2}R$ agrees with experiment for monatomic gases but fails for diatomic and polyatomic gases. Let us try to explain the discrepancy by considering the possibility that molecules with more than one atom can store internal energy in forms other than translational motion.

Figure 20-11 shows common models of helium (a *monatomic* molecule, containing a single atom), oxygen (a *diatomic* molecule, containing two atoms), and methane (a *polyatomic* molecule). From such models, we would assume that all three types of molecules can have translational motions (say, moving left–right and up–down) and rotational motions (spinning about an axis like a top). In addition, we would assume that the diatomic and polyatomic molecules can have oscillatory motions, with the atoms oscillating slightly toward and away from one another, as if attached to opposite ends of a spring.

To keep account of the various ways in which energy can be stored in a gas, James Clerk Maxwell introduced the theorem of the **equipartition of energy:**

▶ Every kind of molecule has a certain number *f* of *degrees of freedom*, which are independent ways in which the molecule can store energy. Each such degree of freedom has associated with it—on average—an energy of $\frac{1}{2}kT$ per molecule (or $\frac{1}{2}RT$ per mole).

Let us apply the theorem to the translational and rotational motions of the molecules in Fig. 20-11. (We discuss oscillatory motion in the next section.) For the translational motion, superimpose an *xyz* coordinate system on any gas. The molecules will, in general, have velocity components along all three axes. Thus, gas molecules of all types have three degrees of translational freedom (three ways to move in translation) and, on average, an associated energy of $3(\frac{1}{2}kT)$ per molecule.

For the rotational motion, imagine the origin of our *xyz* coordinate system at the center of each molecule in Fig. 20-11. In a gas, each molecule should be able to rotate with an angular velocity component along each of the three axes, so each gas should have three degrees of rotational freedom and, on average, an additional energy of $3(\frac{1}{2}kT)$ per molecule. *However,* experiment shows this is true only for the polyatomic molecules. As explained by quantum theory, a monatomic gas molecule does not rotate and so has no rotational energy (a single atom cannot rotate like a top). A diatomic molecule can rotate like a top only about axes perpendicular to the line connecting the atoms (the axes are shown in Fig. 20-11*b*) and not about that line itself. Therefore, a diatomic molecule can have only two degrees of rotational freedom and a rotational energy of only $2(\frac{1}{2}kT)$ per molecule.

To extend our analysis of molar specific heats (C_p and C_V, in Section 20-8) to ideal diatomic and polyatomic gases, it is necessary to retrace the derivations of that analysis in detail. First, we replace Eq. 20-38 ($E_{int} = \frac{3}{2}nRT$) with $E_{int} = (f/2)nRT$, where *f* is the number of degrees of freedom listed in Table 20-3. Doing so leads to

TABLE 20-3 Degrees of Freedom for Various Molecules

| Molecule | Example | Degrees of Freedom | | | Predicted Molar Specific Heats | |
		Translational	Rotational	Total (f)	C_V (Eq. 20-51)	$C_p = C_V + R$
Monatomic	He	3	0	3	$\frac{3}{2}R$	$\frac{5}{2}R$
Diatomic	O_2	3	2	5	$\frac{5}{2}R$	$\frac{7}{2}R$
Polyatomic	CH_4	3	3	6	$3R$	$4R$

the prediction

$$C_V = \left(\frac{f}{2}\right)R = 4.16f \text{ J/mol} \cdot \text{K}, \qquad (20\text{-}51)$$

which agrees—as it must—with Eq. 20-43 for monatomic gases ($f = 3$). As Table 20-2 shows, this prediction also agrees with experiment for diatomic gases ($f = 5$), but it is too low for polyatomic gases.

Sample Problem 20-8

A cabin of volume V is filled with air (which we consider to be an ideal diatomic gas) at an initial low temperature T_1. After you light a wood stove, the air temperature increases to T_2. What is the resulting change ΔE_{int} in the internal energy of the air in the cabin?

SOLUTION: As the air temperature increases, the air pressure p cannot change but must always be equal to the air pressure outside the room. The reason is that, because the room is not air-tight, the air is not confined. As the temperature increases, air molecules leave through various openings and thus the number of moles n of air in the room decreases. Thus, one Key Idea here is that we *cannot* use Eq. 20-45 ($\Delta E_{int} = nC_V \, \Delta T$) to find ΔE_{int}, because it requires constant n.

A second Key Idea is that we *can* relate the internal energy E_{int} at any instant to n and the temperature T with Eq. 20-44 ($E_{int} = nC_VT$). From that equation we can then write

$$\Delta E_{int} = \Delta(nC_VT) = C_V \, \Delta(nT).$$

Next, using Eq. 20-5 ($pV = nRT$), we can replace nT with pV/R, obtaining

$$\Delta E_{int} = C_V \, \Delta\left(\frac{pV}{R}\right). \qquad (20\text{-}52)$$

Now, because p, V, and R are all constants, Eq. 20-52 yields

$$\Delta E_{int} = 0, \qquad \text{(Answer)}$$

even though the temperature changes.

Why does the cabin feel more comfortable at the higher temperature? There are at least two factors involved: (1) You exchange electromagnetic radiation (thermal radiation) with surfaces inside the room, and (2) you exchange energy with air molecules that collide with you. When the room temperature is increased, (1) the amount of thermal radiation emitted by the surfaces and absorbed by you is increased, and (2) the amount of energy you gain through the collisions of air molecules with you is increased.

20-10   A Hint of Quantum Theory

We can improve the agreement of kinetic theory with experiment by including the oscillations of the atoms in a gas of diatomic or polyatomic molecules. For example, the two atoms in the O_2 molecule of Fig. 20-11b can oscillate toward and away from each other, with the interconnecting bond acting like a spring. However, experiment shows that such oscillations occur only at relatively high temperatures of the gas— the motion is "turned on" only when the gas molecules have relatively large energies. Rotational motion is also subject to such "turning on," but at a lower temperature.

Figure 20-12 is of help in seeing this turning on of rotational motion and oscillatory motion. The ratio C_V/R for diatomic hydrogen gas (H_2) is plotted there against temperature, with the temperature scale logarithmic to cover several orders of magnitude. Below about 80 K, we find that $C_V/R = 1.5$. This result implies that only the three translational degrees of freedom of hydrogen are involved in the specific heat.

As the temperature increases, the value of C_V/R gradually increases to 2.5, implying that two additional degrees of freedom have become involved. Quantum theory shows that these two degrees of freedom are associated with the rotational motion of the hydrogen molecules and that this motion requires a certain minimum amount of energy. At very low temperatures (below 80 K), the molecules do not have enough energy to rotate. As the temperature increases from 80 K, first a few molecules and then more and more obtain enough energy to rotate, and C_V/R increases, until all of them are rotating and $C_V/R = 2.5$.

Fig. 20-12 A plot of C_V/R verses temperature for (diatomic) hydrogen gas. Because rotational and oscillatory motions begin at certain energies, only translation is possible at very low temperatures. As the temperature increases, rotational motion can begin. At still higher temperatures, oscillatory motion can begin.

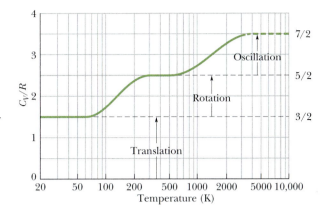

Similarly, quantum theory shows that oscillatory motion of the molecules requires a certain (higher) minimum amount of energy. This minimum amount is not met until the molecules reach a temperature of about 1000 K, as shown in Fig. 20-12. As the temperature increases beyond 1000 K, the number of molecules with enough energy to oscillate increases, and C_V/R increases, until all of them are oscillating and $C_V/R = 3.5$. (In Fig. 20-12, the plotted curve stops at 3200 K because at that temperature, the atoms of a hydrogen molecule oscillate so much that they overwhelm their bond, and the molecule then *dissociates* into two separate atoms.)

20-11 The Adiabatic Expansion of an Ideal Gas

We saw in Section 18-3 that sound waves are propagated through air and other gases as a series of compressions and expansions; these variations in the transmission medium take place so rapidly that there is no time for energy to be transferred from one part of the medium to another as heat. As we saw in Section 19-10, a process for which $Q = 0$ is an *adiabatic process*. We can ensure that $Q = 0$ either by carrying out the process very quickly (as in sound waves) or by doing it (at any rate) in a well-insulated container. Let us see what the kinetic theory has to say about adiabatic processes.

Figure 20-13a shows our usual insulated cylinder, now containing an ideal gas and resting on an insulating stand. By removing mass from the piston, we can allow the gas to expand adiabatically. As the volume increases, both the pressure and the temperature drop. We shall prove next that the relation between the pressure and the volume during such an adiabatic process is

$$pV^\gamma = \text{a constant} \qquad \text{(adiabatic process)}, \qquad (20\text{-}53)$$

in which $\gamma = C_p/C_V$, the ratio of the molar specific heats for the gas. On a p-V diagram such as that in Fig. 20-13b, the process occurs along a line (called an *adiabat*) that has the equation $p = (\text{a constant})/V^\gamma$. Since the gas goes from an initial state i to a final state f, we can rewrite Eq. 20-53 as

$$p_i V_i^\gamma = p_f V_f^\gamma \qquad \text{(adiabatic process)}. \qquad (20\text{-}54)$$

We can also write an equation for an adiabatic process in terms of T and V. To do so, we use the ideal gas equation ($pV = nRT$) to eliminate p from Eq. 20-53, finding

$$\left(\frac{nRT}{V}\right) V^\gamma = \text{a constant}.$$

Fig. 20-13 (a) The volume of an ideal gas is increased by removing mass from the piston. The process is adiabatic ($Q = 0$). (b) The process proceeds from i to f along an adiabat on a p-V diagram.

Because n and R are constants, we can rewrite this in the alternative form

$$TV^{\gamma - 1} = \text{a constant} \qquad \text{(adiabatic process),} \qquad (20\text{-}55)$$

in which the constant is different from that in Eq. 20-53. When the gas goes from an initial state i to a final state f, we can rewrite Eq. 20-55 as

$$T_i V_i^{\gamma - 1} = T_f V_f^{\gamma - 1} \qquad \text{(adiabatic process).} \qquad (20\text{-}56)$$

We can now answer the question that opens this chapter. At the top of an unopened carbonated drink, there is a gas of carbon dioxide and water vapor. Because the pressure of the gas is greater than atmospheric pressure, the gas expands out into the atmosphere when the container is opened. Thus, the gas increases its volume, but that means it must do work to push against the atmosphere. Because the expansion is so rapid, it is adiabatic and the only source of energy for the work is the internal energy of the gas. Because the internal energy decreases, the temperature of the gas must also decrease, which can cause the water vapor in the gas to condense into tiny drops, forming the fog. (Note that Eq. 20-56 also tells us that the temperature must decrease during an adiabatic expansion: Since V_f is greater than V_i, then T_f must be less than T_i.)

Proof of Eq. 20-53

Suppose that you remove some shot from the piston of Fig. 20-13a, allowing the ideal gas to push the piston and the remaining shot upward and thus to increase the volume by a differential amount dV. Since the volume change is tiny, we may assume that the pressure p of the gas on the piston is constant during the change. This assumption allows us to say that the work dW done by the gas during the volume increase is equal to $p\,dV$. From Eq. 19-27, the first law of thermodynamics can then be written as

$$dE_{\text{int}} = Q - p\,dV. \qquad (20\text{-}57)$$

Since the gas is thermally insulated (and thus the expansion is adiabatic), we substitute 0 for Q. Then we use Eq. 20-45 to substitute $nC_V\,dT$ for dE_{int}. With these substitutions, and after some rearranging, we have

$$n\,dT = -\left(\frac{p}{C_V}\right) dV. \qquad (20\text{-}58)$$

Now from the ideal gas law ($pV = nRT$) we have

$$p \, dV + V \, dp = nR \, dT. \quad (20\text{-}59)$$

Replacing R with its equal, $C_p - C_V$, in Eq. 20-59 yields

$$n \, dT = \frac{p \, dV + V \, dp}{C_p - C_V}. \quad (20\text{-}60)$$

Equating Eqs. 20-58 and 20-60 and rearranging then give

$$\frac{dp}{p} + \left(\frac{C_p}{C_V}\right) \frac{dV}{V} = 0.$$

Replacing the ratio of the molar specific heats with γ and integrating (see integral 5 in Appendix E) yield

$$\ln p + \gamma \ln V = \text{a constant.}$$

Rewriting the left side as $\ln pV^\gamma$ and then taking the antilog of both sides, we find

$$pV^\gamma = \text{a constant,} \quad (20\text{-}61)$$

which is what we set out to prove.

Free Expansions

Recall from Section 19-10 that a *free expansion* of a gas is an adiabatic process that involves no work done on or by the gas, and no change in the internal energy of the gas. A free expansion is thus quite different from the type of adiabatic process described by Eqs. 20-53 through 20-61, in which work is done and the internal energy changes. Those equations then do *not* apply to a free expansion, even though such an expansion is adiabatic.

Also recall that in a free expansion, a gas is in equilibrium only at its initial and final points; thus, we can plot only those points, but not the expansion itself, on a p-V diagram. In addition, because $\Delta E_{int} = 0$, the temperature of the final state must be that of the initial state. Thus, the initial and final points on a p-V diagram must be on the same isotherm, and instead of Eq. 20-56 we have

$$T_i = T_f \quad \text{(free expansion).} \quad (20\text{-}62)$$

If we next assume that the gas is ideal (so that $pV = nRT$), then because there is no change in temperature, there can be no change in the product pV. Thus, instead of Eq. 20-53 a free expansion involves the relation

$$p_i V_i = p_f V_f \quad \text{(free expansion).} \quad (20\text{-}63)$$

Sample Problem 20-9

In Sample Problem 20-2, 1 mol of oxygen (assumed to be an ideal gas) expands isothermally (at 310 K) from an initial volume of 12 L to a final volume of 19 L.

(a) What would be the final temperature if the gas had expanded adiabatically to this same final volume? Oxygen (O_2) is diatomic and here has rotation but not oscillation.

SOLUTION: The Key Ideas here are

1. When a gas expands against the pressure of its environment, it must do work.

2. When the process is adiabatic (no energy is transferred as heat), then the energy required for the work can come only from the internal energy of the gas.

3. Because the internal energy decreases, the temperature T must also decrease.

We can relate the initial and final temperatures and volumes with Eq. 20-56:

$$T_i V_i^{\gamma-1} = T_f V_f^{\gamma-1}. \quad (20\text{-}64)$$

Because the molecules are diatomic and have rotation but not oscillation, we can take the molar specific heats from Table 20-3.

Thus,

$$\gamma = \frac{C_p}{C_V} = \frac{\frac{7}{2}R}{\frac{5}{2}R} = 1.40.$$

Solving Eq. 20-64 for T_f and inserting known data then yield

$$T_f = \frac{T_i V_i^{\gamma-1}}{V_f^{\gamma-1}} = \frac{(310 \text{ K})(12 \text{ L})^{1.40-1}}{(19 \text{ L})^{1.40-1}}$$

$$= (310 \text{ K})(\tfrac{12}{19})^{0.40} = 258 \text{ K.} \qquad \text{(Answer)}$$

(b) What would be the final temperature and pressure if, instead, the gas had expanded freely to the new volume, from an initial pressure of 2.0 Pa?

SOLUTION: Here the Key Idea is that the temperature does not change in a free expansion:

$$T_f = T_i = 310 \text{ K.} \qquad \text{(Answer)}$$

We find the new pressure using Eq. 20-63, which gives us

$$p_f = p_i \frac{V_i}{V_f} = (2.0 \text{ Pa}) \frac{12 \text{ L}}{19 \text{ L}} = 1.3 \text{ Pa.} \qquad \text{(Answer)}$$

PROBLEM-SOLVING TACTICS

Tactic 2: *A Graphical Summary of Four Gas Processes*
In this chapter we have discussed four special processes that an ideal gas can undergo. An example of each is shown in Fig. 20-14, and some associated characteristics are given in Table 20-4, including two process names (isobaric and isochoric) that we have not used but that you might see in other courses.

✔CHECKPOINT 5:   Rank paths 1, 2, and 3 in Fig. 20-14 according to the heat transfer to the gas, greatest first.

TABLE 20-4   Four Special Processes

Path in Fig. 20-14	Constant Quantity	Process Type	Some Special Results ($\Delta E_{int} = Q - W$ and $\Delta E_{int} = nC_V \Delta T$ for all paths)
1	p	Isobaric	$Q = nC_p \Delta T;$ $W = p \Delta V$
2	T	Isothermal	$Q = W = nRT \ln(V_f/V_i);$ $\Delta E_{int} = 0$
3	$pV^\gamma,$ $TV^{\gamma-1}$	Adiabatic	$Q = 0;$ $W = -\Delta E_{int}$
4	V	Isochoric	$Q = \Delta E_{int} = nC_V \Delta T;$ $W = 0$

Fig. 20-14 A *p-V* diagram representing four special processes for an ideal gas. Table 20-4 explains the processes.

REVIEW & SUMMARY

Kinetic Theory of Gases   The *kinetic theory of gases* relates the *macroscopic* properties of gases (for example, pressure and temperature) to the *microscopic* properties of gas molecules (for example, speed and kinetic energy).

Avogadro's Number   One mole of a substance contains N_A (*Avogadro's number*) elementary units (usually atoms or molecules), where N_A is found experimentally to be

$$N_A = 6.02 \times 10^{23} \text{ mol}^{-1} \qquad \text{(Avogadro's number).} \quad \text{(20-1)}$$

One molar mass M of any substance is the mass of one mole of the substance. It is related to the mass m of the individual molecules of the substance by

$$M = mN_A. \qquad (20\text{-}4)$$

The number of moles n contained in a sample of mass M_{sam}, con-

sisting of N molecules, is given by

$$n = \frac{N}{N_A} = \frac{M_{sam}}{M} = \frac{M_{sam}}{mN_A}. \qquad (20\text{-}2, 20\text{-}3)$$

Ideal Gas   An *ideal gas* is one for which the pressure p, volume V, and temperature T are related by

$$pV = nRT \qquad \text{(ideal gas law).} \qquad (20\text{-}5)$$

Here n is the number of moles of the gas present and R is a constant (8.31 J/mol · K) called the **gas constant.** The ideal gas law can also be written as

$$pV = NkT, \qquad (20\text{-}9)$$

where the **Boltzmann constant** k is

$$k = \frac{R}{N_A} = 1.38 \times 10^{-23} \text{ J/K}. \qquad (20\text{-}7)$$

Work in an Isothermal Volume Change
The work done *by* an ideal gas during an **isothermal** (constant-temperature) change from volume V_i to volume V_f is

$$W = nRT \ln \frac{V_f}{V_i} \qquad \text{(ideal gas, isothermal process).} \qquad (20\text{-}14)$$

Pressure, Temperature, and Molecular Speed
The pressure exerted by n moles of an ideal gas, in terms of the speed of its molecules, is

$$p = \frac{nMv_{\text{rms}}^2}{3V}, \qquad (20\text{-}21)$$

where $v_{\text{rms}} = \sqrt{(v^2)_{\text{avg}}}$ is the **root-mean-square speed** of the molecules of the gas. With Eq. 20-5 this gives

$$v_{\text{rms}} = \sqrt{\frac{3RT}{M}}. \qquad (20\text{-}22)$$

Temperature and Kinetic Energy
The average translational kinetic energy K_{avg} per molecule of an ideal gas is

$$K_{\text{avg}} = \tfrac{3}{2}kT. \qquad (20\text{-}24)$$

Mean Free Path
The *mean free path* λ of a gas molecule is its average path length between collisions and is given by

$$\lambda = \frac{1}{\sqrt{2}\pi d^2 \, N/V}, \qquad (20\text{-}25)$$

where N/V is the number of molecules per unit volume and d is the molecular diameter.

Maxwell Speed Distribution
The *Maxwell speed distribution* $P(v)$ is a function such that $P(v)\, dv$ gives the *fraction* of molecules with speeds in the interval dv centered on speed v:

$$P(v) = 4\pi \left(\frac{M}{2\pi RT}\right)^{3/2} v^2 e^{-Mv^2/2RT}. \qquad (20\text{-}27)$$

Three measures of the distribution of speeds among the molecules of a gas are

$$v_{\text{avg}} = \sqrt{\frac{8RT}{\pi M}} \qquad \text{(average speed),} \qquad (20\text{-}31)$$

$$v_P = \sqrt{\frac{2RT}{M}} \qquad \text{(most probable speed),} \qquad (20\text{-}35)$$

and the rms speed defined above in Eq. 20-22.

Molar Specific Heats
The molar specific heat C_V of a gas at constant volume is defined as

$$C_V = \frac{1}{n}\frac{Q}{\Delta T} = \frac{1}{n}\frac{\Delta E_{\text{int}}}{\Delta T}, \qquad (20\text{-}39, 20\text{-}41)$$

in which Q is the energy transferred as heat to or from a sample of n moles of the gas, ΔT is the resulting temperature change of the gas, and ΔE_{int} is the resulting change in the internal energy of the gas. For an ideal monatomic gas,

$$C_V = \tfrac{3}{2}R = 12.5 \text{ J/mol} \cdot \text{K}. \qquad (20\text{-}43)$$

The molar specific heat C_p of a gas at constant pressure is defined to be

$$C_p = \frac{1}{n}\frac{Q}{\Delta T}, \qquad (20\text{-}46)$$

in which Q, n, and ΔT are defined as above. C_p is also given by

$$C_p = C_V + R. \qquad (20\text{-}49)$$

For n moles of an ideal gas,

$$E_{\text{int}} = nC_V T \qquad \text{(ideal gas).} \qquad (20\text{-}44)$$

If n moles of a confined ideal gas undergo a temperature change ΔT due to *any* process, the change in the internal energy of the gas is

$$\Delta E_{\text{int}} = nC_V \, \Delta T \qquad \text{(ideal gas, any process),} \qquad (20\text{-}45)$$

in which the appropriate value of C_V must be substituted, according to the type of ideal gas.

Degrees of Freedom and C_V
We find C_V itself by using the *equipartition of energy* theorem, which states that every *degree of freedom* of a molecule (that is, every independent way it can store energy) has associated with it—on average—an energy $\tfrac{1}{2}kT$ per molecule ($= \tfrac{1}{2}RT$ per mole). If f is the number of degrees of freedom, then $E_{\text{int}} = (f/2)nRT$, and

$$C_V = \left(\frac{f}{2}\right)R = 4.16f \text{ J/mol} \cdot \text{K}. \qquad (20\text{-}51)$$

For monatomic gases $f = 3$ (three translational degrees); for diatomic gases $f = 5$ (three translational and two rotational degrees).

Adiabatic Process
When an ideal gas undergoes a slow adiabatic volume change (a change for which $Q = 0$), its pressure and volume are related by

$$pV^\gamma = \text{a constant} \qquad \text{(adiabatic process),} \qquad (20\text{-}53)$$

in which $\gamma \, (= C_p/C_V)$ is the ratio of molar specific heats for the gas. For a free expansion, however, $pV =$ a constant.

QUESTIONS

1. If the temperature of an ideal gas is changed from 20°C to 40°C while the volume is unchanged, is the pressure of the gas doubled, increased but less than doubled, or increased and more than doubled?

2. In Fig. 20-15*a*, three isothermal processes are shown for the same gas and for the same change in volume (V_i to V_f) but at different temperatures. Rank the processes according to (a) the work done by the gas, (b) the change in the internal energy of the gas, and (c) the energy transferred as heat to the gas, greatest first.

In Fig. 20-15*b*, three isothermal processes are shown along a

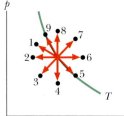

Fig. 20-15 Question 2.

single isotherm, for the same change ΔV in volume. Rank the processes according to (d) the work done by the gas, (e) the change in the internal energy of the gas, and (f) the energy transferred as heat to the gas, greatest first.

3. The volume of a gas and the number of gas molecules within that volume for four situations are (a) $2V_0$ and N_0, (b) $3V_0$ and $3N_0$, (c) $8V_0$ and $4N_0$, and (d) $3V_0$ and $9N_0$. Rank the situations according to the mean free path of the molecules, greatest first.

4. In Sample Problem 20-2, how much energy is transferred as heat during the expansion?

5. Figure 20-16 shows the initial state of an ideal gas and an isotherm through that state. Which of the paths shown result in a decrease in the temperature of the gas?

6. For four situations for an ideal gas, the table gives the heat Q and either the work W done by the gas or the work W_{on} done on the gas, all in joules. Rank the four situations in terms of the temperature change of the gas, most positive first, most negative last.

Fig. 20-16 Question 5.

	a	b	c	d
Q	−50	+35	−15	+20
W	−50	+35		
W_{on}			−40	+40

7. For a temperature increase of ΔT_1, a certain amount of an ideal gas requires 30 J when heated at constant volume and 50 J when heated at constant pressure. How much work is done by the gas in the second situation?

8. An ideal diatomic gas, with molecular rotation but not oscillation, loses energy as heat Q. Is the resulting decrease in the internal energy of the gas greater if the loss occurs in a constant-volume process or in a constant-pressure process?

9. A certain amount of energy is to be transferred as heat to 1 mol of a monatomic gas (a) at constant pressure and (b) at constant volume, and to 1 mol of a diatomic gas (c) at constant pressure and (d) at constant volume. Figure 20-17 shows four paths from an initial point to four final points on a p-V diagram. Which path goes with which process? (e) Are the molecules of the diatomic gas rotating?

Fig. 20-17 Question 9.

10. Does the temperature of an ideal gas increase, decrease, or stay the same during (a) an isothermal expansion, (b) an expansion at constant pressure, (c) an adiabatic expansion, and (d) an increase in pressure at constant volume?

11. (a) Rank the four paths of Fig. 20-14 according to the work done by the gas, greatest first. (b) Rank paths 1, 2, and 3 according to the change in the internal energy of the gas, most positive first and most negative last.

12. In the p-V diagram of Fig. 20-18, the gas does 5 J of work along isotherm ab and 4 J along adiabat bc. What is the change in the internal energy of the gas if the gas traverses the straight path from a to c?

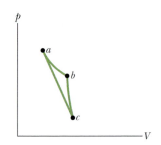

Fig. 20-18 Question 12.

EXERCISES & PROBLEMS

ssm   Solution is in the Student Solutions Manual.
www   Solution is available on the World Wide Web at:
              http://www.wiley.com/college/hrw
ilw   Solution is available on the Interactive LearningWare.

SEC. 20-2   Avogadro's Number

1E. Find the mass in kilograms of 7.50×10^{24} atoms of arsenic, which has a molar mass of 74.9 g/mol.   ssm

2E. Gold has a molar mass of 197 g/mol. (a) How many moles of gold are in a 2.50 g sample of pure gold? (b) How many atoms are in the sample?

3P. If the water molecules in 1.00 g of water were distributed uniformly over the surface of Earth, how many such molecules would there be on 1.00 cm^2 of the surface?

4P. A distinguished scientist has written: "There are enough molecules in the ink that makes one letter of this sentence to provide not only one for every inhabitant of Earth, but one for every creature if each star of our galaxy had a planet as populous as Earth." Check this statement. Assume the ink sample (molar mass = 18 g/mol) to have a mass of 1 μg, the population of Earth to be 5×10^9, and the number of stars in our galaxy to be 10^{11}.

SEC. 20-3   Ideal Gases

5E. Compute (a) the number of moles and (b) the number of molecules in 1.00 cm^3 of an ideal gas at a pressure of 100 Pa and a temperature of 220 K.   ssm

6E. The best laboratory vacuum has a pressure of about 1.00×10^{-18} atm, or 1.01×10^{-13} Pa. How many gas molecules are there per cubic centimeter in such a vacuum at 293 K?

7E. Oxygen gas having a volume of 1000 cm³ at 40.0°C and 1.01×10^5 Pa expands until its volume is 1500 cm³ and its pressure is 1.06×10^5 Pa. Find (a) the number of moles of oxygen present and (b) the final temperature of the sample. ssm

8E. An automobile tire has a volume of 1.64×10^{-2} m³ and contains air at a gauge pressure (pressure above atmospheric pressure) of 165 kPa when the temperature is 0.00°C. What is the gauge pressure of the air in the tires when its temperature rises to 27.0°C and its volume increases to 1.67×10^{-2} m³? Assume atmospheric pressure is 1.01×10^5 Pa.

9E. A quantity of ideal gas at 10.0°C and 100 kPa occupies a volume of 2.50 m³. (a) How many moles of the gas are present? (b) If the pressure is now raised to 300 kPa and the temperature is raised to 30.0°C, how much volume does the gas occupy? Assume no leaks.

10E. Calculate the work done by an external agent during an isothermal compression of 1.00 mol of oxygen from a volume of 22.4 L at 0°C and 1.00 atm pressure to 16.8 L.

11P. Pressure p, volume V, and temperature T for a certain material are related by

$$p = \frac{AT - BT^2}{V},$$

where A and B are constants. Find an expression for the work done by the material if the temperature changes from T_1 to T_2 while the pressure remains constant. ssm

12P. A container encloses two ideal gases. Two moles of the first gas are present, with molar mass M_1. The second gas has molar mass $M_2 = 3M_1$, and 0.5 mol of this gas is present. What fraction of the total pressure on the container wall is attributable to the second gas? (The kinetic theory explanation of pressure leads to the experimentally discovered law of partial pressures for a mixture of gases that do not react chemically: *The total pressure exerted by the mixture is equal to the sum of the pressures that the several gases would exert separately if each were to occupy the vessel alone.*)

13P. Air that initially occupies 0.14 m³ at a gauge pressure of 103.0 kPa is expanded isothermally to a pressure of 101.3 kPa and then cooled at constant pressure until it reaches its initial volume. Compute the work done by the air. (Gauge pressure is the difference between the actual pressure and atmospheric pressure.) ssm ilw www

14P. A sample of an ideal gas is taken through the cyclic process *abca* shown in Fig. 20-19; at point a, $T = 200$ K. (a) How many moles of gas are in the sample? What are (b) the temperature of the gas at point b, (c) the temperature of the gas at point c, and (d) the net energy added to the gas as heat during the cycle?

Fig. 20-19 Problem 14.

15P. An air bubble of 20 cm³ volume is at the bottom of a lake 40 m deep where the temperature is 4.0°C. The bubble rises to the surface, which is at a temperature of 20°C. Take the temperature of the bubble's air to be the same as that of the surrounding water. Just as the bubble reaches the surface, what is its volume? ssm

16P. A pipe of length $L = 25.0$ m that is open at one end contains air at atmospheric pressure. It is thrust vertically into a freshwater lake until the water rises halfway up in the pipe, as shown in Fig. 20-20. What is the depth h of the lower end of the pipe? Assume that the temperature is the same everywhere and does not change.

Fig. 20-20 Problem 16.

17P. Container A in Fig. 20-21 holds an ideal gas at a pressure of 5.0×10^5 Pa and a temperature of 300 K. It is connected by a thin tube (and a closed valve) to container B, with four times the volume of A. Container B holds the same ideal gas at a pressure of 1.0×10^5 Pa and a temperature of 400 K. The valve is opened to allow the pressures to equalize, but the temperature of each container is kept constant at its initial value. What then is the pressure in the two containers? ilw

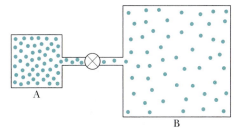

Fig. 20-21 Problem 17.

SEC. 20-4 Pressure, Temperature, and RMS Speed

18E. Calculate the rms speed of helium atoms at 1000 K. See Appendix F for the molar mass of helium atoms.

19E. The lowest possible temperature in outer space is 2.7 K. What is the root-mean-square speed of hydrogen molecules at this temperature? (The molar mass of hydrogen molecules (H_2) is given in Table 20-1.) ssm

20E. Find the rms speed of argon atoms at 313 K. See Appendix F for the molar mass of argon atoms.

21E. The temperature and pressure in the Sun's atmosphere are 2.00×10^6 K and 0.0300 Pa. Calculate the rms speed of free electrons (mass $= 9.11 \times 10^{-31}$ kg) there, assuming they are an ideal gas.

22E. (a) Compute the root-mean-square speed of a nitrogen molecule at 20.0°C. The molar mass of nitrogen molecules (N_2) is given in Table 20-1. At what temperatures will the root-mean-square speed be (b) half that value and (c) twice that value?

23P. A beam of hydrogen molecules (H_2) is directed toward a wall, at an angle of 55° with the normal to the wall. Each molecule in the beam has a speed of 1.0 km/s and a mass of 3.3×10^{-24} g. The beam strikes the wall over an area of 2.0 cm², at the rate of

10^{23} molecules per second. What is the beam's pressure on the wall? ssm

24P. At 273 K and 1.00×10^{-2} atm, the density of a gas is 1.24×10^{-5} g/cm^3. (a) Find v_{rms} for the gas molecules. (b) Find the molar mass of the gas and identify the gas. (*Hint:* The gas is listed in Table 20-1.)

SEC. 20-5   Translational Kinetic Energy

25E. What is the average translational kinetic energy of nitrogen molecules at 1600 K? ssm

26E. Determine the average value of the translational kinetic energy of the molecules of an ideal gas at (a) 0.00°C and (b) 100°C. What is the translational kinetic energy per mole of an ideal gas at (c) 0.00°C and (d) 100°C?

27P. Water standing in the open at 32.0°C evaporates because of the escape of some of the surface molecules. The heat of vaporization (539 cal/g) is approximately equal to εn, where ε is the average energy of the escaping molecules and n is the number of molecules per gram. (a) Find ε. (b) What is the ratio of ε to the average kinetic energy of H_2O molecules, assuming the latter is related to temperature in the same way as it is for gases? ssm www

28P. Show that the ideal gas equation, Eq. 20-5, can be written in the alternative form $p = \rho RT/M$, where ρ is the mass density of the gas and M is the molar mass.

29P. *Avogadro's law* states that under the same conditions of temperature and pressure, equal volumes of gas contain equal numbers of molecules. Is this law equivalent to the ideal gas law? Explain. ssm

SEC. 20-6   Mean Free Path

30E. The mean free path of nitrogen molecules at 0.0°C and 1.0 atm is 0.80×10^{-5} cm. At this temperature and pressure there are 2.7×10^{19} molecules/cm^3. What is the molecular diameter?

31E. At 2500 km above Earth's surface, the density of the atmosphere is about 1 molecule/cm^3. (a) What mean free path is predicted by Eq. 20-25 and (b) what is its significance under these conditions? Assume a molecular diameter of 2.0×10^{-8} cm. ssm

32E. At what frequency would the wavelength of sound in air be equal to the mean free path of oxygen molecules at 1.0 atm pressure and 0.00°C? Take the diameter of an oxygen molecule to be 3.0×10^{-8} cm.

33E. What is the mean free path for 15 spherical jelly beans in a bag that is vigorously shaken? The volume of the bag is 1.0 L, and the diameter of a jelly bean is 1.0 cm. (Consider bean–bean collisions, not bean–bag collisions.) ssm

34P. At 20°C and 750 torr pressure, the mean free paths for argon gas (Ar) and nitrogen gas (N_2) are $\lambda_{Ar} = 9.9 \times 10^{-6}$ cm and $\lambda_{N_2} = 27.5 \times 10^{-6}$ cm. (a) Find the ratio of the effective diameter of argon to that of nitrogen. What is the mean free path of argon at (b) 20°C and 150 torr, and (c) −40°C and 750 torr?

35P. In a certain particle accelerator, protons travel around a circular path of diameter 23.0 m in an evacuated chamber, whose residual gas is at 295 K and 1.00×10^{-6} torr pressure. (a) Calculate the number of gas molecules per cubic centimeter at this pressure. (b) What is the mean free path of the gas molecules if the molecular diameter is 2.00×10^{-8} cm? ssm

SEC. 20-7   The Distribution of Molecular Speeds

36E. Twenty-two particles have speeds as follows (N_i represents the number of particles that have speed v_i):

N_i	2	4	6	8	2
v_i (cm/s)	1.0	2.0	3.0	4.0	5.0

(a) Compute their average speed v_{avg}. (b) Compute their root-mean-square speed v_{rms}. (c) Of the five speeds shown, which is the most probable speed v_P?

37E. The speeds of 10 molecules are 2.0, 3.0, 4.0, . . . , 11 km/s. (a) What is their average speed? (b) What is their root-mean-square speed? ssm

38E. (a) Ten particles are moving with the following speeds: four at 200 m/s, two at 500 m/s, and four at 600 m/s. Calculate their average and root-mean-square speeds. Is $v_{rms} > v_{avg}$? (b) Make up your own speed distribution for the 10 particles and show that $v_{rms} \geq v_{avg}$ for your distribution. (c) Under what condition (if any) does $v_{rms} = v_{avg}$?

39P. (a) Compute the temperatures at which the rms speed for (a) molecular hydrogen and (b) molecular oxygen is equal to the speed of escape from Earth. (c) Do the same for the speed of escape from the Moon, assuming the gravitational acceleration on its surface to be 0.16g. (d) The temperature high in Earth's upper atmosphere is about 1000 K. Would you expect to find much hydrogen there? Much oxygen? Explain. ssm www

40P. It is found that the most probable speed of molecules in a gas when it has (uniform) temperature T_2 is the same as the rms speed of the molecules in this gas when it has (uniform) temperature T_1. Calculate T_2/T_1.

41P. A molecule of hydrogen (diameter 1.0×10^{-8} cm), traveling with the rms speed, escapes from a furnace ($T = 4000$ K) into a chamber containing atoms of *cold* argon (diameter 3.0×10^{-8} cm) at a density of 4.0×10^{19} atoms/cm^3. (a) What is the speed of the hydrogen molecule? (b) If the H_2 molecule collides with an argon atom, what is the closest their centers can be, considering each as spherical? (c) What is the initial number of collisions per second experienced by the hydrogen molecule? (*Hint:* Assume that the cold argon atoms are stationary. Then the mean free path of the hydrogen molecule is given by Eq. 20-26 and not Eq. 20-25.) ssm

42P. Two containers are at the same temperature. The first contains gas with pressure p_1, molecular mass m_1, and root-mean-square speed v_{rms1}. The second contains gas with pressure $2p_1$, molecular mass m_2, and average speed $v_{avg2} = 2v_{rms1}$. Find the mass ratio m_1/m_2.

43P. Figure 20-22 shows a hypothetical speed distribution for a sample of N gas particles (note that $P(v) = 0$ for $v > 2v_0$). (a) Express a in terms of N and v_0. (b) How many of the particles have speeds between $1.5v_0$ and $2.0v_0$? (c) Express the average speed of the particles in terms of v_0. (d) Find v_{rms}. ssm

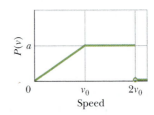

Fig. 20-22 Problem 43.

SEC. 20-8 The Molar Specific Heats of an Ideal Gas

44E. What is the internal energy of 1.0 mol of an ideal monatomic gas at 273 K?

45E. One mole of an ideal gas undergoes an isothermal expansion. Find the energy added to the gas as heat in terms of the initial and final volumes and the temperature. (*Hint:* Use the first law of thermodynamics.) **ssm**

46P. When 20.9 J was added as heat to a particular ideal gas, the volume of the gas changed from 50.0 cm³ to 100 cm³ while the pressure remained constant at 1.00 atm. (a) By how much did the internal energy of the gas change? If the quantity of gas present is 2.00×10^{-3} mol, find the molar specific heat of the gas at (b) constant pressure and (c) constant volume.

47P. A container holds a mixture of three nonreacting gases: n_1 moles of the first gas with molar specific heat at constant volume C_1, and so on. Find the molar specific heat at constant volume of the mixture, in terms of the molar specific heats and quantities of the separate gases. **ssm**

48P. One mole of an ideal diatomic gas goes from *a* to *c* along the diagonal path in Fig. 20-23. During the transition, (a) what is the change in internal energy of the gas, and (b) how much energy is added to the gas as heat? (c) How much heat is required if the gas goes from *a* to *c* along the indirect path *abc*?

Fig. 20-23 Problem 48.

49P. The mass of a gas molecule can be computed from its specific heat at constant volume c_V. Take $c_V = 0.075$ cal/g · C° for argon and calculate (a) the mass of an argon atom and (b) the molar mass of argon. **ilw**

SEC. 20-9 Degrees of Freedom and Molar Specific Heats

50E. We give 70 J as heat to a diatomic gas, which then expands at constant pressure. The gas molecules rotate but do not oscillate. By how much does the internal energy of the gas increase?

51E. One mole of oxygen (O_2) is heated at constant pressure starting at 0°C. How much energy must be added to the gas as heat to double its volume? (The molecules rotate but do not oscillate.) **ilw**

52E. Suppose 12.0 g of oxygen (O_2) is heated at constant atmospheric pressure from 25.0°C to 125°C. (a) How many moles of oxygen are present? (See Table 20-1 for the molar mass.) (b) How much energy is transferred to the oxygen as heat? (The molecules rotate but do not oscillate.) (c) What fraction of the heat is used to raise the internal energy of the oxygen?

53P. Suppose 4.00 mol of an ideal diatomic gas, with molecular rotation but not oscillation, experienced a temperature increase of 60.0 K under constant-pressure conditions. (a) How much energy was transferred to the gas as heat? (b) How much did the internal energy of the gas increase? (c) How much work was done by the gas? (d) How much did the translational kinetic energy of the gas increase? **ssm** **www**

SEC. 20-11 The Adiabatic Expansion of an Ideal Gas

54E. (a) One liter of a gas with $\gamma = 1.3$ is at 273 K and 1.0 atm pressure. It is suddenly compressed adiabatically to half its original volume. Find its final pressure and temperature. (b) The gas is now cooled back to 273 K at constant pressure. What is its final volume?

55E. A certain gas occupies a volume of 4.3 L at a pressure of 1.2 atm and a temperature of 310 K. It is compressed adiabatically to a volume of 0.76 L. Determine (a) the final pressure and (b) the final temperature, assuming the gas to be an ideal gas for which $\gamma = 1.4$. **ssm**

56E. We know that for an adiabatic process $pV^\gamma = $ a constant. Evaluate "a constant" for an adiabatic process involving exactly 2.0 mol of an ideal gas passing through the state having exactly $p = 1.0$ atm and $T = 300$ K. Assume a diatomic gas whose molecules have rotation but not oscillation.

57E. Let *n* moles of an ideal gas expand adiabatically from an initial temperature T_1 to a final temperature T_2. Prove that the work done by the gas is $nC_V(T_1 - T_2)$, where C_V is the molar specific heat at constant volume. (*Hint:* Use the first law of thermodynamics.) **ssm**

58E. For adiabatic processes in an ideal gas, show that (a) the bulk modulus is given by

$$B = -V\frac{dp}{dV} = \gamma p,$$

and therefore (b) the speed of sound in the gas is

$$v_s = \sqrt{\frac{\gamma p}{\rho}} = \sqrt{\frac{\gamma RT}{M}}.$$

See Eqs. 18-2 and 18-3.

59E. Air at 0.000°C and 1.00 atm pressure has a density of 1.29×10^{-3} g/cm³, and the speed of sound in air is 331 m/s at that temperature. Use those data to compute the ratio γ of the molar specific heats of air. (*Hint:* See Exercise 58.)

60P. (a) An ideal gas initially at pressure p_0 undergoes a free expansion until its volume is 3.00 times its initial volume. What then is its pressure? (b) The gas is next slowly and adiabatically compressed back to its original volume. The pressure after compression is $(3.00)^{1/3}p_0$. Is the gas monatomic, diatomic, or polyatomic? (c) How does the average kinetic energy per molecule in this final state compare with that in the initial state?

61P. One mole of an ideal monatomic gas traverses the cycle of Fig. 20-24. Process $1 \rightarrow 2$ occurs at constant volume, process $2 \rightarrow 3$ is adiabatic, and process $3 \rightarrow 1$ occurs at constant pressure. (a) Compute the heat Q, the change in internal energy ΔE_{int}, and the work done W, for each of the three processes and for the cycle as a whole. (b) The initial pressure at point 1 is 1.00 atm. Find the pressure and the volume at points 2 and 3. Use 1.00 atm = 1.013×10^5 Pa and $R = 8.314$ J/mol · K. **ssm**

Fig. 20-24 Problem 61.

NEW PROBLEMS

N1. At what temperature is the average translational kinetic energy of a molecule in a gas equal to 4.0×10^{-19} J?

N2. An ideal gas with 0.825 mol is to undergo an isothermal expansion as energy is added to it as heat Q. Figure 20N-1 shows the final volume V_f versus heat Q. What is the temperature of the gas?

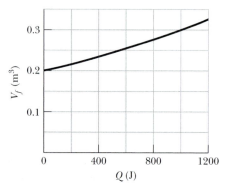

Fig. 20N-1 Problem N2.

N3. (a) Determine the work done by any ideal gas (monatomic, diatomic, or polyatomic) during an adiabatic expansion from volume V_i and pressure p_i to volume V_f and pressure p_f by applying Eq. 19-25 ($W = \int p \, dV$) to Eq. 20-53 (pV^γ = a constant). (b) Show that the result is equivalent to $-\Delta E_{int}$, where the change in internal energy is given by Eq. 20-45 ($\Delta E_{int} = nC_V \, \Delta T$).

N4. Figure 20N-2 gives the probability distribution for nitrogen gas. What are (a) the gas temperature and (b) the rms speed of the molecules?

Fig. 20N-2 Problem N4.

N5. An ideal gas is taken through a complete cycle in three steps:

1. adiabatic expansion, with work equal to 125 J,

2. isothermal contraction at temperature 325 K,

3. increase in pressure at constant volume.

(a) Draw a p-V diagram for the three steps. (b) How much energy is transferred as heat in step 3, and (c) is it transferred to or from the gas?

N6. An ideal gas with 3.00 mol is initially in state 1 with a pressure of $p_1 = 20.0$ atm and a volume of $V_1 = 1500$ cm³. First it is taken to state 2 with a pressure of $p_2 = 1.50p_1$ and a volume of $V_2 = 2.00V_1$. Then it is taken to state 3 with a pressure of $p_3 = 2.00p_1$ and a volume of $V_3 = 0.500V_1$. What are the temperatures of the gas in (a) state 1 and (b) state 2? (c) What is the net change in internal energy from state 1 to state 3?

N7. Oxygen (O_2) gas at 273 K and 1.0 atm pressure is confined to a cubical container 10 cm on a side. Calculate the ratio of (1) the change in gravitational potential energy of an oxygen molecule falling the height of the box to (2) the molecule's average translational kinetic energy.

N8. The temperature of an ideal monatomic gas with 2.00 moles is to be increased by 15.0 K. If the increase is done with the volume held constant, what are (a) the work done by the gas, (b) the energy transferred as heat, (c) the change in the internal energy of the gas, and (d) the change in the average kinetic energy of the atoms? Repeat the question (in the same order of parts) if the increase in temperature is done as (e) a constant-pressure process and (f) an adiabatic process.

N9. The envelope and basket of a hot-air balloon have a combined weight of 2.45 kN, and the envelope has a capacity (volume) of 2.18×10^3 m³. When it is fully inflated, what should be the temperature of the enclosed air to give the balloon a *lifting capacity* (force) of 2.67 kN (in addition to the balloon's weight)? Assume that the surrounding air, at 20.0°C, has a weight density (weight per volume) of 11.9 N/m³ and a molecular mass of 0.028 kg/mole, and is at a pressure of 1.0 atm.

N10. An ideal gas of 1.8 moles is taken from a volume of 3.00 m³ to a volume of 1.50 m³ via an isothermal compression at temperature 30°C. (a) How much energy is transferred as heat during the compression, and (b) is the transfer to or from the gas?

N11. A container holds a gas of molecular hydrogen (H_2) at a temperature of 250 K. What are (a) the most probable speed v_P of the molecules and (b) the maximum value P_{max} of the probability distribution function $P(v)$? (c) With a graphing calculator or computer math package, determine what percentage of the molecules have speeds between $0.500v_P$ and $1.50v_P$. The temperature is then increased to 500 K. What are (d) the most probable speed v_P of the molecules and (e) the maximum value P_{max} of the probability distribution function $P(v)$? Did (f) v_P and (g) P_{max} increase, decrease, or remain the same because of the temperature increase?

N12. An ideal diatomic gas, with rotation but no oscillation, undergoes an adiabatic expansion. Its initial pressure and volume are 1.20 atm and 0.200 m³. Its final pressure is 2.40 atm. How much work is done by the gas?

N13. At what frequency would the wavelength of sound in air be equal to the mean free path of oxygen molecules at 1.0 atm pressure and 0.0°C? Take 3.0×10^{-8} cm to be the diameter of an oxygen molecule.

N14. During a compression at a constant pressure of 250 Pa, the volume of an ideal gas decreases from 0.80 m³ to 0.20 m³. The initial temperature is 360 K, and the gas loses 210 J as heat. What are (a) the change in the internal energy of the gas and (b) the final temperature of the gas?

N15. At what temperature is the average translational kinetic energy of a molecule equal to 1.6×10^{-19} J?

N16. A gas is to be expanded from initial state i to final state f along either path 1 or path 2 on a p-V diagram. Path 1 consists of three steps:

1. an isothermal expansion (work is 40 J in magnitude),
2. an adiabatic expansion (work is 20 J in magnitude),
3. another isothermal expansion (work is 30 J in magnitude).

Path 2 consists of two steps:

1. a pressure reduction at constant volume,
2. an expansion at constant pressure.

What would be the change in the internal energy of the gas along path 2?

N17. Consider this sentence: A _____ of water contains about as many molecules as there are _____ s of water in all the oceans. What single word best fits both blank spaces: drop, teaspoon (4.8 mL), tablespoon (14.5 mL), cup (250 mL), quart ($\approx$ 1 L), barrel (119 L) or ton? The oceans cover 75% of Earth's surface and have an average depth of about 5 km. (After Edward M. Purcell.)

N18. An ideal gas consists of 1.50 moles of diatomic molecules having rotation but not oscillation. The molecular diameter is 250 pm. The gas is expanded at a constant pressure of 1.50×10^5 Pa, with a transfer of 200 J as heat. What is the change in the mean free path of the molecules?

N19. An ideal gas is adiabatically compressed from a volume of 200 L to a volume of 74.3 L. The initial pressure and temperature are 1.00 atm and 300 K. The final pressure is 4.00 atm. (a) Is the

gas monatomic, diatomic, or polyatomic? (b) What is the final temperature? (c) How many moles are in the gas?

N20. At what frequency do molecules collide in oxygen gas (O_2) at temperature 400 K and pressure 2.00 atm? Assume that the molecular diameter is 290 pm and that the gas is ideal.

N21. In an industrial process the volume of 25.0 moles of a monatomic ideal gas is reduced at a uniform rate from 0.616 m^3 to 0.308 m^3 in 2.00 h while its temperature is increased at a uniform rate from 27.0°C to 450°C. Throughout the process, the gas passes through thermodynamic equilibrium states. (a) Plot the cumulative work done on the gas and the cumulative energy absorbed by the gas as heat as a function of time for the duration of the process. (b) What are the values of these quantities for the entire process? (c) What is the molar specific heat for the process? (*Hint:* To evaluate the integral for the work, you might use

$$\int \frac{a + bx}{A + Bx} \, dx = \frac{bx}{B} + \frac{aB - bA}{B^2} \ln(A + Bx),$$

an indefinite integral.) (d) Compare these values with those that would be obtained if the volume is first reduced at constant temperature, then the temperature is increased at constant volume, with the gas reaching the same final state.

N22. An ideal monatomic gas initially has a temperature of 330 K and a pressure of 6.0 atm. It is to expand from volume 500 cm^3 to volume 1500 cm^3. If the expansion is isothermal, what are (a) the final pressure and (b) the work done by the gas? If, instead, the expansion is adiabatic, what are (c) the final pressure and (d) the work done by the gas?

N23. (a) What is the difference between the average value n_{avg} and the rms value n_{rms} of the numbers 1, 7, 84, and 201? Does the difference increase, decrease, or remain the same (b) if the first number is increased by 10 and (c) if the last number is increased by 10?

21 Entropy and the Second Law of Thermodynamics

An anonymous graffito on a wall of the Pecan Street Cafe in Austin, Texas, read: "Time is God's way of keeping things from happening all at once." Time also has direction—some things happen in a certain sequence and could never happen on their own in a reverse sequence. As an example, an accidentally dropped egg splatters in a cup. The reverse process, a splattered egg reforming into a whole egg and jumping up to an outstretched hand, will never happen on its own—but why not? Why can't that process be reversed, like a videotape run backward?

What in the world gives direction to time?

The answer is in this chapter.

21-1 Some One-Way Processes

Suppose you come indoors on a very cold day and wrap your cold hands around a warm mug of cocoa. Then your hands get warmer and the mug gets cooler. However, it never happens the other way around; that is, your cold hands never get still colder while the warm mug gets still warmer.

The system consisting of your hands and the mug is a *closed system,* one that is isolated from its environment. Here are some other one-way processes that occur in closed systems: (1) A crate sliding over an ordinary surface eventually stops—but you never see an initially stationary crate start to move all by itself. (2) If you drop a glob of putty, it falls to the floor—but an initially motionless glob of putty never leaps spontaneously into the air. (3) If you puncture a helium-filled balloon in a closed room, the helium gas spreads throughout the room—but the individual helium atoms will never clump up again into the shape of the balloon. We say that such one-way processes are **irreversible,** meaning that they cannot be reversed by means of only small changes in their environment.

The one-way character of such thermodynamic processes is so pervasive that we take it for granted. If these processes were to occur *spontaneously* (on their own) in the "wrong" direction, we would be astonished beyond belief. *Yet none of these "wrong-way" events would violate the law of conservation of energy.* In the cocoa mug example, that law would be obeyed even for a wrong-way transfer of energy as heat between hands and mug. It would be obeyed even if a stationary crate or a stationary glob of putty suddenly were to transfer some of its thermal energy to kinetic energy and begin to move. It would also be obeyed even if the helium atoms released from a balloon were, on their own, to clump together again.

Thus, changes in energy within a closed system do not set the direction of irreversible processes. Rather, that direction is set by another property that we shall discuss in this chapter—the *change in entropy* ΔS of the system. The change in entropy of a system is defined in the next section, but we can here state its central property, often called the *entropy postulate:*

> ▶ If an irreversible process occurs in a *closed* system, the entropy S of the system always increases; it never decreases.

Entropy differs from energy in that it does *not* obey a conservation law. The *energy* of a closed system is conserved; it always remains constant. For irreversible processes, the *entropy* of a closed system always increases. Because of this property, the change in entropy is sometimes called "the arrow of time." For example, we associate the egg of our opening photograph, breaking irreversibly as it drops into a cup, with the forward direction of time and with an increase in entropy. The backward direction of time (a videotape run backward) would correspond to the broken egg re-forming into a whole egg and rising into the air. This backward process would result in an entropy decrease, so it never happens.

There are two equivalent ways to define the change in entropy of a system: (1) in terms of the system's temperature and the energy it gains or loses as heat, and (2) by counting the ways in which the atoms or molecules that make up the system can be arranged. We use the first approach in the next section, and the second in Section 21-7.

21-2 Change in Entropy

Let's approach this definition of *change in entropy* by looking again at a process that we described in Sections 19-10 and 20-11: the free expansion of an ideal gas.

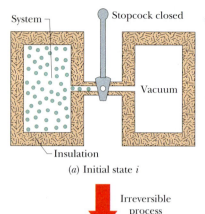

System — Stopcock closed

Vacuum

Insulation

(a) Initial state *i*

Irreversible process

Stopcock open

(b) Final state *f*

Fig. 21-1 The free expansion of an ideal gas. (*a*) The gas is confined to the left half of an insulated container by a closed stopcock. (*b*) When the stopcock is opened, the gas rushes to fill the entire container. This process is irreversible; that is, it does not occur in reverse, with the gas spontaneously collecting itself in the left half of the container.

Figure 21-1*a* shows the gas in its initial equilibrium state *i*, confined by a closed stopcock to the left half of a thermally insulated container. If we open the stopcock, the gas rushes to fill the entire container, eventually reaching the final equilibrium state *f* shown in Fig. 21-1*b*. This is an irreversible process; all the molecules of the gas will never return, by themselves, to the left half of the container.

The *p-V* plot of the process, in Fig. 21-2, shows the pressure and volume of the gas in its initial state *i* and final state *f*. Pressure and volume are *state properties*, properties that depend only on the state of the gas and not on how it reached that state. Other state properties are temperature and energy. We now assume that the gas has still another state property—its entropy. Furthermore, we define the **change in entropy** $S_f - S_i$ of a system during a process that takes the system from an initial state *i* to a final state *f* as

$$\Delta S = S_f - S_i = \int_i^f \frac{dQ}{T} \qquad \text{(change in entropy defined).} \qquad (21\text{-}1)$$

Here *Q* is the energy transferred as heat to or from the system during the process, and *T* is the temperature of the system in kelvins. Thus, an entropy change depends not only on the energy transferred as heat but also on the temperature at which the transfer takes place. Because *T* is always positive, the sign of ΔS is the same as that of *Q*. We see from Eq. 21-1 that the SI unit for entropy and entropy change is the joule per kelvin.

There is a problem, however, in applying Eq. 21-1 to the free expansion of Fig. 21-1. As the gas rushes to fill the entire container, the pressure, temperature, and volume of the gas fluctuate unpredictably. In other words, they do not have a sequence of well-defined equilibrium values during the intermediate stages of the change from initial equilibrium state *i* to final equilibrium state *f*. Thus, we cannot trace a pressure–volume path for the free expansion on the *p-V* plot of Fig. 21-2 and, more important, we cannot find a relation between *Q* and *T* that allows us to integrate as Eq. 21-1 requires.

However, if entropy is truly a state property, the difference in entropy between states *i* and *f* must depend *only on those states* and not at all on the way the system went from one state to the other. Suppose, then, that we replace the irreversible free expansion of Fig. 21-1 with a *reversible* process that connects states *i* and *f*. With a reversible process we can trace a pressure–volume path on a *p-V* plot, and we can find a relation between *Q* and *T* that allows us to use Eq. 21-1 to obtain the entropy change.

We saw in Section 20-11 that the temperature of an ideal gas does not change during a free expansion: $T_i = T_f = T$. Thus, points *i* and *f* in Fig. 21-2 must be on the same isotherm. A convenient replacement process is then a reversible isothermal expansion from state *i* to state *f*, which actually proceeds *along* that isotherm. Furthermore, because *T* is constant throughout a reversible isothermal expansion, the integral of Eq. 21-1 is greatly simplified.

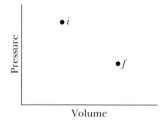

Fig. 21-2 A *p-V* diagram showing the initial state *i* and the final state *f* of the free expansion of Fig. 21-1. The intermediate states of the gas cannot be shown because they are not equilibrium states.

Insulation

Lead shot

Q

Thermal reservoir   Control knob

T

(a) Initial state i

Reversible
process

Lead shot

T

(b) Final state f

Fig. 21-3 The isothermal expansion of
an ideal gas, done in a reversible way.
The gas has the same initial state i
and same final state f as in the irreversible process of Figs. 21-1 and 21-2.

Figure 21-3 shows how to produce such a reversible isothermal expansion. We
confine the gas to an insulated cylinder that rests on a thermal reservoir maintained
at the temperature T. We begin by placing just enough lead shot on the movable
piston so that the pressure and volume of the gas are those of the initial state i of
Fig. 21-1a. We then remove shot slowly (piece by piece) until the pressure and
volume of the gas are those of the final state f of Fig. 21-1b. The temperature of the
gas does not change because the gas remains in thermal contact with the reservoir
throughout the process.

The reversible isothermal expansion of Fig. 21-3 is physically quite different
from the irreversible free expansion of Fig. 21-1. However, *both processes have the
same initial state and the same final state and thus must have the same change in
entropy*. Because we removed the lead shot slowly, the intermediate states of the
gas are equilibrium states, so we can plot them on a p-V diagram (Fig. 21-4).

To apply Eq. 21-1 to the isothermal expansion, we take the constant temperature
T outside the integral, obtaining

$$\Delta S = S_f - S_i = \frac{1}{T} \int_i^f dQ.$$

Because $\int dQ = Q$, where Q is the total energy transferred as heat during the process,
we have

$$\Delta S = S_f - S_i = \frac{Q}{T} \qquad \text{(change in entropy, isothermal process)}. \qquad (21\text{-}2)$$

To keep the temperature T of the gas constant during the isothermal expansion of
Fig. 21-3, heat Q must have been energy transferred *from* the reservoir *to* the gas.
Thus, Q is positive and the entropy of the gas *increases* during the isothermal process
and during the free expansion of Fig. 21-1.

To summarize:

> To find the entropy change for an irreversible process occurring in a *closed* system,
> replace that process with any reversible process that connects the same initial and final
> states. Calculate the entropy change for this reversible process with Eq. 21-1.

When the temperature change ΔT of a system is small relative to the temperature (in kelvins) before and after the process, the entropy change can be approximated as

$$\Delta S = S_f - S_i \approx \frac{Q}{T_{avg}}, \qquad (21\text{-}3)$$

where T_{avg} is the average temperature of the system in kelvins during the process.

✓**CHECKPOINT 1:**   Water is heated on a stove. Rank the entropy changes of the water as
its temperature rises (a) from 20°C to 30°C, (b) from 30°C to 35°C, and (c) from 80°C to
85°C, greatest first.

Fig. 21-4 A p-V diagram for the reversible isothermal expansion of Fig. 21-3. The intermediate states, which are
now equilibrium states, are shown.

Sample Problem 21-1

One mole of nitrogen gas is confined to the left side of the container of Fig. 21-1a. You open the stopcock and the volume of the gas doubles. What is the entropy change of the gas for this irreversible process? Treat the gas as ideal.

SOLUTION: We need two **Key Ideas** here. One is that we can determine the entropy change for the irreversible process by calculating it for a reversible process that provides the same change in volume. The other is that the temperature of the gas does not change in the free expansion. Thus, the reversible process should be an isothermal expansion—namely, the one of Figs. 21-3 and 21-4.

From Table 20-4, the energy Q added as heat to the gas as it expands isothermally at temperature T from an initial volume V_i to a final volume V_f is

$$Q = nRT \ln \frac{V_f}{V_i},$$

in which n is the number of moles of gas present. From Eq. 21-2 the entropy change for this reversible process is

$$\Delta S_{rev} = \frac{Q}{T} = \frac{nRT \ln(V_f/V_i)}{T} = nR \ln \frac{V_f}{V_i}.$$

Substituting $n = 1.00$ mol and $V_f/V_i = 2$, we find

$$\Delta S_{rev} = nR \ln \frac{V_f}{V_i} = (1.00 \text{ mol})(8.31 \text{ J/mol} \cdot \text{K})(\ln 2)$$

$$= +5.76 \text{ J/K}.$$

Thus, the entropy change for the free expansion (and for all other processes that connect the initial and final states shown in Fig. 21-2) is

$$\Delta S_{irrev} = \Delta S_{rev} = +5.76 \text{ J/K.} \qquad \text{(Answer)}$$

ΔS is positive, so the entropy increases, in accordance with the entropy postulate of Section 21-1.

✔**CHECKPOINT 2:**   An ideal gas has temperature T_1 at the initial state i shown in the p-V diagram here. The gas has a higher temperature T_2 at final states a and b, which it can reach along the paths shown. Is the entropy change along the path to state a larger than, smaller than, or the same as that along the path to state b?

Sample Problem 21-2

Figure 21-5a shows two identical copper blocks of mass $m = 1.5$ kg: block L at temperature $T_{iL} = 60°C$ and block R at temperature $T_{iR} = 20°C$. The blocks are in a thermally insulated box and are separated by an insulating shutter. When we lift the shutter, the blocks eventually come to the equilibrium temperature $T_f = 40°C$ (Fig. 21-5b). What is the net entropy change of the two-block system during this irreversible process? The specific heat of copper is 386 J/kg · K.

SOLUTION: The **Key Idea** here is that to calculate the entropy change, we must find a reversible process that takes the system from the initial state of Fig. 21-5a to the final state of Fig. 21-5b. We can calculate the net entropy change ΔS_{rev} of the reversible process using Eq. 21-1, and then the entropy change for the irreversible process is equal to ΔS_{rev}. For such a reversible process we need a thermal reservoir whose temperature can be changed slowly (say, by turning a knob). We then take the blocks through the following two steps, illustrated in Fig. 21-6.

Step 1.   With the reservoir's temperature set at 60°C, put block L on the reservoir. (Since block and reservoir are at the same temperature, they are already in thermal equilibrium.) Then slowly lower the temperature of the reservoir and the block to 40°C. As the block's temperature changes by each increment dT during this process, energy dQ is transferred as heat *from* the block to the reservoir. Using Eq. 19-14, we can write this transferred energy as $dQ = mc \, dT$, where c is the specific heat of copper. According to Eq. 21-1, the entropy change

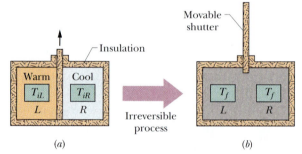

Fig. 21-5 Sample Problem 21-2. (a) In the initial state, two copper blocks L and R, identical except for their temperatures, are in an insulating box and are separated by an insulating shutter. (b) When the shutter is removed, the blocks exchange energy as heat and come to a final state, both with the same temperature T_f.

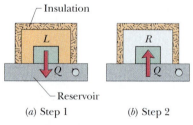

Fig. 21-6 The blocks of Fig. 21-5 can proceed from their initial state to their final state in a reversible way if we use a reservoir with a controllable temperature (a) to extract heat reversibly from block L and (b) to add heat reversibly to block R.

ΔS_L of block L during the full temperature change from initial temperature T_{iL} ($= 60°C = 333$ K) to final temperature T_f ($= 40°C = 313$ K) is

$$\Delta S_L = \int_i^f \frac{dQ}{T} = \int_{T_{iL}}^{T_f} \frac{mc\ dT}{T} = mc \int_{T_{iL}}^{T_f} \frac{dT}{T}$$

$$= mc \ln \frac{T_f}{T_{iL}}.$$

Inserting the given data yields

$$\Delta S_L = (1.5\text{ kg})(386\text{ J/kg} \cdot \text{K}) \ln \frac{313\text{ K}}{333\text{ K}}$$

$$= -35.86\text{ J/K}.$$

Step 2. With the reservoir's temperature now set at 20°C, put block R on the reservoir. Then slowly raise the temperature of the reservoir and the block to 40°C. With the same reasoning used to find ΔS_L, you can show that the entropy change ΔS_R of block R during this process is

$$\Delta S_R = (1.5\text{ kg})(386\text{ J/kg} \cdot \text{K}) \ln \frac{313\text{ K}}{293\text{ K}}$$

$$= +38.23\text{ J/K}.$$

The net entropy change ΔS_{rev} of the two-block system undergoing this two-step reversible process is then

$$\Delta S_{\text{rev}} = \Delta S_L + \Delta S_R$$

$$= -35.86\text{ J/K} + 38.23\text{ J/K} = 2.4\text{ J/K}.$$

Thus, the net entropy change ΔS_{irrev} for the two-block system undergoing the actual irreversible process is

$$\Delta S_{\text{irrev}} = \Delta S_{\text{rev}} = 2.4\text{ J/K.} \qquad \text{(Answer)}$$

This result is positive, in accordance with the entropy postulate of Section 21-1.

Entropy as a State Function

We have assumed that entropy, like pressure, energy, and temperature, is a property of the state of a system and is independent of how that state is reached. That entropy is indeed a *state function* (as state properties are usually called) can only be deduced by experiment. However, we can prove it is a state function for the special and important case in which an ideal gas is taken through a reversible process.

To make the process reversible, it is done slowly in a series of small steps, with the gas in an equilibrium state at the end of each step. For each small step, the energy transferred as heat to or from the gas is dQ, the work done by the gas is dW, and the change in internal energy is dE_{int}. These are related by the first law of thermodynamics in differential form (Eq. 19-27):

$$dE_{\text{int}} = dQ - dW.$$

Because the steps are reversible, with the gas in equilibrium states, we can use Eq. 19-24 to replace dW with $p\ dV$ and Eq. 20-45 to replace dE_{int} with $nC_V\ dT$. Solving for dQ then leads to

$$dQ = p\ dV + nC_V\ dT.$$

Using the ideal gas law, we replace p in this equation with nRT/V. Then we divide each term in the resulting equation by T, obtaining

$$\frac{dQ}{T} = nR \frac{dV}{V} + nC_V \frac{dT}{T}.$$

Now let us integrate each term of this equation between an arbitrary initial state i and an arbitrary final state f to get

$$\int_i^f \frac{dQ}{T} = \int_i^f nR \frac{dV}{V} + \int_i^f nC_V \frac{dT}{T}.$$

The quantity on the left is the entropy change ΔS ($= S_f - S_i$) defined by Eq. 21-1. Substituting this and integrating the quantities on the right yield

$$\Delta S = S_f - S_i = nR \ln \frac{V_f}{V_i} + nC_V \ln \frac{T_f}{T_i}. \qquad (21\text{-}4)$$

Note that we did not have to specify a particular reversible process when we integrated. Therefore, the integration must hold for all reversible processes that take the gas from state i to state f. Thus, the change in entropy ΔS between the initial and final states of an ideal gas depends only on properties of the initial state (V_i and T_i) and properties of the final state (V_f and T_f); ΔS does not depend on how the gas changes between the two states.

21-3 The Second Law of Thermodynamics

Here is a puzzle. We saw in Sample Problem 21-1 that if we cause the reversible process of Fig. 21-3 to proceed from (a) to (b) in that figure, the change in entropy of the gas—which we take as our system—is positive. However, because the process is reversible, we can just as easily make it proceed from (b) to (a), simply by slowly adding lead shot to the piston of Fig. 21-3b until the original volume of the gas is restored. In this reverse process, energy must be extracted as heat *from the gas* to keep its temperature from rising. Hence Q is negative and so, from Eq. 21-2, the entropy of the gas must decrease.

Doesn't this decrease in the entropy of the gas violate the entropy postulate of Section 21-1, which states that entropy always increases? No, because that postulate holds only for *irreversible* processes occurring in closed systems. The procedure suggested here does not meet these requirements. The process is *not* irreversible and (because energy is transferred as heat from the gas to the reservoir) the system—which is the gas alone—is *not* closed.

However, if we include the reservoir, along with the gas, as part of the system, then we do have a closed system. Let's check the change in entropy of the enlarged system *gas + reservoir* for the process that takes it from (b) to (a) in Fig. 21-3. During this reversible process, energy is transferred as heat from the gas to the reservoir—that is, from one part of the enlarged system to another. Let $|Q|$ represent the absolute value (or magnitude) of this heat. With Eq. 21-2, we can then calculate separately the entropy changes for the gas (which loses $|Q|$) and the reservoir (which gains $|Q|$). We get

$$\Delta S_{\text{gas}} = -\frac{|Q|}{T}$$

and

$$\Delta S_{\text{res}} = +\frac{|Q|}{T}.$$

The entropy change of the closed system is the sum of these two quantities, *which is zero*.

With this result, we can modify the entropy postulate of Section 21-1 to include both reversible and irreversible processes:

▶ If a process occurs in a *closed* system, the entropy of the system increases for irreversible processes and remains constant for reversible processes. It never decreases.

Although entropy may decrease in part of a closed system, there will always be an equal or larger entropy increase in another part of the system, so that the entropy of the system as a whole never decreases. This fact is one form of the **second law of thermodynamics** and can be written as

$$\Delta S \geq 0 \qquad \text{(second law of thermodynamics)}, \qquad (21\text{-}5)$$

where the greater-than sign applies to irreversible processes, and the equals sign to reversible processes. Equation 21-5 applies only to closed systems.

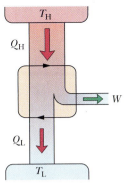

Fig. 21-7 The elements of an engine. The two black arrowheads on the central loop suggest the working substance operating in a cycle, as if on a p-V plot. Energy $|Q_H|$ is transferred as heat from the high-temperature reservoir at temperature T_H to the working substance. Energy $|Q_L|$ is transferred as heat from the working substance to the low-temperature reservoir at temperature T_L. Work W is done by the engine (actually by the working substance) on something in the environment.

In the real world almost all processes are irreversible to some extent because of friction, turbulence, and other factors, so the entropy of real closed systems undergoing real processes always increases. Processes in which the system's entropy remains constant are always idealizations.

21-4 Entropy in the Real World: Engines

A **heat engine,** or more simply, an **engine,** is a device that extracts energy from its environment in the form of heat and does useful work. At the heart of every engine is a *working substance.* In a steam engine, the working substance is water, in both its vapor and its liquid form. In an automobile engine the working substance is a gasoline–air mixture. If an engine is to do work on a sustained basis, the working substance must operate in a *cycle*; that is, the working substance must pass through a closed series of thermodynamic processes, called *strokes,* returning again and again to each state in its cycle. Let us see what the laws of thermodynamics can tell us about the operation of engines.

A Carnot Engine

We have seen that we can learn much about real gases by analyzing an ideal gas, which obeys the simple law $pV = nRT$. This is a useful plan because, although an ideal gas does not exist, any real gas approaches ideal behavior as closely as you wish if its density is low enough. In much the same spirit we choose to study real engines by analyzing the behavior of an **ideal engine.**

> In an ideal engine, all processes are reversible and no wasteful energy transfers occur due to, say, friction and turbulence.

We shall focus on a particular ideal engine called a **Carnot engine** after the French scientist and engineer N. L. Sadi Carnot (pronounced "car-no"), who first proposed the engine's concept in 1824. This ideal engine turns out to be the best (in principle) at using energy as heat to do useful work. Surprisingly, Carnot was able to analyze the performance of this engine before the first law of thermodynamics and the concept of entropy had been discovered.

Figure 21-7 shows schematically the operation of a Carnot engine. During each cycle of the engine, the working substance absorbs energy $|Q_H|$ as heat from a thermal reservoir at constant temperature T_H and discharges energy $|Q_L|$ as heat to a second thermal reservoir at a constant lower temperature T_L.

Figure 21-8 shows a p-V plot of the *Carnot cycle*—the cycle followed by the working substance. As indicated by the arrows, the cycle is traversed in the clockwise direction. Imagine the working substance to be a gas, confined to an insulating cylinder with a weighted, movable piston. The cylinder may be placed at will on either of the two thermal reservoirs, as in Fig. 21-3, or on an insulating slab. Figure 21-8 shows that, if we place the cylinder in contact with the high-temperature reservoir at temperature T_H, heat $|Q_H|$ is transferred *to* the working substance *from* this reservoir as the gas undergoes an isothermal *expansion* from volume V_a to volume V_b. Similarly, with the working substance in contact with the low-temperature reservoir at temperature T_L, heat $|Q_L|$ is transferred *from* the working substance *to* the low-temperature reservoir, as the gas undergoes an isothermal *compression* from volume V_c to volume V_d.

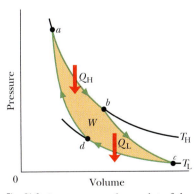

Fig. 21-8 A pressure–volume plot of the cycle followed by the working substance of the Carnot engine in Fig. 21-7. The cycle consists of two isotherm (*ab* and *cd*) and two adiabatic processes (*bc* and *da*). The shaded area enclosed by the cycle is equal to the work W per cycle done by the Carnot engine.

In the engine of Fig. 21-7, we assume that heat transfers to or from the working substance can take place *only* during the isothermal processes *ab* and *cd* of Fig. 21-8. Therefore, processes *bc* and *da* in that figure, which connect the two isotherms at temperatures T_H and T_L, must be (reversible) adiabatic processes; that is, they must be processes in which no energy is transferred as heat. To ensure this, during processes *bc* and *da* the cylinder is placed on an insulating slab as the volume of the working substance is changed.

During the consecutive processes *ab* and *bc* of Fig. 21-8, the working substance is expanding and thus doing positive work as it raises the weighted piston. This work is represented in Fig. 21-8 by the area under curve *abc*. During the consecutive processes *cd* and *da*, the working substance is being compressed, which means that it is doing negative work on its environment or, equivalently, that its environment is doing work on it as the loaded piston descends. This work is represented by the area under curve *cda*. The *net work per cycle,* which is represented by *W* in both Figs. 21-7 and 21-8, is the difference between these two areas and is a positive quantity equal to the area enclosed by cycle *abcda* in Fig. 21-8. This work *W* is performed on some outside object, such as a load to be lifted.

Equation 21-1 ($\Delta S = \int dQ/T$) tells us that any energy transfer as heat must involve a change in entropy. To illustrate the entropy changes for a Carnot engine, we can plot the Carnot cycle on a temperature-entropy (*T-S*) diagram as shown in Fig. 21-9. The lettered points *a, b, c,* and *d* in Fig. 21-9 correspond to the lettered points in the *p-V* diagram in Fig. 21-8. The two horizontal lines in Fig. 21-9 correspond to the two isothermal processes of the Carnot cycle (because the temperature is constant). Process *ab* is the isothermal expansion of the cycle. As the working substance (reversibly) absorbs energy $|Q_H|$ as heat at constant temperature T_H during the expansion, its entropy increases. Similarly, during the isothermal compression *cd*, the working substance (reversibly) loses energy $|Q_L|$ as heat at constant temperature T_L, and its entropy decreases.

The two vertical lines in Fig. 21-9 correspond to the two adiabatic processes of the Carnot cycle. Because no energy is transferred as heat during the two processes, the entropy of the working substance is constant during them.

Fig. 21-9 The Carnot cycle of Fig. 21-8 plotted on a temperature–entropy diagram. During processes *ab* and *cd* the temperature remains constant. During processes *bc* and *da* the entropy remains constant.

The Work: To calculate the net work done by a Carnot engine during a cycle, let us apply Eq. 19-26, the first law of thermodynamics ($\Delta E_{int} = Q - W$), to the working substance. That substance must return again and again to any arbitrarily selected state in the cycle. Thus, if *X* represents any state property of the working substance, such as pressure, temperature, volume, internal energy, or entropy, we must have $\Delta X = 0$ for every cycle. It follows that $\Delta E_{int} = 0$ for a complete cycle of the working substance. Recalling that *Q* in Eq. 19-26 is the *net* heat transfer per cycle and *W* is the *net* work, we can write the first law of thermodynamics for the Carnot cycle as

$$W = |Q_H| - |Q_L|. \tag{21-6}$$

Entropy Changes: In a Carnot engine, there are *two* (and only two) reversible energy transfers as heat, and thus two changes in the entropy of the working substance— one at temperature T_H and one at T_L. The net entropy change per cycle is then

$$\Delta S = \Delta S_H + \Delta S_L = \frac{|Q_H|}{T_H} - \frac{|Q_L|}{T_L}. \tag{21-7}$$

Here ΔS_H is positive because energy $|Q_H|$ is *added to* the working substance as heat (an increase in entropy) and ΔS_L is negative because energy $|Q_L|$ is *removed from* the working substance as heat (a decrease in entropy). Because entropy is a state function, we must have $\Delta S = 0$ for a complete cycle. Putting $\Delta S = 0$ in Eq. 21-7 requires that

$$\frac{|Q_H|}{T_H} = \frac{|Q_L|}{T_L}. \tag{21-8}$$

Note that, because $T_H > T_L$, we must have $|Q_H| > |Q_L|$; that is, more energy is extracted as heat from the high-temperature reservoir than is delivered to the low-temperature reservoir.

We shall now use Eqs. 21-6 and 21-8 to derive an expression for the efficiency of a Carnot engine.

Efficiency of a Carnot Engine

The purpose of any engine is to transform as much of the extracted energy Q_H into work as possible. We measure its success in doing so by its **thermal efficiency** ε, defined as the work the engine does per cycle ("energy we get") divided by the energy it absorbs as heat per cycle ("energy we pay for"):

$$\varepsilon = \frac{\text{energy we get}}{\text{energy we pay for}} = \frac{|W|}{|Q_H|} \qquad \text{(efficiency, any engine)}. \qquad (21\text{-}9)$$

For a Carnot engine we can substitute for W from Eq. 21-6 to write Eq. 21-9 as

$$\varepsilon_C = \frac{|Q_H| - |Q_L|}{|Q_H|} = 1 - \frac{|Q_L|}{|Q_H|}. \qquad (21\text{-}10)$$

Using Eq. 21-8 we can write this as

$$\varepsilon_C = 1 - \frac{T_L}{T_H} \qquad \text{(efficiency, Carnot engine)}, \qquad (21\text{-}11)$$

where the temperatures T_L and T_H are in kelvins. Because $T_L < T_H$, the Carnot engine necessarily has a thermal efficiency less than unity—that is, less than 100%. This is indicated in Fig. 21-7, which shows that only part of the energy extracted as heat from the high-temperature reservoir is available to do work, and the rest is delivered to the low-temperature reservoir. We will show in Section 21-6 that no real engine can have a thermal efficiency greater than that calculated from Eq. 21-11.

Inventors continually try to improve engine efficiency by reducing the energy $|Q_L|$ that is "thrown away" during each cycle. The inventor's dream is to produce the *perfect engine,* diagrammed in Fig. 21-10, in which $|Q_L|$ is reduced to zero and $|Q_H|$ is converted completely into work. Such an engine on an ocean liner, for example, could extract energy as heat from the water and use it to drive the propellers, with no fuel cost. An automobile, fitted with such an engine, could extract energy as heat from the surrounding air and use it to drive the car, again with no fuel cost. Alas, a perfect engine is only a dream: Inspection of Eq. 21-11 shows that we can achieve 100% engine efficiency (that is, $\varepsilon = 1$) only if $T_L = 0$ or $T_H \to \infty$, requirements that are impossible to meet. Instead, decades of practical engineering experience have led to the following alternative version of the second law of thermodynamics:

> No series of processes is possible whose sole result is the transfer of energy as heat from a thermal reservoir and the complete conversion of this energy to work.

In short, *there are no perfect engines.*

To summarize: The thermal efficiency given by Eq. 21-11 applies only to Carnot engines. Real engines, in which the processes that form the engine cycle are not reversible, have lower efficiencies. If your car were powered by a Carnot engine, it

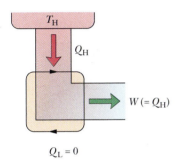

$Q_L = 0$

Fig. 21-10 The elements of a perfect engine—that is, one that converts heat Q_H from a high-temperature reservoir directly to work W with 100% efficiency.

Fig. 21-11 The North Anna nuclear power plant near Charlottesville, Virginia, which generates electric energy at the rate of 900 MW. At the same time, by design, it discards energy into the nearby river at the rate of 2100 MW. This plant—and all others like it—throws away more energy than it delivers in useful form. It is a real counterpart to the ideal engine of Fig. 21-7.

would have an efficiency of about 55% according to Eq. 21-11; its actual efficiency is probably about 25%. A nuclear power plant (Fig. 21-11), taken in its entirety, is an engine. It extracts energy as heat from a reactor core, does work by means of a turbine, and discharges energy as heat to a nearby river. If the power plant operated as a Carnot engine, its efficiency would be about 40%; its actual efficiency is about 30%. In designing engines of any type, there is simply no way to beat the efficiency limitation imposed by Eq. 21-11.

Stirling Engine

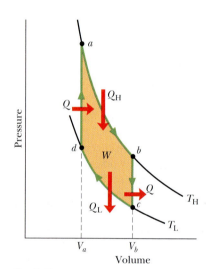

Fig. 21-12 A p-V plot for the working substance of an ideal Stirling engine, assumed for convenience to be an ideal gas.

Equation 21-11 does not apply to all ideal engines, but only to engines that can be represented as in Fig. 21-8—that is, to Carnot engines. For example, Fig. 21-12 shows the operating cycle of an ideal **Stirling engine.** Comparison with the Carnot cycle of Fig. 21-8 shows that each engine has isothermal heat transfers at temperatures T_H and T_L. However, the two isotherms of the Stirling engine cycle are connected, not by adiabatic processes as for the Carnot engine, but by constant-volume processes (Fig. 21-12). To increase the temperature of a gas at constant volume reversibly from T_L to T_H (process da of Fig. 21-12) requires a transfer of energy as heat to the working substance from a thermal reservoir whose temperature can be varied smoothly between those limits. Also, a reverse transfer is required in process bc. Thus, reversible heat transfers (and corresponding entropy changes) occur in all four of the processes that form the cycle of a Stirling engine, not just two processes as in a Carnot engine. Thus, the derivation that led to Eq. 21-11 does not apply to an ideal Stirling engine. More important, the efficiency of an ideal Stirling engine is lower than that of a Carnot engine operating between the same two temperatures. Real Stirling engines have even lower efficiencies.

The Stirling engine was developed in 1816 by Robert Stirling. This engine, long neglected, is now being developed for use in automobiles and spacecraft. A Stirling engine delivering 5000 hp (3.7 MW) has been built.

✔**CHECKPOINT 3:** Three Carnot engines operate between reservoir temperatures of (a) 400 and 500 K, (b) 600 and 800 K, and (c) 400 and 600 K. Rank the engines according to their thermal efficiencies, greatest first.

Sample Problem 21-3

Imagine a Carnot engine that operates between the temperatures $T_H = 850$ K and $T_L = 300$ K. The engine performs 1200 J of work each cycle, which takes 0.25 s.

(a) What is the efficiency of this engine?

SOLUTION: The Key Idea here is that the efficiency ε of a Carnot engine depends only on the ratio T_L/T_H of the temperatures (in kelvins) of the thermal reservoirs to which it is connected. Thus, from Eq. 21-11, we have

$$\varepsilon = 1 - \frac{T_L}{T_H} = 1 - \frac{300 \text{ K}}{850 \text{ K}} = 0.647 \approx 65\%. \quad \text{(Answer)}$$

(b) What is the average power of this engine?

SOLUTION: Here the Key Idea is that the average power P of an engine is the ratio of the work W it does per cycle to the time t that each cycle takes. For this Carnot engine, we find

$$P = \frac{W}{t} = \frac{1200 \text{ J}}{0.25 \text{ s}} = 4800 \text{ W} = 4.8 \text{ kW}. \quad \text{(Answer)}$$

(c) How much energy $|Q_H|$ is extracted as heat from the high-temperature reservoir every cycle?

SOLUTION: Now the Key Idea is that for any engine, including a Carnot engine, the efficiency ε is the ratio of the work W that is done per cycle to the energy $|Q_H|$ that is extracted as heat from the high-temperature reservoir per cycle ($\varepsilon = W/|Q_H|$). Thus,

$$|Q_H| = \frac{W}{\varepsilon} = \frac{1200 \text{ J}}{0.647} = 1855 \text{ J}. \quad \text{(Answer)}$$

(d) How much energy $|Q_L|$ is delivered as heat to the low-temperature reservoir every cycle?

SOLUTION: The Key Idea here is that for a Carnot engine, the work W done per cycle is equal to the difference in the energy transfers as heat: $|Q_H| - |Q_L|$, as in Eq. 21-6. Thus, we have

$$|Q_L| = |Q_H| - W$$
$$= 1855 \text{ J} - 1200 \text{ J} = 655 \text{ J}. \quad \text{(Answer)}$$

(e) What is the entropy change of the working substance for the energy transfer to it from the high-temperature reservoir? From it to the low-temperature reservoir?

SOLUTION: The Key Idea here is that the entropy change ΔS during a transfer of energy Q as heat at constant temperature T is given by Eq. 21-2 ($\Delta S = Q/T$). Thus, for the *positive* transfer of energy Q_H from the high-temperature reservoir at T_H, the change in the entropy of the working substance is

$$\Delta S_H = \frac{Q_H}{T_H} = \frac{1855 \text{ J}}{850 \text{ K}} = +2.18 \text{ J/K}. \quad \text{(Answer)}$$

Similarly, for the *negative* transfer of energy Q_L to the low-temperature reservoir at T_L, we have

$$\Delta S_L = \frac{Q_L}{T_L} = \frac{-655 \text{ J}}{300 \text{ K}} = -2.18 \text{ J/K}. \quad \text{(Answer)}$$

Note that the net entropy change of the working substance for one cycle is zero, as we discussed in deriving Eq. 21-8..

Sample Problem 21-4

An inventor claims to have constructed an engine that has an efficiency of 75% when operated between the boiling and freezing points of water. Is this possible?

SOLUTION: The Key Idea here is that the efficiency of a real engine (with its irreversible processes and wasteful energy transfers) must be less than the efficiency of a Carnot engine operating between the same two temperatures. From Eq. 21-11, we find that the effi-

ciency of a Carnot engine operating between the boiling and freezing points of water is

$$\varepsilon = 1 - \frac{T_L}{T_H} = 1 - \frac{(0 + 273) \text{ K}}{(100 + 273) \text{ K}} = 0.268 \approx 27\%.$$

Thus, the claimed efficiency of 75% for a real engine operating between the given temperatures is impossible.

Tactic 1: *The Language of Thermodynamics*
A rich, but sometimes misleading, language is used in scientific and engineering studies of thermodynamics. You may see statements that say heat is added, absorbed, subtracted, extracted, rejected, discharged, discarded, withdrawn, delivered, gained, lost, transferred, or expelled, or that it flows from one body to another (as if it were a liquid). You may also see statements that describe a body as *having* heat (as if heat can be held or possessed), or that its heat is increased or decreased. You should always keep in mind what is meant by the term *heat:*

▶ Heat is energy that is transferred from one body to another body owing to a difference in the temperatures of the bodies.

When we identify one of the bodies as being our system of interest, any such transfer of energy into the system is positive heat Q, and any such transfer out of the system is negative heat Q.

The term *work* also requires close attention. You may see statements that say work is produced or generated, or combined with heat or changed from heat. Here is what is meant by the term *work:*

▶ Work is energy that is transferred from one body to another body owing to a force that acts between them.

When we identify one of the bodies as being our system of interest, any such transfer of energy out of the system is either positive work W done *by* the system or negative work W done *on* the system. Any such transfer of energy into the system is negative work done *by* the system or positive work done *on* the system. (The preposition that is used is important.) Obviously, this can be confusing—whenever you see the term *work,* you should read carefully to determine the intent.

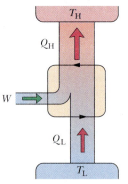

Fig. 21-13 The elements of a refrigerator. The two black arrowheads on the central loop suggest the working substance operating in a cycle, as if on a p-V plot. Energy Q_L is transferred as heat to the working substance from the low-temperature reservoir. Energy Q_H is transferred as heat to the high-temperature reservoir from the working substance. Work W is done on the refrigerator (on the working substance) by something in the environment.

21-5 Entropy in the Real World: Refrigerators

A **refrigerator** is a device that uses work to transfer energy from a low-temperature reservoir to a high-temperature reservoir as it continuously repeats a set series of thermodynamic processes. In a household refrigerator, for example, work is done by an electrical compressor to transfer energy from the food storage compartment (a low-temperature reservoir) to the room (a high-temperature reservoir).

Air conditioners and heat pumps are also refrigerators. The differences are only in the nature of the high- and low-temperature reservoirs. For an air conditioner, the low-temperature reservoir is the room that is to be cooled, and the high-temperature reservoir is the (presumably warmer) outdoors. A heat pump is an air conditioner that can be operated in reverse to heat a room; the room is the high-temperature reservoir and heat is transferred to it from the (presumably cooler) outdoors.

Let us consider an *ideal refrigerator:*

> In an ideal refrigerator, all processes are reversible and no wasteful energy transfers occur due to, say, friction and turbulence.

Figure 21-13 shows the basic elements of an ideal refrigerator that operates in the reverse of the Carnot engine of Fig. 21-7. In other words, all the energy transfers, as either heat or work, are reversed from those of a Carnot engine. We can call such an ideal refrigerator a **Carnot refrigerator.**

The designer of a refrigerator would like to extract as much energy $|Q_L|$ as possible from the low-temperature reservoir (what we want) for the least amount of work $|W|$ (what we pay for). A measure of the efficiency of a refrigerator, then, is

$$K = \frac{\text{what we want}}{\text{what we pay for}} = \frac{|Q_L|}{|W|} \qquad \text{(coefficient of performance, any refrigerator),} \qquad (21\text{-}12)$$

where K is called the *coefficient of performance.* For a Carnot refrigerator, the first law of thermodynamics gives $|W| = |Q_H| - |Q_L|$, where $|Q_H|$ is the magnitude of the energy transferred as heat to the high-temperature reservoir. Equation 21-12 then becomes

$$K_C = \frac{|Q_L|}{|Q_H| - |Q_L|}. \qquad (21\text{-}13)$$

Because a Carnot refrigerator is a Carnot engine operating in reverse, we can combine Eq. 21-8 with Eq. 21-13; after some algebra we find

$$K_C = \frac{T_L}{T_H - T_L} \qquad \text{(coefficient of performance, Carnot refrigerator).} \qquad (21\text{-}14)$$

For typical room air conditioners, $K \approx 2.5$. For household refrigerators, $K \approx 5$. Perversely, the value of K is higher the closer the temperatures of the two reservoirs are to each other. That is why heat pumps are more effective in temperate climates than in climates where the outside temperature varies widely.

It would be nice to own a refrigerator that did not require some input of work—that is, one that would run without being plugged in. Figure 21-14 represents another "inventor's dream," a *perfect refrigerator* that transfers heat Q from a cold reservoir to a warm reservoir without the need for work. Because the unit operates in cycles, the entropy of the working substance does not change during a complete cycle. The

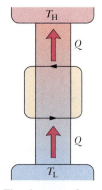

Fig. 21-14 The elements of a perfect refrigerator—that is, one that transfers energy from a low-temperature reservoir to a high-temperature reservoir without any input of work.

entropies of the two reservoirs, however, do change: The entropy change for the cold reservoir is $-|Q|/T_L$, and that for the warm reservoir is $+|Q|/T_H$. Thus, the net entropy change for the entire system is

$$\Delta S = -\frac{|Q|}{T_L} + \frac{|Q|}{T_H}.$$

Because $T_H > T_L$, the right side of this equation is negative and thus the net change in entropy per cycle for the closed system *refrigerator + reservoirs* is also negative. Because such a decrease in entropy violates the second law of thermodynamics (Eq. 21-5), a perfect refrigerator does not exist. (If you want your refrigerator to operate, you must plug it in.)

This result leads us to another (equivalent) formulation of the second law of thermodynamics:

> No series of processes is possible whose sole result is the transfer of energy as heat from a reservoir at a given temperature to a reservoir at a higher temperature.

In short, *there are no perfect refrigerators.*

✔**CHECKPOINT 4:** You wish to increase the coefficient of performance of an ideal refrigerator. You can do so by (a) running the cold chamber at a slightly higher temperature, (b) running the cold chamber at a slightly lower temperature, (c) moving the unit to a slightly warmer room, or (d) moving it to a slightly cooler room. The magnitudes of the temperature changes are to be the same in all four cases. List the changes according to the resulting coefficients of performance, greatest first.

21-6 The Efficiencies of Real Engines

Let ε_C be the efficiency of a Carnot engine operating between two given temperatures. In this section we prove that no real engine operating between those temperatures can have an efficiency greater than ε_C. If it could, the engine would violate the second law of thermodynamics.

Let us assume that an inventor, working in her garage, has constructed an engine X, which she claims has an efficiency ε_X that is greater than ε_C:

$$\varepsilon_X > \varepsilon_C \qquad \text{(a claim)}. \qquad (21\text{-}15)$$

Let us couple engine X to a Carnot refrigerator, as in Fig. 21-15a. We adjust the strokes of the Carnot refrigerator so that the work it requires per cycle is just equal

Fig. 21-15 (*a*) Engine X drives a Carnot refrigerator. (*b*) If, as claimed, engine X is more efficient than a Carnot engine, then the combination shown in (*a*) is equivalent to the perfect refrigerator shown here. This violates the second law of thermodynamics, so we conclude that engine X *cannot* be more efficient than a Carnot engine.

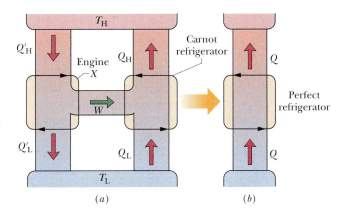

to that provided by engine X. Thus, no (external) work is performed on or by the combination *engine + refrigerator* of Fig. 21-15a, which we take as our system.

If Eq. 21-15 is true, from the definition of efficiency (Eq. 21-9), we must have

$$\frac{|W|}{|Q'_H|} > \frac{|W|}{|Q_H|},$$

where the prime refers to engine X and the right side of the inequality is the efficiency of the Carnot refrigerator when it operates as an engine. This inequality requires that

$$|Q_H| > |Q'_H|. \tag{21-16}$$

Because the work done by engine X is equal to the work done on the Carnot refrigerator, we have, from the first law of thermodynamics (see Eq. 21-6),

$$|Q_H| - |Q_L| = |Q'_H| - |Q'_L|,$$

which we can write as

$$|Q_H| - |Q'_H| = |Q_L| - |Q'_L| = Q. \tag{21-17}$$

Because of Eq. 21-16, the quantity Q in Eq. 21-17 must be positive.

Comparison of Eq. 21-17 with Fig. 21-15 shows that the net effect of engine X and the Carnot refrigerator, working in combination, is to transfer energy Q as heat from a low-temperature reservoir to a high-temperature reservoir without the requirement of work. Thus, the combination acts like the perfect refrigerator of Fig. 21-14, whose existence is a violation of the second law of thermodynamics.

Something must be wrong with one or more of our assumptions, and it can only be Eq. 21-15. We conclude that *no real engine can have an efficiency greater than that of a Carnot engine when both engines work between the same two temperatures.* At most, it can have an efficiency equal to that of a Carnot engine. In that case, engine X *is* a Carnot engine.

21-7 A Statistical View of Entropy

In Chapter 20 we saw that the macroscopic properties of gases can be explained in terms of their microscopic, or molecular, behavior. For one example, recall that we were able to account for the pressure exerted by a gas on the walls of its container in terms of the momentum transferred to those walls by rebounding gas molecules. Such explanations are part of a study called **statistical mechanics.**

Here we shall focus our attention on a single problem, involving the distribution of gas molecules between the two halves of an insulated box. This problem is reasonably simple to analyze, and it allows us to use statistical mechanics to calculate the entropy change for the free expansion of an ideal gas. You will see in Sample Problem 21-6 that statistical mechanics leads to the same entropy change we obtained in Sample Problem 21-1 using thermodynamics.

Figure 21-16 shows a box that contains six identical (and thus indistinguishable) molecules of a gas. At any instant, a given molecule will be in either the left or the right half of the box; because the two halves have equal volumes, the molecule has the same likelihood, or probability, of being in either half.

Table 21-1 shows four of the seven possible *configurations* of the six molecules, each configuration labeled with a Roman numeral. For example, in configuration I, all six molecules are in the left half of the box ($n_1 = 6$) and none are in the right half ($n_2 = 0$). The three configurations not shown are V with a (2, 4) split, VI with a (1, 5) split, and VII with a (0, 6) split. We see that, in general, a given configuration can be achieved in a number of different ways. We call these different

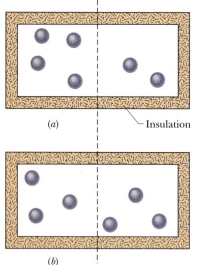

(a)   Insulation

(b)

Fig. 21-16 An insulated box contains six gas molecules. Each molecule has the same probability of being in the left half of the box as in the right half. The arrangement in (a) corresponds to configuration III in Table 21-1, and that in (b) corresponds to configuration IV.

Table 21-1   Six Molecules in a Box

Configuration			Multiplicity W	Calculation of W	Entropy 10^{-23} J/K
Label	n_1	n_2	(number of microstates)	(Eq. 21-18)	(Eq. 21-19)
I	6	0	1	$6!/(6!\,0!) = 1$	0
II	5	1	6	$6!/(5!\,1!) = 6$	2.47
III	4	2	15	$6!/(4!\,2!) = 15$	3.74
IV	3	3	20	$6!/(3!\,3!) = 20$	4.13

Total number
of microstates = 64

arrangements of the molecules *microstates*. Let us see how to calculate the number of microstates that correspond to a given configuration.

Suppose we have N molecules, distributed with n_1 molecules in one half of the box and n_2 in the other. (Thus $n_1 + n_2 = N$.) Let us imagine that we distribute the molecules "by hand," one at a time. If $N = 6$, we can select the first molecule in six independent ways; that is, we can pick any one of the six molecules. We can pick the second molecule in five ways, by picking any one of the remaining five molecules; and so on. The total number of ways in which we can select all six molecules is the product of these independent ways, or $6 \times 5 \times 4 \times 3 \times 2 \times 1 = 720$. In mathematical shorthand we write this product as $6! = 720$, where 6! is pronounced "six factorial." Your hand calculator can probably calculate factorials. For later use you will need to know that $0! = 1$. (Check this on your calculator.)

However, because the molecules are indistinguishable, these 720 arrangements are not all different. In the case that $n_1 = 4$ and $n_2 = 2$ (which is configuration III in Table 21-1), for example, the order in which you put four molecules in one half of the box does not matter, because after you have put all four in, there is no way that you can tell the order in which you did so. The number of ways in which you can order the four molecules is 4! or 24. Similarly, the number of ways in which you can order two molecules for the other half of the box is simply 2! or 2. To get the number of *different* arrangements that lead to the (4, 2) split of configuration III, we must divide 720 by 24 and also by 2. We call the resulting quantity, which is the number of microstates that correspond to a given configuration, the *multiplicity* W of that configuration. Thus, for configuration III,

$$W_{\text{III}} = \frac{6!}{4!\,2!} = \frac{720}{24 \times 2} = 15.$$

Thus, Table 21-1 tells us there are 15 independent microstates that correspond to configuration III. Note that, as the table also tells us, the total number of microstates for six molecules distributed over the seven configurations is 64.

Extrapolating from six molecules to the general case of N molecules, we have

$$W = \frac{N!}{n_1!\,n_2!} \qquad \text{(multiplicity of configuration).} \qquad (21\text{-}18)$$

You should verify that Eq. 21-18 gives the multiplicities for all the configurations listed in Table 21-1.

The basic assumption of statistical mechanics is:

> All microstates are equally probable.

Fig. 21-17 For a *large* number of molecules in a box, a plot of the number of microstates that require various percentages of the molecules to be in the left half of the box. Nearly all the microstates correspond to an approximately equal sharing of the molecules between the two halves of the box; those microstates form the *central configuration peak* on the plot. For $N \approx 10^{22}$, the central configuration peak is much too narrow to be drawn on this plot.

In other words, if we were to take a great many snapshots of the six molecules as they jostle around in the box of Fig. 21-16 and then count the number of times each microstate occurred, we would find that all 64 microstates would occur equally often. In other words, the system will spend, on average, the same amount of time in each of the 64 microstates.

Because the microstates are equally probable, but different configurations have different numbers of microstates, the configurations are *not* equally probable. In Table 21-1 configuration IV, with 20 microstates, is the *most probable configuration,* with a probability of 20/64 = 0.313. This result means that the system is in configuration IV 31.3% of the time. Configurations I and VII, in which all the molecules are in one half of the box, are the least probable, each with a probability of 1/64 = 0.016 or 1.6%. It is not surprising that the most probable configuration is the one in which the molecules are evenly divided between the two halves of the box, because that is what we expect at thermal equilibrium. However, it *is* surprising that there is *any* probability, however small, of finding all six molecules clustered in half of the box, with the other half empty. In Sample Problem 21-5 we show that this state can occur because six molecules is an extremely small number.

For large values of N there are extremely large numbers of microstates, but nearly all the microstates belong to the configuration in which the molecules are divided equally between the two halves of the box, as Fig. 21-17 indicates. Even though the measured temperature and pressure of the gas remain constant, the gas is churning away endlessly as its molecules "visit" all probable microstates with equal probability. However, because so few microstates lie outside the very narrow central configuration peak of Fig. 21-17, we might as well assume that the gas molecules are always divided equally between the two halves of the box. As we shall see, this is the configuration with the greatest entropy.

Sample Problem 21-5

Suppose that there are 100 indistinguishable molecules in the box of Fig. 21-16. How many microstates are associated with the configuration $n_1 = 50$ and $n_2 = 50$? How many are associated with the configuration $n_1 = 100$ and $n_2 = 0$? Interpret the results in terms of the relative probabilities of the two configurations.

SOLUTION: The **Key Idea** here is that the multiplicity W of a configuration of indistinguishable molecules in a closed box is the number of independent microstates with that configuration, as given by Eq. 21-18. For the (n_1, n_2) configuration (50, 50), that equation yields

$$W = \frac{N!}{n_1! \, n_2!} = \frac{100!}{50! \, 50!}$$

$$= \frac{9.33 \times 10^{157}}{(3.04 \times 10^{64})(3.04 \times 10^{64})}$$

$$= 1.01 \times 10^{29}. \qquad \text{(Answer)}$$

Similarly, for the configuration of (100, 0), we have

$$W = \frac{N!}{n_1! \, n_2!} = \frac{100!}{100! \, 0!} = \frac{1}{0!} = \frac{1}{1} = 1. \qquad \text{(Answer)}$$

Thus, a 50–50 distribution is more likely than a 100–0 distribution by the enormous factor of about 1×10^{29}. If you could count, at one per nanosecond, the number of microstates that correspond to the 50–50 distribution, it would take you about 3×10^{12} years, which is about 750 times longer than the age of the universe. Even 100 molecules is *still* a very small number. Imagine what these calculated probabilities would be like for a mole of molecules, say about $N = 10^{24}$. Thus, you need never worry about suddenly finding all the air molecules clustering in one corner of your room!

Probability and Entropy

In 1877, Austrian physicist Ludwig Boltzmann (the Boltzmann of Boltzmann's constant k) derived a relationship between the entropy S of a configuration of a gas and the multiplicity W of that configuration. That relationship is

$$S = k \ln W \qquad \text{(Boltzmann's entropy equation).} \qquad (21\text{-}19)$$

This famous formula is engraved on Boltzmann's tombstone.

It is natural that S and W should be related by a logarithmic function. The total entropy of two systems is the *sum* of their separate entropies. The probability of occurrence of two independent systems is the *product* of their separate probabilities. Because $\ln ab = \ln a + \ln b$, the logarithm seems the logical way to connect these quantities.

Table 21-1 displays the entropies of the configurations of the six-molecule system of Fig. 21-16, computed using Eq. 21-19. Configuration IV, which has the greatest multiplicity, also has the greatest entropy.

When you use Eq. 21-18 to calculate W, your calculator may signal "OVERFLOW" if you try to find the factorial of a number greater than a few hundred. Fortunately, there is a very good approximation, known as **Stirling's approximation,** not for $N!$ but for $\ln N!$, which is exactly what is needed in Eq. 21-19. Stirling's approximation is

$$\ln N! \approx N(\ln N) - N \qquad \text{(Stirling's approximation).} \qquad (21\text{-}20)$$

The Stirling of this approximation is not the Stirling of the Stirling engine.

CHECKPOINT 5: A box contains one mole of a gas. Consider two configurations: (a) each half of the box contains half the molecules, and (b) each third of the box contains one-third of the molecules. Which configuration has more microstates?

Sample Problem 21-6

In Sample Problem 21-1 we showed that when n moles of an ideal gas doubles its volume in a free expansion, the entropy increase from the initial state i to the final state f is $S_f - S_i = nR \ln 2$. Derive this result with statistical mechanics.

SOLUTION: One Key Idea here is that we can relate the entropy S of any given configuration of the molecules in the gas to the multiplicity W of microstates for that configuration, using Eq. 21-19 ($S = k \ln W$). We are interested in two configurations: the final configuration f (with the molecules occupying the full volume of their container in Fig. 21-1b) and the initial configuration i (with the molecules occupying the left half of the container).

A second Key Idea is that, because the molecules are in a closed container, we can calculate the multiplicity W of their microstates with Eq. 21-18. Here we have N molecules in the n moles of the gas. Initially, with the molecules all in the left half of the container, their (n_1, n_2) configuration is $(N, 0)$. Then, Eq. 21-18 gives their multiplicity as

$$W_i = \frac{N!}{N! \, 0!} = 1.$$

Finally, with the molecules spread through the full volume, their (n_1, n_2) configuration is $(N/2, N/2)$. Then, Eq. 21-18 gives their multiplicity as

$$W_f = \frac{N!}{(N/2)! \, (N/2)!}.$$

From Eq. 21-19, the initial and final entropies are

$$S_i = k \ln W_i = k \ln 1 = 0$$

and

$$S_f = k \ln W_f = k \ln(N!) - 2k \ln[(N/2)!]. \qquad (21\text{-}21)$$

In writing Eq. 21-21, we have used the relation

$$\ln \frac{a}{b^2} = \ln a - 2 \ln b.$$

Now, applying Eq. 21-20 to evaluate Eq. 21-21, we find that

$$S_f = k \ln(N!) - 2k \ln[(N/2)!]$$
$$= k[N(\ln N) - N] - 2k[(N/2) \ln(N/2) - (N/2)]$$
$$= k[N(\ln N) - N - N \ln(N/2) + N]$$
$$= k[N(\ln N) - N(\ln N - \ln 2)] = Nk \ln 2. \qquad (21\text{-}22)$$

From Section 20-3 we can substitute nR for Nk, where R is the universal gas constant. Equation 21-22 then becomes

$$S_f = nR \ln 2.$$

The change in entropy from the initial state to the final is thus

$$S_f - S_i = nR \ln 2 - 0$$
$$= nR \ln 2, \qquad \text{(Answer)}$$

which is what we set out to show. In Sample Problem 21-1 we calculated this entropy increase for a free expansion with thermodynamics by finding an equivalent reversible process and calculating the entropy change for *that* process in terms of temperature and heat transfer. In this sample problem, we calculate the same increase in entropy with statistical mechanics using the fact that the system consists of molecules.

REVIEW & SUMMARY

One-Way Processes   An **irreversible process** is one that cannot be reversed by means of small changes in the environment. The direction in which an irreversible process proceeds is set by the *change in entropy* ΔS of the system undergoing the process. Entropy S is a *state property* (or *state function*) of the system; that is, it depends only on the state of the system and not on how the system reached that state. The *entropy postulate* states (in part): *If an irreversible process occurs in a closed system, the entropy of the system always increases.*

Calculating Entropy Change   The **entropy change** ΔS for an irreversible process that takes a system from an initial state i to a final state f is exactly equal to the entropy change ΔS for *any reversible process* that takes the system between those same two states. We can compute the latter (but not the former) with

$$\Delta S = S_f - S_i = \int_i^f \frac{dQ}{T}. \qquad (21\text{-}1)$$

Here Q is the energy transferred as heat to or from the system during the process, and T is the temperature of the system in kelvins during the process.

For a reversible isothermal process, Eq. 21-1 reduces to

$$\Delta S = S_f - S_i = \frac{Q}{T}. \qquad (21\text{-}2)$$

When the temperature change ΔT of a system is small relative to the temperature (in kelvins) before and after the process, the entropy change can be approximated as

$$\Delta S = S_f - S_i \approx \frac{Q}{T_{\text{avg}}}, \qquad (21\text{-}3)$$

where T_{avg} is the system's average temperature during the process.

When an ideal gas changes reversibly from an initial state with temperature T_i and volume V_i to a final state with temperature T_f and volume V_f, the change ΔS in the entropy of the gas is

$$\Delta S = S_f - S_i = nR \ln \frac{V_f}{V_i} + nC_V \ln \frac{T_f}{T_i}. \qquad (21\text{-}4)$$

The Second Law of Thermodynamics   This law, which is an extension of the entropy postulate, states: *If a process occurs in a closed system, the entropy of the system increases for irreversible processes and remains constant for reversible processes. It never decreases.* In equation form,

$$\Delta S \geq 0. \qquad (21\text{-}5)$$

Engines   An **engine** is a device that, operating in a cycle, extracts energy $|Q_H|$ as heat from a high-temperature reservoir and does a certain amount of work $|W|$. The *efficiency* ε of any engine is defined as

$$\varepsilon = \frac{\text{energy we get}}{\text{energy we pay for}} = \frac{|W|}{|Q_H|}. \qquad (21\text{-}9)$$

In an **ideal engine,** all processes are reversible and no wasteful energy transfers occur due to, say, friction and turbulence. A **Carnot engine** is an ideal engine that follows the cycle of Fig. 21-8.

Its efficiency is

$$\varepsilon_C = 1 - \frac{|Q_L|}{|Q_H|} = 1 - \frac{T_L}{T_H}, \qquad (21\text{-}10, 21\text{-}11)$$

in which T_H and T_L are the temperatures of the high- and low-temperature reservoirs, respectively. Real engines always have an efficiency lower than that given by Eq. 21-11. Ideal engines that are not Carnot engines also have efficiencies lower than that given by Eq. 21-11.

A *perfect engine* is an imaginary engine in which energy extracted as heat from the high-temperature reservoir is converted completely to work. Such would violate the second law of thermodynamics, which can be restated as follows: No series of processes is possible whose sole result is the absorption of energy as heat from a thermal reservoir and the complete conversion of this energy to work.

Refrigerators   A refrigerator is a device that, operating in a cycle, has work W done on it as it extracts energy $|Q_L|$ as heat from a low-temperature reservoir. The coefficient of performance K of a refrigerator is defined as

$$K = \frac{\text{what we want}}{\text{what we pay for}} = \frac{|Q_L|}{|W|}. \qquad (21\text{-}12)$$

A **Carnot refrigerator** is a Carnot engine operating in reverse. For a Carnot refrigerator, Eq. 21-12 becomes

$$K_C = \frac{|Q_L|}{|Q_H| - |Q_L|} = \frac{T_L}{T_H - T_L}. \qquad (21\text{-}13, 21\text{-}14)$$

A *perfect refrigerator* is an imaginary refrigerator in which energy extracted as heat from the low-temperature reservoir is converted completely to heat discharged to the high-temperature reservoir, without any need for work. Such would violate the second law of thermodynamics, which can be restated as follows: No series of processes is possible whose sole result is the transfer of energy as heat from a reservoir at a given temperature to a reservoir at a higher temperature.

Entropy from a Statistical View   The entropy of a system can be defined in terms of the possible distributions of its molecules. For identical molecules, each possible distribution of molecules is called a **microstate** of the system. All equivalent microstates are grouped into a **configuration** of the system. The number of microstates in a configuration is the **multiplicity** W of the configuration.

For a system of N molecules that may be distributed between the two halves of a box, the multiplicity is given by

$$W = \frac{N!}{n_1! \, n_2!}, \qquad (21\text{-}18)$$

in which n_1 is the number of molecules in one half of the box and n_2 is the number in the other half. A basic assumption of *statistical mechanics* is that all the microstates are equally probable. Thus, configurations with a large multiplicity occur most often. When N is very large (say, $N = 10^{22}$ molecules or more), the molecules are nearly always in the configuration in which $n_1 = n_2$.

The multiplicity W of a configuration of a system and the entropy S of the system in that configuration are related by Boltzmann's entropy equation:

$$S = k \ln W, \qquad (21\text{-}19)$$

where $k = 1.38 \times 10^{-23}$ J/K is the Boltzmann constant.

When N is very large (the usual case), we can approximate $\ln N!$ with *Stirling's approximation:*

$$\ln N! \approx N(\ln N) - N. \qquad (21\text{-}20)$$

QUESTIONS

1. A gas, confined to an insulated cylinder, is compressed adiabatically to half its volume. Does the entropy of the gas increase, decrease, or remain unchanged during this process?

2. In four experiments, blocks A and B, starting at different initial temperatures, were brought together in an insulating box (as in Sample Problem 21-2) and allowed to reach a common final temperature. The entropy changes for the blocks in the four experiments had the following values (in Joules per kelvin), but not necessarily in the order given. Determine which values for A go with which values for B.

Block		Values		
A	8	5	3	9
B	−3	−8	−5	−2

3. Point i in Fig. 21-18 represents the initial state of an ideal gas at temperature T. Taking algebraic signs into account, rank the entropy changes that the gas undergoes as it moves, successively and reversibly, from point i to points a, b, c, and d, greatest first.

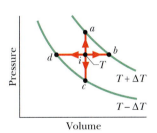

Fig. 21-18 Question 3.

4. An ideal gas, in contact with a controllable thermal reservoir, can be taken from initial state i to final state f along the four reversible paths in Fig. 21-19. Rank the paths according to the magnitudes of the resulting entropy changes of (a) the gas, (b) the reservoir, and (c) the gas–reservoir system, greatest first.

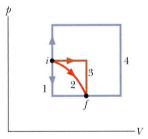

Fig. 21-19 Question 4.

5. You allow a gas to expand freely from volume V to volume $2V$. Later you allow that gas to expand freely from volume $2V$ to volume $3V$. Is the net entropy change for these two expansions greater than, less than, or equal to the entropy change that would occur if you allowed the gas to expand freely from volume V directly to volume $3V$?

6. Three Carnot engines operate between temperature limits of (a) 400 and 500 K, (b) 500 and 600 K, and (c) 400 and 600 K. Each engine extracts the same amount of energy per cycle from the high-temperature reservoir. Rank the magnitudes of the work done by the engines per cycle, greatest first.

7. Does the entropy per cycle increase, decrease, or remain the same for (a) a Carnot engine, (b) a real engine, and (c) a perfect engine (which is, of course, impossible to build)?

8. If you leave the door of your kitchen refrigerator open for several hours, does the temperature of the kitchen increase, decrease, or remain the same? Assume that the kitchen is closed and well insulated.

9. Does the entropy per cycle increase, decrease, or remain the same for (a) a Carnot refrigerator, (b) a real refrigerator, and (c) a perfect refrigerator (which is, of course, impossible to build)?

10. A box contains 100 atoms in a configuration, with 50 atoms in each half of the box. Suppose that you could count the different microstates associated with this configuration at the rate of 100 billion states per second, using a supercomputer. Without written calculation, guess how much computing time you would need: a day, a year, or much more than a year.

11. Figure 21-20 shows a snapshot at time $t = 0$ of molecules a and b in a box (similar to that of Fig. 21-16). The molecules have the same mass and speed v, and collisions between the molecules and the walls are elastic. What is the probability that snapshots taken at times (a) $t = 0.10L/v$ and (b) $t = 10L/v$ will show that a is in the left side of the box and b is in the right side of the box? (c) What is the probability that at some later time only half of the kinetic energy of the molecules will be in the right side of the box?

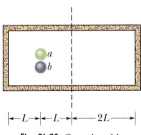

Fig. 21-20 Question 11.

EXERCISES & PROBLEMS

ssm Solution is in the Student Solutions Manual.
www Solution is available on the World Wide Web at:
        http://www.wiley.com/college/hrw
ilw Solution is available on the Interactive LearningWare.

SEC. 21-2 Change in Entropy

1E. A 2.50 mol sample of an ideal gas expands reversibly and isothermally at 360 K until its volume is doubled. What is the increase in entropy of the gas? ilw

2E. How much heat is required for a reversible isothermal expansion of an ideal gas at 132°C if the entropy of the gas increases by 46.0 J/K?

3E. Four moles of an ideal gas undergo a reversible isothermal expansion from volume V_1 to volume $V_2 = 2V_1$ at temperature $T = 400$ K. Find (a) the work done by the gas and (b) the entropy change of the gas. (c) If the expansion is reversible and adiabatic instead of isothermal, what is the entropy change of the gas? ssm

4E. An ideal gas undergoes a reversible isothermal expansion at 77.0°C, increasing its volume from 1.30 L to 3.40 L. The entropy change of the gas is 22.0 J/K. How many moles of gas are present?

5E. Find (a) the energy absorbed as heat and (b) the change in entropy of a 2.00 kg block of copper whose temperature is increased reversibly from 25°C to 100°C. The specific heat of copper is 386 J/kg·K. ilw

6E. An ideal monatomic gas at initial temperature T_0 (in kelvins) expands from initial volume V_0 to volume $2V_0$ by each of the five

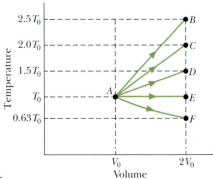

Fig. 21-21 Exercise 6.

processes indicated in the T-V diagram of Fig. 21-21. In which process is the expansion (a) isothermal, (b) isobaric (constant pressure), and (c) adiabatic? Explain your answers. (d) In which processes does the entropy of the gas decrease?

7E. (a) What is the entropy change of a 12.0 g ice cube that melts completely in a bucket of water whose temperature is just above the freezing point of water? (b) What is the entropy change of a 5.00 g spoonful of water that evaporates completely on a hot plate whose temperature is slightly above the boiling point of water?

8P. A 2.0 mol sample of an ideal monatomic gas undergoes the reversible process shown in Fig. 21-22. (a) How much energy is absorbed as heat by the gas? (b) What is the change in the internal energy of the gas? (c) How much work is done by the gas?

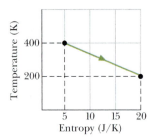

Fig. 21-22 Problem 8.

9P. In an experiment, 200 g of aluminum (with a specific heat of 900 J/kg·K) at 100°C is mixed with 50.0 g of water at 20.0°C, with the mixture thermally isolated. (a) What is the equilibrium temperature? What are the entropy changes of (b) the aluminum, (c) the water, and (d) the aluminum–water system? ssm www

10P. In the irreversible process of Fig. 21-5, let the initial temper-

atures of identical blocks L and R be 305.5 and 294.5 K, respectively, and let 215 J be the energy transfer between the blocks required to reach equilibrium. Then for the reversible processes of Fig. 21-6, what are the entropy changes of (a) block L, (b) its reservoir, (c) block R, (d) its reservoir, (e) the two-block system, and (f) the system of the two blocks and the two reservoirs?

11P. Use the reversible apparatus of Fig. 21-6 to show that, if the process of Fig. 21-5 happened in reverse, the entropy of the system would decrease, a violation of the second law of thermodynamics.

12P. An ideal diatomic gas, whose molecules are rotating but not oscillating, is taken through the cycle in Fig. 21-23. Determine for all three processes, in terms of p_1, V_1, T_1, and R: (a) p_2, p_3, and T_3 and (b) W, Q, ΔE_{int}, and ΔS per mole.

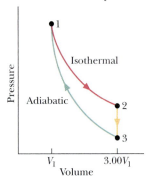

Fig. 21-23 Problem 12.

13P. A 50.0 g block of copper whose temperature is 400 K is placed in an insulating box with a 100 g block of lead whose temperature is 200 K. (a) What is the equilibrium temperature of the two-block system? (b) What is the change in the internal energy of the two-block system between the initial state and the equilibrium state? (c) What is the change in the entropy of the two-block system? (See Table 19-3.) ilw

14P. One mole of a monatomic ideal gas is taken from an initial pressure p and volume V to a final pressure $2p$ and volume $2V$ by two different processes: (I) It expands isothermally until its volume is doubled, and then its pressure is increased at constant volume to the final pressure. (II) It is compressed isothermally until its pressure is doubled, and then its volume is increased at constant pressure to the final volume. (a) Show the path of each process on a p-V diagram. For each process calculate, in terms of p and V, (b) the energy absorbed by the gas as heat in each part of the process, (c) the work done by the gas in each part of the process, (d) the change in internal energy of the gas, $E_{int,f} - E_{int,i}$, and (e) the change in entropy of the gas, $S_f - S_i$.

15P. A 10 g ice cube at -10°C is placed in a lake whose temperature is 15°C. Calculate the change in entropy of the cube–lake system as the ice cube comes to thermal equilibrium with the lake. The specific heat of ice is 2220 J/kg·K. (*Hint:* Will the ice cube affect the temperature of the lake?) ssm

16P. An 8.0 g ice cube at -10°C is put into a Thermos flask containing 100 cm^3 of water at 20°C. By how much has the entropy of the cube–water system changed when a final equilibrium state is reached? The specific heat of ice is 2220 J/kg·K.

17P. A mixture of 1773 g of water and 227 g of ice is in an initial equilibrium state at 0.00°C. The mixture is then, in a reversible process, brought to a second equilibrium state where the water–ice ratio, by mass, is 1:1 at 0.00°C. (a) Calculate the entropy change of the system during this process. (The heat of fusion for water is 333 kJ/kg.) (b) The system is then returned to the initial equilibrium state in an irreversible process (say, by using a Bunsen burner). Calculate the entropy change of the system during this process. (c) Are your answers consistent with the second law of thermodynamics? ssm

18P. A cylinder contains n moles of a monatomic ideal gas. If the gas undergoes a reversible isothermal expansion from initial volume V_i to final volume V_f along path I in Fig. 21-24, its change in entropy is $\Delta S = nR \ln (V_f/V_i)$. (See Sample Problem 21-1.) Now consider path II in Fig. 21-24, which takes the gas from the same initial state i to state x by a reversible adiabatic expansion, and then from that state x to the same final state f by a reversible constant-volume process. (a) Describe how you would carry out the two reversible processes for path II. (b) Show that the temperature of the gas in state x is

$$T_x = T_i(V_i/V_f)^{2/3}.$$

(c) What are the energy Q_I transferred as heat along path I and the energy Q_{II} transferred as heat along path II? Are they equal? (d) What is the entropy change ΔS for path II? Is the entropy change for path I equal to it? (e) Evaluate T_x, Q_I, Q_{II}, and ΔS for $n = 1$, $T_i = 500$ K, and $V_f/V_i = 2$.

Fig. 21-24 Problem 18.

19P. One mole of an ideal monatomic gas is taken through the cycle in Fig. 21-25. (a) How much work is done by the gas in going from state a to state c along path abc? What are the changes in internal energy and entropy in going (b) from b to c and (c) through one complete cycle? Express all answers in terms of the pressure p_0, volume V_0, and temperature T_0 of state a. **ssm**

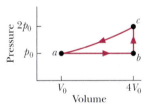

Fig. 21-25 Problem 19.

20P. One mole of an ideal monatomic gas, at an initial pressure of 5.00 kPa and initial temperature of 600 K, expands from initial volume $V_i = 1.00$ m^3 to final volume $V_f = 2.00$ m^3. During the expansion, the pressure p and volume V of the gas are related by $p = 5.00 \exp[(V_i - V)/a]$, where p is in kilopascals, V_i and V are in cubic meters, and $a = 1.00$ m^3. What are (a) the final pressure and (b) the final temperature of the gas? (c) How much work is done by the gas during the expansion? (d) What is the change in entropy of the gas during the expansion? (*Hint:* Use two simple reversible processes to find the entropy change.)

SEC. 21-4 Entropy in the Real World: Engines

21E. A Carnot engine absorbs 52 kJ as heat and exhausts 36 kJ as heat in each cycle. Calculate (a) the engine's efficiency and (b) the work done per cycle in kilojoules.

22E. A Carnot engine whose low-temperature reservoir is at 17°C has an efficiency of 40%. By how much should the temperature of the high-temperature reservoir be increased to increase the efficiency to 50%?

23E. A Carnot engine operates between 235°C and 115°C, absorbing 6.30×10^4 J per cycle at the higher temperature. (a) What is the efficiency of the engine? (b) How much work per cycle is this engine capable of performing? **ssm**

24E. In a hypothetical nuclear fusion reactor, the fuel is deuterium gas at a temperature of about 7×10^8 K. If this gas could be used

to operate a Carnot engine with $T_L = 100$°C, what would be the engine's efficiency?

25E. A Carnot engine has an efficiency of 22.0%. It operates between constant-temperature reservoirs differing in temperature by 75.0 C°. What are the temperatures of the two reservoirs? **ssm**

26P. A Carnot engine has a power of 500 W. It operates between constant-temperature reservoirs at 100°C and 60.0°C. What are (a) the rate of heat input and (b) the rate of exhaust heat output, in kilojoules per second?

27P. One mole of a monatomic ideal gas is taken through the reversible cycle shown in Fig. 21-26. Process bc is an adiabatic expansion, with $p_b = 10.0$ atm and $V_b = 1.00 \times 10^{-3}$ m^3. Find (a) the energy added to the gas as heat, (b) the energy leaving the gas as heat, (c) the net work done by the gas, and (d) the efficiency of the cycle. **ssm ilw**

Fig. 21-26 Problem 27.

28P. Show that the area enclosed by the Carnot cycle on the temperature-entropy plot of Fig. 21-9 represents the net energy transfer per cycle as heat to the working substance.

29P. One mole of an ideal monatomic gas is taken through the cycle shown in Fig. 21-27. Assume that $p = 2p_0$, $V = 2V_0$, $p_0 = 1.01 \times 10^5$ Pa, and $V_0 = 0.0225$ m^3. Calculate (a) the work done during the cycle, (b) the energy added as heat during stroke abc, and (c) the efficiency of the cycle. (d) What is the efficiency of a Carnot engine operating between the highest and lowest temperatures that occur in the cycle? How does this compare to the efficiency calculated in (c)?

Fig. 21-27 Problem 29.

30P. In the first stage of a two-stage Carnot engine, energy Q_1 is absorbed as heat at temperature T_1, work W_1 is done, and energy Q_2 is expelled as heat at a lower temperature T_2. The second stage absorbs that energy Q_2, does work W_2, and expels energy Q_3 at a still lower temperature T_3. Prove that the efficiency of the two-stage engine is $(T_1 - T_3)/T_1$.

31P. Suppose that a deep shaft were drilled in Earth's crust near one of the poles, where the surface temperature is -40°C, to a depth where the temperature is 800°C. (a) What is the theoretical limit to the efficiency of an engine operating between these temperatures? (b) If all the energy released as heat into the low-temperature reservoir were used to melt ice that was initially at -40°C, at what rate could liquid water at 0°C be produced by a 100 MW power plant (treat it as an engine)? The specific heat of ice is 2220 J/kg·K; water's heat of fusion is 333 kJ/kg. (Note that the engine can operate only between 0°C and 800°C in this case. Energy exhausted at -40°C cannot be used to raise the temperature of anything above -40°C.) **ssm www**

32P. One mole of an ideal gas is used as the working substance of an engine that operates on the cycle shown in Fig. 21-28. *BC* and *DA* are reversible adiabatic processes. (a) Is the gas monatomic, diatomic, or polyatomic? (b) What is the efficiency of the engine?

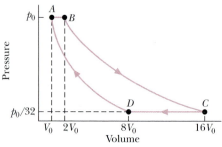

Fig. 21-28 Problem 32.

33P. The operation of a gasoline internal combustion engine is represented by the cycle in Fig. 21-29. Assume the gasoline–air intake mixture is an ideal gas and use a compression ratio of 4:1 ($V_4 = 4V_1$). Assume that $p_2 = 3p_1$. (a) Determine the pressure and temperature at each of the vertex points of the *p-V* diagram in terms of p_1, T_1, and the ratio γ of the molar specific heats of the gas. (b) What is the efficiency of the cycle? ssm

Fig. 21-29 Problem 33.

SEC. 21-5 Entropy in the Real World: Refrigerators

34E. A Carnot refrigerator does 200 J of work to remove 600 J from its cold compartment. (a) What is the refrigerator's coefficient of performance? (b) How much energy per cycle is exhausted to the kitchen as heat?

35E. A Carnot air conditioner takes energy from the thermal energy of a room at 70°F and transfers it to the outdoors, which is at 96°F. For each joule of electric energy required to operate the air conditioner, how many joules are removed from the room? ssm

36E. The electric motor of a heat pump transfers energy as heat from the outdoors, which is at −5.0°C, to a room, which is at 17°C. If the heat pump were a Carnot heat pump (a Carnot engine working in reverse), how many joules of heat would be transferred to the thermal energy of the room for each joule of electric energy consumed?

37E. A heat pump is used to heat a building. The outside temperature is −5.0°C, and the temperature inside the building is to be maintained at 22°C. The pump's coefficient of performance is 3.8, and the heat pump delivers 7.54 MJ as heat to the building each hour. If the heat pump is a Carnot engine working in reverse, at what rate must work be done to run the heat pump? ssm

38E. How much work must be done by a Carnot refrigerator to transfer 1.0 J as heat (a) from a reservoir at 7.0°C to one at 27°C, (b) from a reservoir at −73°C to one at 27°C, (c) from a reservoir at −173°C to one at 27°C, and (d) from a reservoir at −223°C to one at 27°C?

39P. An air conditioner operating between 93°F and 70°F is rated at 4000 Btu/h cooling capacity. Its coefficient of performance is 27% of that of a Carnot refrigerator operating between the same two temperatures. What horsepower is required of the air conditioner motor? ilw

40P. The motor in a refrigerator has a power of 200 W. If the freezing compartment is at 270 K and the outside air is at 300 K, and assuming the efficiency of a Carnot refrigerator, what is the maximum amount of energy that can be extracted as heat from the freezing compartment in 10.0 min?

41P. A Carnot engine works between temperatures T_1 and T_2. It drives a Carnot refrigerator that works between temperatures T_3 and T_4 (Fig. 21-30). Find the ratio Q_3/Q_1 in terms of T_1, T_2, T_3, and T_4. ssm

Fig. 21-30 Problem 41.   Engine

SEC. 21-7 A Statistical View of Entropy

42E. Construct a table like Table 21-1 for eight molecules.

43E. Show that for *N* molecules in a box, the number of possible microstates is 2^N when microstates are defined by whether a given molecule is in the left half of the box or the right half. Check this for the situation of Table 21-1.

44P. A box contains *N* gas molecules, equally divided between its two halves. For *N* = 50: (a) What is the multiplicity of this central configuration? (b) What is the total number of microstates for the system? (*Hint:* See Exercise 43.) (c) What percentage of the time does the system spend in its central configuration? (d) Repeat (a) through (c) for *N* = 100. (e) Repeat (a) through (c) for *N* = 200. (f) As *N* increases, you will find that the system spends *less* time (not more) in its central configuration. Explain why this is so.

45P. A box contains *N* gas molecules. Consider the box to be divided into three equal parts. (a) By extension of Eq. 21-18, write a formula for the multiplicity of any given configuration. (b) Consider two configurations: configuration *A* with equal numbers of molecules in all three thirds of the box, and configuration *B* with equal numbers of molecules in both halves of the box. What is the ratio W_A/W_B of the multiplicity of configuration *A* to that of configuration *B*? (c) Evaluate W_A/W_B for *N* = 100. (Because 100 is not evenly divisible by 3, put 34 molecules into one of the three box parts and 33 in each of the other parts for configuration *A*.) ssm www

NEW PROBLEMS

N1. A brass rod is in thermal contact with a constant-temperature reservoir at 130°C at one end and another reservoir at 24.0°C at the other end. (a) Compute the total change in entropy of the rod–reservoirs system that occurs when 5030 J of heat is conducted through the rod, from one reservoir to the other. (b) Does the entropy of the rod change in the process?

N2. A gas sample undergoes a reversible isothermal expansion. Figure 21N-1 gives the change ΔS in entropy of the gas versus the final volume V_f of the gas. How many moles are in the gas?

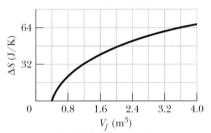

Fig. 21N-1 Problem N2.

N3. At very low temperatures, the molar specific heat C_V of many solids is approximately $C_V = AT^3$, where A depends on the particular substance. For aluminum, $A = 3.15 \times 10^{-5}$ J/mol·K⁴. Find the entropy change for 4.00 mol of aluminum when its temperature is raised from 5.00 K to 10.0 K.

N4. A Carnot engine is set up to produce a certain work W per cycle. In each cycle, energy Q_H in the form of heat is transferred to the working substance of the engine from the higher-temperature thermal reservoir, which is at an adjustable temperature T_H. The lower-temperature thermal reservoir is maintained at temperature $T_L = 250$ K. Figure 21N-2 gives Q_H for a range of T_H. If T_H is set at 550 K, what is Q_H?

Fig. 21N-2 Problem N4.

N5. An object of (constant) heat capacity C is heated from an initial temperature T_i to a final temperature T_f by a constant-temperature reservoir at T_f. (a) Represent the process on a graph of C/T versus T, and show graphically that the total change in entropy ΔS of the object–reservoir system is positive. (b) Explain how the use of reservoirs at intermediate temperatures would allow the process to be carried out in a way that makes ΔS as small as desired.

N6. A sample of nitrogen gas (N_2) undergoes a temperature increase at constant volume. As a result, the distribution of molecular speeds increases. That is, the probability distribution function $P(v)$ for the nitrogen molecules spreads to higher values of speed, as suggested in Fig. 20-7b. One way to report the spread in $P(v)$ is to measure the difference Δv between the most probable speed v_P and the rms speed v_{rms}. When $P(v)$ spreads to higher speeds, Δv increases. (a) Write an equation relating the change ΔS in the entropy of the nitrogen gas to the initial difference Δv_i and the final difference Δv_f. Assume that the gas is an ideal diatomic gas with rotation but not oscillation of its molecules. Let the number of moles be 1.5 mol, the initial temperature be 250 K, and the final temperature be 500 K. What are (b) the initial difference Δv_i, (c) the final difference Δv_f, and (d) the entropy change ΔS for the gas?

N7. An athlete dives from a platform and comes to rest in the water; this process is irreversible, and the diver–water–Earth system is closed. Devise a reversible way of transforming the system from its initial state (diver on platform) to its final state (diver in water, with the temperature of the diver and water slightly increased due to frictional heating). You can use an ideal engine and a reservoir whose temperature can be controlled.

N8. A 364 g block is put in contact with a thermal reservoir. The block is initially at a lower temperature than the reservoir. Assume that the consequent transfer of energy as heat from the reservoir to the block is reversible. Figure 21N-3 gives the change in entropy of the block ΔS until thermal equilibrium is reached. What is the specific heat of the block?

Fig. 21N-3 Problem N8.

N9. A moving block is slowed to a stop by friction. Both the block and the surface along which it slides are in an insulated enclosure. (a) Devise a reversible process to change the system from its initial state (block moving) to its final state (block stationary; temperature of the block and surface slightly increased). Show that this reversible process results in an entropy increase for the closed block–surface system. For the process, you can use an ideal engine and a constant-temperature reservoir. (b) Show, using the same process, but in reverse, that if the temperature of the system were to decrease spontaneously and the block were to start moving again, the entropy of the system would decrease (a violation of the second law of thermodynamics).

N10. (a) During each cycle, a Carnot engine absorbs 750 J as heat from a high-temperature reservoir at 360 K, with the low-temperature reservoir at 280 K. How much work is done per cycle? (b) The engine is then made to work in reverse to function as a Carnot refrigerator between those same two reservoirs. During each cycle, how much work is required to remove 1200 J as heat from the low-temperature reservoir?

N11. An ideal engine has a power of 500 W. It operates between constant-temperature reservoirs at 100°C and 60.0°C. What are (a) the rate of heat input and (b) the rate of exhaust heat output, in kilojoules per second?

N12. An ideal gas with 0.550 mol is isothermally and reversibly expanded in four situations. The temperature, initial volume, and final volume for each situation are given in the table. What is the change in the entropy of the gas for each situation?

Situation	(a)	(b)	(c)	(d)
Temperature (K)	250	350	400	450
Initial volume (cm³)	0.200	0.200	0.300	0.300
Final volume (cm³)	0.800	0.800	1.20	1.20

N13. An inventor has built an engine (engine X) and claims that its efficiency ε_X is greater than the efficiency ε of an ideal engine operating between the same two temperatures. Suppose that you couple engine X to an ideal refrigerator (Fig. 21N-4a) and adjust the cycle of engine X so that the work per cycle that it provides equals the work per cycle required by the ideal refrigerator. Treat this combination as a single unit and show that if the inventor's claim were true (if $\varepsilon_X > \varepsilon$), the combined unit would act as a perfect refrigerator (Fig. 21N-4b), transferring heat from the low-temperature reservoir to the high-temperature reservoir without the need of work.

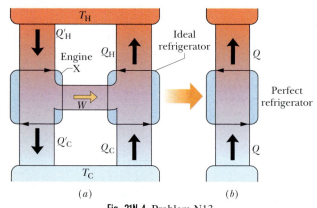

Fig. 21N-4 Problem N13.

N14. A Carnot refrigerator extracts an energy of 35.0 kJ as heat during each cycle, operating with a coefficient of performance of 4.60. What are (a) the energy per cycle transferred as heat to the room and (b) the work done per cycle?

N15. An ideal engine has efficiency ε. Show that if you run it backward as an ideal refrigerator, the coefficient of performance will be $K = (1 - \varepsilon)/\varepsilon$.

N16. An ideal monatomic gas with 3.2 mol undergoes a reversible increase in temperature from 380 K to 425 K at constant volume. What is the entropy change of the gas?

N17. An insulated thermos contains 130 g of water at 80.0°C. You put in a 12.0 g ice cube at 0°C to form a system of *ice + original water*. (a) What is the equilibrium temperature of the system? What are the entropy changes of the water that was originally the ice

cube (b) for its melting and then (c) for its warming to the equilibrium temperature? (d) What is the entropy change of the original water to reach the equilibrium temperature? (e) What is the net entropy change of the *ice + original water* system to reach the equilibrium temperature?

N18. A diatomic gas of 2 mol is taken reversibly around the cycle shown in the *T-S* diagram of Fig. 21N-5. The molecules do not rotate or oscillate. What are the heats Q for (a) the path from point 1 to point 2, (b) the path from point 2 to point 3, and (c) the full cycle? (d) What is the work W for the isothermal process? The volume V_1 at point 1 is 0.200 m³. What are the volumes at (e) point 2 and (f) point 3?

What are the changes ΔE_{int} in internal energy for (g) the path from point 1 to point 2, (h) the path from point 2 to point 3, and (i) the full cycle? (*Hint:* Part (h) can be done with one or two lines of calculation using Section 20-8 or with a page of calculation using Section 20-11.) (j) What is the work W for the adiabatic process?

Fig. 21N-5 Problem N18.

N19. An ideal refrigerator does 150 J of work to remove 560 J as heat from its cold compartment. (a) What is the refrigerator's coefficient of performance? (b) How much heat per cycle is exhausted to the kitchen?

N20. An ideal diatomic gas with 3.4 mol is taken through a cycle of three processes:

1. its temperature is increased from 200 K to 500 K at constant volume;

2. it is then isothermally expanded to its original pressure;

3. it is then contracted at constant pressure back to its original volume.

Throughout the cycle, the molecules rotate but do not oscillate. What is the efficiency of the cycle?

N21. A cylindrical copper rod of length 1.50 m and radius 2.00 cm is insulated to prevent heat loss through its curved surface. One end is attached to a thermal reservoir fixed at 300°C; the other is attached to a thermal reservoir fixed at 30.0°C. What is the rate at which entropy increases for the rod–reservoirs system?

N22. A 0.600 kg sample of water is initially ice at temperature −20°C. What is the sample's entropy change if its temperature is increased to 40°C?

N23. Four particles are in the insulated box of Fig. 21-16. What are (a) the least multiplicity, (b) the greatest multiplicity, (c) the least entropy, and (d) the greatest entropy of the four-particle system?

The International System of Units (SI)*

1. The SI Base Units

Quantity	Name	Symbol	Definition
length	meter	m	". . . the length of the path traveled by light in vacuum in 1/299,792,458 of a second." (1983)
mass	kilogram	kg	". . . this prototype [a certain platinum–iridium cylinder] shall henceforth be considered to be the unit of mass." (1889)
time	second	s	". . . the duration of 9,192,631,770 periods of the radiation corresponding to the transition between the two hyperfine levels of the ground state of the cesium-133 atom." (1967)
electric current	ampere	A	". . . that constant current which, if maintained in two straight parallel conductors of infinite length, of negligible circular cross section, and placed 1 meter apart in vacuum, would produce between these conductors a force equal to 2×10^{-7} newton per meter of length." (1946)
thermodynamic temperature	kelvin	K	". . . the fraction 1/273.16 of the thermodynamic temperature of the triple point of water." (1967)
amount of substance	mole	mol	". . . the amount of substance of a system which contains as many elementary entities as there are atoms in 0.012 kilogram of carbon-12." (1971)
luminous intensity	candela	cd	". . . the luminous intensity, in the perpendicular direction, of a surface of 1/600,000 square meter of a blackbody at the temperature of freezing platinum under a pressure of 101.325 newtons per square meter." (1967)

*Adapted from "The International System of Units (SI)," National Bureau of Standards Special Publication 330, 1972 edition. The definitions above were adopted by the General Conference of Weights and Measures, an international body, on the dates shown. In this book we do not use the candela.

2.   Some SI Derived Units

Quantity	Name of Unit	Symbol	
area	square meter	m^2	
volume	cubic meter	m^3	
frequency	hertz	Hz	s^{-1}
mass density (density)	kilogram per cubic meter	kg/m^3	
speed, velocity	meter per second	m/s	
angular velocity	radian per second	rad/s	
acceleration	meter per second per second	m/s^2	
angular acceleration	radian per second per second	rad/s^2	
force	newton	N	$kg \cdot m/s^2$
pressure	pascal	Pa	N/m^2
work, energy, quantity of heat	joule	J	$N \cdot m$
power	watt	W	J/s
quantity of electric charge	coulomb	C	$A \cdot s$
potential difference, electromotive force	volt	V	W/A
electric field strength	volt per meter (or newton per coulomb)	V/m	N/C
electric resistance	ohm	Ω	V/A
capacitance	farad	F	$A \cdot s/V$
magnetic flux	weber	Wb	$V \cdot s$
inductance	henry	H	$V \cdot s/A$
magnetic flux density	tesla	T	Wb/m^2
magnetic field strength	ampere per meter	A/m	
entropy	joule per kelvin	J/K	
specific heat	joule per kilogram kelvin	$J/(kg \cdot K)$	
thermal conductivity	watt per meter kelvin	$W/(m \cdot K)$	
radiant intensity	watt per steradian	W/sr	

3.   The SI Supplementary Units

Quantity	Name of Unit	Symbol
plane angle	radian	rad
solid angle	steradian	sr

APPENDIX B
Some Fundamental Constants of Physics*

Constant	Symbol	Computational Value	Best (1998) Value Value[a]	Best (1998) Value Uncertainty[b]
Speed of light in a vacuum	c	3.00×10^8 m/s	2.997 924 58	exact
Elementary charge	e	1.60×10^{-19} C	1.602 176 462	0.039
Gravitational constant	G	6.67×10^{-11} m³/s² · kg	6.673	1500
Universal gas constant	R	8.31 J/mol · K	8.314 472	1.7
Avogadro constant	N_A	6.02×10^{23} mol⁻¹	6.022 141 99	0.079
Boltzmann constant	k	1.38×10^{-23} J/K	1.380 650 3	1.7
Stefan–Boltzmann constant	σ	5.67×10^{-8} W/m² · K⁴	5.670 400	7.0
Molar volume of ideal gas at STP[d]	V_m	2.27×10^{-2} m³/mol	2.271 098 1	1.7
Permittivity constant	ϵ_0	8.85×10^{-12} F/m	8.854 187 817 62	exact
Permeability constant	μ_0	1.26×10^{-6} H/m	1.256 637 061 43	exact
Planck constant	h	6.63×10^{-34} J · s	6.626 068 76	0.078
Electron mass[c]	m_e	9.11×10^{-31} kg	9.109 381 88	0.079
		5.49×10^{-4} u	5.485 799 110	0.0021
Proton mass[c]	m_p	1.67×10^{-27} kg	1.672 621 58	0.079
		1.0073 u	1.007 276 466 88	1.3×10^{-4}
Ratio of proton mass to electron mass	m_p/m_e	1840	1836.152 667 5	0.0021
Electron charge-to-mass ratio	e/m_e	1.76×10^{11} C/kg	1.758 820 174	0.040
Neutron mass[c]	m_n	1.68×10^{-27} kg	1.674 927 16	0.079
		1.0087 u	1.008 664 915 78	5.4×10^{-4}
Hydrogen atom mass[c]	m_{1H}	1.0078 u	1.007 825 031 6	0.0005
Deuterium atom mass[c]	m_{2H}	2.0141 u	2.014 101 777 9	0.0005
Helium atom mass[c]	m_{4He}	4.0026 u	4.002 603 2	0.067
Muon mass	m_μ	1.88×10^{-28} kg	1.883 531 09	0.084
Electron magnetic moment	μ_e	9.28×10^{-24} J/T	9.284 763 62	0.040
Proton magnetic moment	μ_p	1.41×10^{-26} J/T	1.410 606 663	0.041
Bohr magneton	μ_B	9.27×10^{-24} J/T	9.274 008 99	0.040
Nuclear magneton	μ_N	5.05×10^{-27} J/T	5.050 783 17	0.040
Bohr radius	r_B	5.29×10^{-11} m	5.291 772 083	0.0037
Rydberg constant	R	1.10×10^7 m⁻¹	1.097 373 156 854 8	7.6×10^{-6}
Electron Compton wavelength	λ_C	2.43×10^{-12} m	2.426 310 215	0.0073

[a]Values given in this column should be given the same unit and power of 10 as the computational value.

[b]Parts per million.

[c]Masses given in u are in unified atomic mass units, where 1 u = 1.660 538 73 $\times 10^{-27}$ kg.

[d]STP means standard temperature and pressure: 0°C and 1.0 atm (0.1 MPa).

*The values in this table were selected from the 1998 CODATA recommended values (www.physics.nist.gov).

APPENDIX C
Some Astronomical Data

Some Distances from Earth

To the Moon*	3.82×10^8 m	To the center of our galaxy	2.2×10^{20} m
To the Sun*	1.50×10^{11} m	To the Andromeda Galaxy	2.1×10^{22} m
To the nearest star (Proxima Centauri)	4.04×10^{16} m	To the edge of the observable universe	$\sim 10^{26}$ m

*Mean distance.

The Sun, Earth, and the Moon

Property	Unit	Sun		Earth	Moon
Mass	kg	1.99×10^{30}		5.98×10^{24}	7.36×10^{22}
Mean radius	m	6.96×10^8		6.37×10^6	1.74×10^6
Mean density	kg/m^3	1410		5520	3340
Free-fall acceleration at the surface	m/s^2	274		9.81	1.67
Escape velocity	km/s	618		11.2	2.38
Period of rotation[a]	—	37 d at poles[b]	26 d at equator[b]	23 h 56 min	27.3 d
Radiation power[c]	W	3.90×10^{26}			

[a]Measured with respect to the distant stars.
[b]The Sun, a ball of gas, does not rotate as a rigid body.
[c]Just outside Earth's atmosphere solar energy is received, assuming normal incidence, at the rate of 1340 W/m^2.

Some Properties of the Planets

	Mercury	Venus	Earth	Mars	Jupiter	Saturn	Uranus	Neptune	Pluto
Mean distance from Sun, 10^6 km	57.9	108	150	228	778	1430	2870	4500	5900
Period of revolution, y	0.241	0.615	1.00	1.88	11.9	29.5	84.0	165	248
Period of rotation,[a] d	58.7	-243^b	0.997	1.03	0.409	0.426	-0.451^b	0.658	6.39
Orbital speed, km/s	47.9	35.0	29.8	24.1	13.1	9.64	6.81	5.43	4.74
Inclination of axis to orbit	<28°	≈3°	23.4°	25.0°	3.08°	26.7°	97.9°	29.6°	57.5°
Inclination of orbit to Earth's orbit	7.00°	3.39°		1.85°	1.30°	2.49°	0.77°	1.77°	17.2°
Eccentricity of orbit	0.206	0.0068	0.0167	0.0934	0.0485	0.0556	0.0472	0.0086	0.250
Equatorial diameter, km	4880	12 100	12 800	6790	143 000	120 000	51 800	49 500	2300
Mass (Earth = 1)	0.0558	0.815	1.000	0.107	318	95.1	14.5	17.2	0.002
Density (water = 1)	5.60	5.20	5.52	3.95	1.31	0.704	1.21	1.67	2.03
Surface value of g,[c] m/s^2	3.78	8.60	9.78	3.72	22.9	9.05	7.77	11.0	0.5
Escape velocity,[c] km/s	4.3	10.3	11.2	5.0	59.5	35.6	21.2	23.6	1.1
Known satellites	0	0	1	2	16 + ring	18 + rings	17 + rings	8 + rings	1

[a]Measured with respect to the distant stars.
[b]Venus and Uranus rotate opposite their orbital motion.
[c]Gravitational acceleration measured at the planet's equator.

APPENDIX D
Conversion Factors

Conversion factors may be read directly from these tables. For example, 1 degree = 2.778×10^{-3} revolutions, so $16.7° = 16.7 \times 2.778 \times 10^{-3}$ rev. The SI units are fully capitalized. Adapted in part from G. Shortley and D. Williams, *Elements of Physics*, 1971, Prentice-Hall, Englewood Cliffs, NJ.

Plane Angle

	°	′	″	RADIAN	rev
1 degree =	1	60	3600	1.745×10^{-2}	2.778×10^{-3}
1 minute =	1.667×10^{-2}	1	60	2.909×10^{-4}	4.630×10^{-5}
1 second =	2.778×10^{-4}	1.667×10^{-2}	1	4.848×10^{-6}	7.716×10^{-7}
1 RADIAN =	57.30	3438	2.063×10^{5}	1	0.1592
1 revolution =	360	2.16×10^{4}	1.296×10^{6}	6.283	1

Solid Angle

1 sphere = 4π steradians = 12.57 steradians

Length

	cm	METER	km	in.	ft	mi
1 centimeter =	1	10^{-2}	10^{-5}	0.3937	3.281×10^{-2}	6.214×10^{-6}
1 METER =	100	1	10^{-3}	39.37	3.281	6.214×10^{-4}
1 kilometer =	10^{5}	1000	1	3.937×10^{4}	3281	0.6214
1 inch =	2.540	2.540×10^{-2}	2.540×10^{-5}	1	8.333×10^{-2}	1.578×10^{-5}
1 foot =	30.48	0.3048	3.048×10^{-4}	12	1	1.894×10^{-4}
1 mile =	1.609×10^{5}	1609	1.609	6.336×10^{4}	5280	1

1 angström = 10^{-10} m
1 nautical mile = 1852 m
    = 1.151 miles = 6076 ft

1 fermi = 10^{-15} m
1 light-year = 9.460×10^{12} km
1 parsec = 3.084×10^{13} km

1 fathom = 6 ft
1 Bohr radius = 5.292×10^{-11} m
1 yard = 3 ft

1 rod = 16.5 ft
1 mil = 10^{-3} in.
1 nm = 10^{-9} m

Area

	METER^2	cm^2	ft^2	in.^{2}
1 SQUARE METER =	1	10^{4}	10.76	1550
1 square centimeter =	10^{-4}	1	1.076×10^{-3}	0.1550
1 square foot =	9.290×10^{-2}	929.0	1	144
1 square inch =	6.452×10^{-4}	6.452	6.944×10^{-3}	1

1 square mile = 2.788×10^{7} ft^2 = 640 acres
1 barn = 10^{-28} m^2

1 acre = 43 560 ft^2
1 hectare = 10^{4} m^2 = 2.471 acres

Volume

	METER^3	cm^3	L	ft^3	in.^{3}
1 CUBIC METER = 1		10^6	1000	35.31	6.102×10^4
1 cubic centimeter = 10^{-6}		1	1.000×10^{-3}	3.531×10^{-5}	6.102×10^{-2}
1 liter = 1.000×10^{-3}		1000	1	3.531×10^{-2}	61.02
1 cubic foot = 2.832×10^{-2}		2.832×10^4	28.32	1	1728
1 cubic inch = 1.639×10^{-5}		16.39	1.639×10^{-2}	5.787×10^{-4}	1

1 U.S. fluid gallon = 4 U.S. fluid quarts = 8 U.S. pints = 128 U.S. fluid ounces = 231 in.^{3}
1 British imperial gallon = 277.4 in.^{3} = 1.201 U.S. fluid gallons

Mass

Quantities in the colored areas are not mass units but are often used as such. When we write, for example, 1 kg ''='' 2.205 lb, this means that a kilogram is a *mass* that *weighs* 2.205 pounds at a location where g has the standard value of 9.80665 m/s^2.

	g	KILOGRAM	slug	u	oz	lb	ton
1 gram = 1		0.001	6.852×10^{-5}	6.022×10^{23}	3.527×10^{-2}	2.205×10^{-3}	1.102×10^{-6}
1 KILOGRAM = 1000		1	6.852×10^{-2}	6.022×10^{26}	35.27	2.205	1.102×10^{-3}
1 slug = 1.459×10^4	14.59		1	8.786×10^{27}	514.8	32.17	1.609×10^{-2}
1 atomic mass unit = 1.661×10^{-24}	1.661×10^{-27}	1.138×10^{-28}	1		5.857×10^{-26}	3.662×10^{-27}	1.830×10^{-30}
1 ounce = 28.35		2.835×10^{-2}	1.943×10^{-3}	1.718×10^{25}	1	6.250×10^{-2}	3.125×10^{-5}
1 pound = 453.6		0.4536	3.108×10^{-2}	2.732×10^{26}	16	1	0.0005
1 ton = 9.072×10^5	907.2		62.16	5.463×10^{29}	3.2×10^4	2000	1

1 metric ton = 1000 kg

Density

Quantities in the colored areas are weight densities and, as such, are dimensionally different from mass densities. See note for mass table.

	slug/ft^3	KILOGRAM/ METER^3	g/cm^3	lb/ft^3	lb/in.^{3}
1 slug per foot^3 = 1		515.4	0.5154	32.17	1.862×10^{-2}
1 KILOGRAM per METER^3 = 1.940×10^{-3}		1	0.001	6.243×10^{-2}	3.613×10^{-5}
1 gram per centimeter^3 = 1.940		1000	1	62.43	3.613×10^{-2}
1 pound per foot^3 = 3.108×10^{-2}		16.02	16.02×10^{-2}	1	5.787×10^{-4}
1 pound per inch^3 = 53.71		2.768×10^4	27.68	1728	1

Time

	y	d	h	min	SECOND
1 year = 1		365.25	8.766×10^3	5.259×10^5	3.156×10^7
1 day = 2.738×10^{-3}		1	24	1440	8.640×10^4
1 hour = 1.141×10^{-4}		4.167×10^{-2}	1	60	3600
1 minute = 1.901×10^{-6}		6.944×10^{-4}	1.667×10^{-2}	1	60
1 SECOND = 3.169×10^{-8}		1.157×10^{-5}	2.778×10^{-4}	1.667×10^{-2}	1

Speed

	ft/s	km/h	METER/SECOND	mi/h	cm/s
1 foot per second = 1		1.097	0.3048	0.6818	30.48
1 kilometer per hour = 0.9113		1	0.2778	0.6214	27.78
1 METER per SECOND = 3.281		3.6	1	2.237	100
1 mile per hour = 1.467		1.609	0.4470	1	44.70
1 centimeter per second = 3.281×10^{-2}		3.6×10^{-2}	0.01	2.237×10^{-2}	1

1 knot = 1 nautical mi/h = 1.688 ft/s   1 mi/min = 88.00 ft/s = 60.00 mi/h

Force

Force units in the colored areas are now little used. To clarify: 1 gram-force (= 1 gf) is the force of gravity that would act on an object whose mass is 1 gram at a location where g has the standard value of 9.80665 m/s^2.

	dyne	NEWTON	lb	pdl	gf	kgf
1 dyne = 1		10^{-5}	2.248×10^{-6}	7.233×10^{-5}	1.020×10^{-3}	1.020×10^{-6}
1 NEWTON = 10^5		1	0.2248	7.233	102.0	0.1020
1 pound = 4.448×10^5		4.448	1	32.17	453.6	0.4536
1 poundal = 1.383×10^4		0.1383	3.108×10^{-2}	1	14.10	1.410×10^2
1 gram-force = 980.7		9.807×10^{-3}	2.205×10^{-3}	7.093×10^{-2}	1	0.001
1 kilogram-force = 9.807×10^5		9.807	2.205	70.93	1000	1

1 ton = 2000 lb

Pressure

	atm	dyne/cm^2	inch of water	cm Hg	PASCAL	lb/in.^{2}	lb/ft^2
1 atmosphere = 1		1.013×10^6	406.8	76	1.013×10^5	14.70	2116
1 dyne per centimeter^2 = 9.869×10^{-7}		1	4.015×10^{-4}	7.501×10^{-5}	0.1	1.405×10^{-5}	2.089×10^{-3}
1 inch of water[a] at 4°C = 2.458×10^{-3}		2491	1	0.1868	249.1	3.613×10^{-2}	5.202
1 centimeter of mercury[a] at 0°C = 1.316×10^{-2}		1.333×10^4	5.353	1	1333	0.1934	27.85
1 PASCAL = 9.869×10^{-6}		10	4.015×10^{-3}	7.501×10^{-4}	1	1.450×10^{-4}	2.089×10^{-2}
1 pound per inch^2 = 6.805×10^{-2}		6.895×10^4	27.68	5.171	6.895×10^3	1	144
1 pound per foot^2 = 4.725×10^{-4}		478.8	0.1922	3.591×10^{-2}	47.88	6.944×10^{-3}	1

[a]Where the acceleration of gravity has the standard value of 9.80665 m/s^2.

1 bar = 10^6 dyne/cm^2 = 0.1 MPa   1 millibar = 10^3 dyne/cm^2 = 10^2 Pa   1 torr = 1 mm Hg

Energy, Work, Heat

Quantities in the colored areas are not energy units but are included for convenience. They arise from the relativistic mass–energy equivalence formula $E = mc^2$ and represent the energy released if a kilogram or unified atomic mass unit (u) is completely converted to energy (bottom two rows) or the mass that would be completely converted to one unit of energy (rightmost two columns).

	Btu	erg	ft·lb	hp·h	JOULE	cal	kW·h	eV	MeV	kg	u
1 British thermal unit =	1	1.055×10^{10}	777.9	3.929×10^{-4}	1055	252.0	2.930×10^{-4}	6.585×10^{21}	6.585×10^{15}	1.174×10^{-14}	7.070×10^{12}
1 erg =	9.481×10^{-11}	1	7.376×10^{-8}	3.725×10^{-14}	10^{-7}	2.389×10^{-8}	2.778×10^{-14}	6.242×10^{11}	6.242×10^{5}	1.113×10^{-24}	670.2
1 foot-pound =	1.285×10^{-3}	1.356×10^{7}	1	5.051×10^{-7}	1.356	0.3238	3.766×10^{-7}	8.464×10^{18}	8.464×10^{12}	1.509×10^{-17}	9.037×10^{9}
1 horsepower-hour =	2545	2.685×10^{13}	1.980×10^{6}	1	2.685×10^{6}	6.413×10^{5}	0.7457	1.676×10^{25}	1.676×10^{19}	2.988×10^{-11}	1.799×10^{16}
1 JOULE =	9.481×10^{-4}	10^{7}	0.7376	3.725×10^{-7}	1	0.2389	2.778×10^{-7}	6.242×10^{18}	6.242×10^{12}	1.113×10^{-17}	6.702×10^{9}
1 calorie =	3.969×10^{-3}	4.186×10^{7}	3.088	1.560×10^{-6}	4.186	1	1.163×10^{-6}	2.613×10^{19}	2.613×10^{13}	4.660×10^{-17}	2.806×10^{10}
1 kilowatt-hour =	3413	3.600×10^{13}	2.655×10^{6}	1.341	3.600×10^{6}	8.600×10^{5}	1	2.247×10^{25}	2.247×10^{19}	4.007×10^{-11}	2.413×10^{16}
1 electron-volt =	1.519×10^{-22}	1.602×10^{-12}	1.182×10^{-19}	5.967×10^{-26}	1.602×10^{-19}	3.827×10^{-20}	4.450×10^{-26}	1	10^{-6}	1.783×10^{-36}	1.074×10^{-9}
1 million electron-volts =	1.519×10^{-16}	1.602×10^{-6}	1.182×10^{-13}	5.967×10^{-20}	1.602×10^{-13}	3.827×10^{-14}	4.450×10^{-20}	10^{-6}	1	1.783×10^{-30}	1.074×10^{-3}
1 kilogram =	8.521×10^{13}	8.987×10^{23}	6.629×10^{16}	3.348×10^{10}	8.987×10^{16}	2.146×10^{16}	2.497×10^{10}	5.610×10^{35}	5.610×10^{29}	1	6.022×10^{26}
1 unified atomic mass unit =	1.415×10^{-13}	1.492×10^{-3}	1.101×10^{-10}	5.559×10^{-17}	1.492×10^{-10}	3.564×10^{-11}	4.146×10^{-17}	9.320×10^{8}	932.0	1.661×10^{-27}	1

Power

	Btu/h	ft·lb/s	hp	cal/s	kW	WATT
1 British thermal unit per hour =	1	0.2161	3.929×10^{-4}	6.998×10^{-2}	2.930×10^{-4}	0.2930
1 foot-pound per second =	4.628	1	1.818×10^{-3}	0.3239	1.356×10^{-3}	1.356
1 horsepower =	2545	550	1	178.1	0.7457	745.7
1 calorie per second =	14.29	3.088	5.615×10^{-3}	1	4.186×10^{-3}	4.186
1 kilowatt =	3413	737.6	1.341	238.9	1	1000
1 WATT =	3.413	0.7376	1.341×10^{-3}	0.2389	0.001	1

Magnetic Field

	gauss	TESLA	milligauss
1 gauss =	1	10^{-4}	1000
1 TESLA =	10^{4}	1	10^{7}
1 milligauss =	0.001	10^{-7}	1

1 tesla = 1 weber/meter^2

Magnetic Flux

	maxwell	WEBER
1 maxwell =	1	10^{-8}
1 WEBER =	10^{8}	1

Mathematical Formulas

Geometry

Circle of radius r: circumference $= 2\pi r$; area $= \pi r^2$.

Sphere of radius r: area $= 4\pi r^2$; volume $= \frac{4}{3}\pi r^3$.

Right circular cylinder of radius r and height h:
area $= 2\pi r^2 + 2\pi rh$; volume $= \pi r^2 h$.

Triangle of base a and altitude h: area $= \frac{1}{2}ah$.

Quadratic Formula

If $ax^2 + bx + c = 0$, then $x = \dfrac{-b \pm \sqrt{b^2 - 4ac}}{2a}$.

Trigonometric Functions of Angle θ

$$\sin \theta = \frac{y}{r} \quad \cos \theta = \frac{x}{r}$$

$$\tan \theta = \frac{y}{x} \quad \cot \theta = \frac{x}{y}$$

$$\sec \theta = \frac{r}{x} \quad \csc \theta = \frac{r}{y}$$

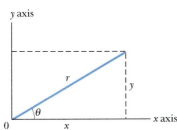

Pythagorean Theorem

In this right triangle,
$$a^2 + b^2 = c^2$$

Triangles

Angles are A, B, C

Opposite sides are a, b, c

Angles $A + B + C = 180°$

$$\frac{\sin A}{a} = \frac{\sin B}{b} = \frac{\sin C}{c}$$

$$c^2 = a^2 + b^2 - 2ab \cos C$$

Exterior angle $D = A + C$

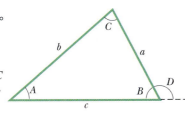

Mathematical Signs and Symbols

$=$ equals

$\approx$ equals approximately

$\sim$ is the order of magnitude of

$\neq$ is not equal to

$\equiv$ is identical to, is defined as

$>$ is greater than ($\gg$ is much greater than)

$<$ is less than ($\ll$ is much less than)

$\geq$ is greater than or equal to (or, is no less than)

$\leq$ is less than or equal to (or, is no more than)

$\pm$ plus or minus

$\propto$ is proportional to

Σ the sum of

x_{avg} the average value of x

Trigonometric Identities

$$\sin(90° - \theta) = \cos \theta$$

$$\cos(90° - \theta) = \sin \theta$$

$$\sin \theta / \cos \theta = \tan \theta$$

$$\sin^2 \theta + \cos^2 \theta = 1$$

$$\sec^2 \theta - \tan^2 \theta = 1$$

$$\csc^2 \theta - \cot^2 \theta = 1$$

$$\sin 2\theta = 2 \sin \theta \cos \theta$$

$$\cos 2\theta = \cos^2 \theta - \sin^2 \theta = 2 \cos^2 \theta - 1 = 1 - 2 \sin^2 \theta$$

$$\sin(\alpha \pm \beta) = \sin \alpha \cos \beta \pm \cos \alpha \sin \beta$$

$$\cos(\alpha \pm \beta) = \cos \alpha \cos \beta \mp \sin \alpha \sin \beta$$

$$\tan(\alpha \pm \beta) = \frac{\tan \alpha \pm \tan \beta}{1 \mp \tan \alpha \tan \beta}$$

$$\sin \alpha \pm \sin \beta = 2 \sin \tfrac{1}{2}(\alpha \pm \beta) \cos \tfrac{1}{2}(\alpha \mp \beta)$$

$$\cos \alpha + \cos \beta = 2 \cos \tfrac{1}{2}(\alpha + \beta) \cos \tfrac{1}{2}(\alpha - \beta)$$

$$\cos \alpha - \cos \beta = -2 \sin \tfrac{1}{2}(\alpha + \beta) \sin \tfrac{1}{2}(\alpha - \beta)$$

Binomial Theorem

$$(1 + x)^n = 1 + \frac{nx}{1!} + \frac{n(n-1)x^2}{2!} + \cdots \qquad (x^2 < 1)$$

Exponential Expansion

$$e^x = 1 + x + \frac{x^2}{2!} + \frac{x^3}{3!} + \cdots$$

Logarithmic Expansion

$$\ln(1 + x) = x - \tfrac{1}{2}x^2 + \tfrac{1}{3}x^3 - \cdots \qquad (|x| < 1)$$

Trigonometric Expansions
(θ in radians)

$$\sin \theta = \theta - \frac{\theta^3}{3!} + \frac{\theta^5}{5!} - \cdots$$

$$\cos \theta = 1 - \frac{\theta^2}{2!} + \frac{\theta^4}{4!} - \cdots$$

$$\tan \theta = \theta + \frac{\theta^3}{3} + \frac{2\theta^5}{15} + \cdots$$

Cramer's Rule

Two simultaneous equations in unknowns x and y,

$$a_1x + b_1y = c_1 \quad \text{and} \quad a_2x + b_2y = c_2,$$

have the solutions

$$x = \frac{\begin{vmatrix} c_1 & b_1 \\ c_2 & b_2 \end{vmatrix}}{\begin{vmatrix} a_1 & b_1 \\ a_2 & b_2 \end{vmatrix}} = \frac{c_1b_2 - c_2b_1}{a_1b_2 - a_2b_1}$$

and

$$y = \frac{\begin{vmatrix} a_1 & c_1 \\ a_2 & c_2 \end{vmatrix}}{\begin{vmatrix} a_1 & b_1 \\ a_2 & b_2 \end{vmatrix}} = \frac{a_1c_2 - a_2c_1}{a_1b_2 - a_2b_1}.$$

Products of Vectors

Let $\hat{i}$, $\hat{j}$, and $\hat{k}$ be unit vectors in the x, y, and z directions. Then

$$\hat{i} \cdot \hat{i} = \hat{j} \cdot \hat{j} = \hat{k} \cdot \hat{k} = 1, \qquad \hat{i} \cdot \hat{j} = \hat{j} \cdot \hat{k} = \hat{k} \cdot \hat{i} = 0,$$

$$\hat{i} \times \hat{i} = \hat{j} \times \hat{j} = \hat{k} \times \hat{k} = 0,$$

$$\hat{i} \times \hat{j} = \hat{k}, \qquad \hat{j} \times \hat{k} = \hat{i}, \qquad \hat{k} \times \hat{i} = \hat{j}.$$

Any vector $\vec{a}$ with components a_x, a_y, and a_z along the x, y, and z axes can be written as

$$\vec{a} = a_x\hat{i} + a_y\hat{j} + a_z\hat{k}.$$

Let $\vec{a}$, $\vec{b}$, and $\vec{c}$ be arbitrary vectors with magnitudes a, b, and c. Then

$$\vec{a} \times (\vec{b} + \vec{c}) = (\vec{a} \times \vec{b}) + (\vec{a} \times \vec{c})$$

$$(s\vec{a}) \times \vec{b} = \vec{a} \times (s\vec{b}) = s(\vec{a} \times \vec{b}) \quad (s = \text{a scalar}).$$

Let θ be the smaller of the two angles between $\vec{a}$ and $\vec{b}$. Then

$$\vec{a} \cdot \vec{b} = \vec{b} \cdot \vec{a} = a_xb_x + a_yb_y + a_zb_z = ab \cos \theta$$

$$\vec{a} \times \vec{b} = -\vec{b} \times \vec{a} = \begin{vmatrix} \hat{i} & \hat{j} & \hat{k} \\ a_x & a_y & a_z \\ b_x & b_y & b_z \end{vmatrix}$$

$$= \hat{i} \begin{vmatrix} a_y & a_z \\ b_y & b_z \end{vmatrix} - \hat{j} \begin{vmatrix} a_x & a_z \\ b_x & b_z \end{vmatrix} + \hat{k} \begin{vmatrix} a_x & a_y \\ b_x & b_y \end{vmatrix}$$

$$= (a_yb_z - b_ya_z)\hat{i} + (a_zb_x - b_za_x)\hat{j} + (a_xb_y - b_xa_y)\hat{k}$$

$$|\vec{a} \times \vec{b}| = ab \sin \theta$$

$$\vec{a} \cdot (\vec{b} \times \vec{c}) = \vec{b} \cdot (\vec{c} \times \vec{a}) = \vec{c} \cdot (\vec{a} \times \vec{b})$$

$$\vec{a} \times (\vec{b} \times \vec{c}) = (\vec{a} \cdot \vec{c})\vec{b} - (\vec{a} \cdot \vec{b})\vec{c}$$

Derivatives and Integrals

In what follows, the letters u and v stand for any functions of x, and a and m are constants. To each of the indefinite integrals should be added an arbitrary constant of integration. The *Handbook of Chemistry and Physics* (CRC Press Inc.) gives a more extensive tabulation.

1. $\dfrac{dx}{dx} = 1$

2. $\dfrac{d}{dx}(au) = a\dfrac{du}{dx}$

3. $\dfrac{d}{dx}(u + v) = \dfrac{du}{dx} + \dfrac{dv}{dx}$

4. $\dfrac{d}{dx}x^m = mx^{m-1}$

5. $\dfrac{d}{dx}\ln x = \dfrac{1}{x}$

6. $\dfrac{d}{dx}(uv) = u\dfrac{dv}{dx} + v\dfrac{du}{dx}$

7. $\dfrac{d}{dx}e^x = e^x$

8. $\dfrac{d}{dx}\sin x = \cos x$

9. $\dfrac{d}{dx}\cos x = -\sin x$

10. $\dfrac{d}{dx}\tan x = \sec^2 x$

11. $\dfrac{d}{dx}\cot x = -\csc^2 x$

12. $\dfrac{d}{dx}\sec x = \tan x \sec x$

13. $\dfrac{d}{dx}\csc x = -\cot x \csc x$

14. $\dfrac{d}{dx}e^u = e^u\dfrac{du}{dx}$

15. $\dfrac{d}{dx}\sin u = \cos u\dfrac{du}{dx}$

16. $\dfrac{d}{dx}\cos u = -\sin u\dfrac{du}{dx}$

1. $\displaystyle\int dx = x$

2. $\displaystyle\int au\,dx = a\int u\,dx$

3. $\displaystyle\int (u + v)\,dx = \int u\,dx + \int v\,dx$

4. $\displaystyle\int x^m\,dx = \dfrac{x^{m+1}}{m + 1} \quad (m \neq -1)$

5. $\displaystyle\int \dfrac{dx}{x} = \ln |x|$

6. $\displaystyle\int u\dfrac{dv}{dx}\,dx = uv - \int v\dfrac{du}{dx}\,dx$

7. $\displaystyle\int e^x\,dx = e^x$

8. $\displaystyle\int \sin x\,dx = -\cos x$

9. $\displaystyle\int \cos x\,dx = \sin x$

10. $\displaystyle\int \tan x\,dx = \ln |\sec x|$

11. $\displaystyle\int \sin^2 x\,dx = \tfrac{1}{2}x - \tfrac{1}{4}\sin 2x$

12. $\displaystyle\int e^{-ax}\,dx = -\dfrac{1}{a}e^{-ax}$

13. $\displaystyle\int xe^{-ax}\,dx = -\dfrac{1}{a^2}(ax + 1)e^{-ax}$

14. $\displaystyle\int x^2 e^{-ax}\,dx = -\dfrac{1}{a^3}(a^2x^2 + 2ax + 2)e^{-ax}$

15. $\displaystyle\int_0^\infty x^n e^{-ax}\,dx = \dfrac{n!}{a^{n+1}}$

16. $\displaystyle\int_0^\infty x^{2n} e^{-ax^2}\,dx = \dfrac{1 \cdot 3 \cdot 5 \cdots (2n - 1)}{2^{n+1}a^n}\sqrt{\dfrac{\pi}{a}}$

17. $\displaystyle\int \dfrac{dx}{\sqrt{x^2 + a^2}} = \ln(x + \sqrt{x^2 + a^2})$

18. $\displaystyle\int \dfrac{x\,dx}{(x^2 + a^2)^{3/2}} = -\dfrac{1}{(x^2 + a^2)^{1/2}}$

19. $\displaystyle\int \dfrac{dx}{(x^2 + a^2)^{3/2}} = \dfrac{x}{a^2(x^2 + a^2)^{1/2}}$

20. $\displaystyle\int_0^\infty x^{2n+1} e^{-ax^2}\,dx = \dfrac{n!}{2a^{n+1}} \quad (a > 0)$

21. $\displaystyle\int \dfrac{x\,dx}{x + d} = x - d\ln(x + d)$

APPENDIX F
Properties of the Elements

All physical properties are for a pressure of 1 atm unless otherwise specified.

Element	Symbol	Atomic Number Z	Molar Mass, g/mol	Density, g/cm^3 at 20°C	Melting Point, °C	Boiling Point, °C	Specific Heat, J/(g · °C) at 25°C
Actinium	Ac	89	(227)	10.06	1323	(3473)	0.092
Aluminum	Al	13	26.9815	2.699	660	2450	0.900
Americium	Am	95	(243)	13.67	1541	—	—
Antimony	Sb	51	121.75	6.691	630.5	1380	0.205
Argon	Ar	18	39.948	1.6626×10^{-3}	−189.4	−185.8	0.523
Arsenic	As	33	74.9216	5.78	817 (28 atm)	613	0.331
Astatine	At	85	(210)	—	(302)	—	—
Barium	Ba	56	137.34	3.594	729	1640	0.205
Berkelium	Bk	97	(247)	14.79	—	—	—
Beryllium	Be	4	9.0122	1.848	1287	2770	1.83
Bismuth	Bi	83	208.980	9.747	271.37	1560	0.122
Bohrium	Bh	107	262.12	—	—	—	—
Boron	B	5	10.811	2.34	2030	—	1.11
Bromine	Br	35	79.909	3.12 (liquid)	−7.2	58	0.293
Cadmium	Cd	48	112.40	8.65	321.03	765	0.226
Calcium	Ca	20	40.08	1.55	838	1440	0.624
Californium	Cf	98	(251)	—	—	—	—
Carbon	C	6	12.01115	2.26	3727	4830	0.691
Cerium	Ce	58	140.12	6.768	804	3470	0.188
Cesium	Cs	55	132.905	1.873	28.40	690	0.243
Chlorine	Cl	17	35.453	3.214×10^{-3} (0°C)	−101	−34.7	0.486
Chromium	Cr	24	51.996	7.19	1857	2665	0.448
Cobalt	Co	27	58.9332	8.85	1495	2900	0.423
Copper	Cu	29	63.54	8.96	1083.40	2595	0.385
Curium	Cm	96	(247)	13.3	—	—	—
Dubnium	Db	105	262.114	—	—	—	—
Dysprosium	Dy	66	162.50	8.55	1409	2330	0.172
Einsteinium	Es	99	(254)	—	—	—	—
Erbium	Er	68	167.26	9.15	1522	2630	0.167
Europium	Eu	63	151.96	5.243	817	1490	0.163
Fermium	Fm	100	(237)	—	—	—	—
Fluorine	F	9	18.9984	1.696×10^{-3} (0°C)	−219.6	−188.2	0.753
Francium	Fr	87	(223)	—	(27)	—	—
Gadolinium	Gd	64	157.25	7.90	1312	2730	0.234
Gallium	Ga	31	69.72	5.907	29.75	2237	0.377

Element	Symbol	Atomic Number Z	Molar Mass, g/mol	Density, g/cm³ at 20°C	Melting Point, °C	Boiling Point, °C	Specific Heat, J/(g · °C) at 25°C
Germanium	Ge	32	72.59	5.323	937.25	2830	0.322
Gold	Au	79	196.967	19.32	1064.43	2970	0.131
Hafnium	Hf	72	178.49	13.31	2227	5400	0.144
Hassium	Hs	108	(265)	—	—	—	—
Helium	He	2	4.0026	0.1664×10^{-3}	−269.7	−268.9	5.23
Holmium	Ho	67	164.930	8.79	1470	2330	0.165
Hydrogen	H	1	1.00797	0.08375×10^{-3}	−259.19	−252.7	14.4
Indium	In	49	114.82	7.31	156.634	2000	0.233
Iodine	I	53	126.9044	4.93	113.7	183	0.218
Iridium	Ir	77	192.2	22.5	2447	(5300)	0.130
Iron	Fe	26	55.847	7.874	1536.5	3000	0.447
Krypton	Kr	36	83.80	3.488×10^{-3}	−157.37	−152	0.247
Lanthanum	La	57	138.91	6.189	920	3470	0.195
Lawrencium	Lr	103	(257)	—	—	—	—
Lead	Pb	82	207.19	11.35	327.45	1725	0.129
Lithium	Li	3	6.939	0.534	180.55	1300	3.58
Lutetium	Lu	71	174.97	9.849	1663	1930	0.155
Magnesium	Mg	12	24.312	1.738	650	1107	1.03
Manganese	Mn	25	54.9380	7.44	1244	2150	0.481
Meitnerium	Mt	109	(266)	—	—	—	—
Mendelevium	Md	101	(256)	—	—	—	—
Mercury	Hg	80	200.59	13.55	−38.87	357	0.138
Molybdenum	Mo	42	95.94	10.22	2617	5560	0.251
Neodymium	Nd	60	144.24	7.007	1016	3180	0.188
Neon	Ne	10	20.183	0.8387×10^{-3}	−248.597	−246.0	1.03
Neptunium	Np	93	(237)	20.25	637	—	1.26
Nickel	Ni	28	58.71	8.902	1453	2730	0.444
Niobium	Nb	41	92.906	8.57	2468	4927	0.264
Nitrogen	N	7	14.0067	1.1649×10^{-3}	−210	−195.8	1.03
Nobelium	No	102	(255)	—	—	—	—
Osmium	Os	76	190.2	22.59	3027	5500	0.130
Oxygen	O	8	15.9994	1.3318×10^{-3}	−218.80	−183.0	0.913
Palladium	Pd	46	106.4	12.02	1552	3980	0.243
Phosphorus	P	15	30.9738	1.83	44.25	280	0.741
Platinum	Pt	78	195.09	21.45	1769	4530	0.134
Plutonium	Pu	94	(244)	19.8	640	3235	0.130
Polonium	Po	84	(210)	9.32	254	—	—
Potassium	K	19	39.102	0.862	63.20	760	0.758
Praseodymium	Pr	59	140.907	6.773	931	3020	0.197
Promethium	Pm	61	(145)	7.22	(1027)	—	—
Protactinium	Pa	91	(231)	15.37 (estimated)	(1230)	—	—
Radium	Ra	88	(226)	5.0	700	—	—
Radon	Rn	86	(222)	9.96×10^{-3} (0°C)	(−71)	−61.8	0.092
Rhenium	Re	75	186.2	21.02	3180	5900	0.134

Element	Symbol	Atomic Number Z	Molar Mass, g/mol	Density, g/cm^3 at 20°C	Melting Point, °C	Boiling Point, °C	Specific Heat, J/(g · °C) at 25°C
Rhodium	Rh	45	102.905	12.41	1963	4500	0.243
Rubidium	Rb	37	85.47	1.532	39.49	688	0.364
Ruthenium	Ru	44	101.107	12.37	2250	4900	0.239
Rutherfordium	Rf	104	261.11	—	—	—	—
Samarium	Sm	62	150.35	7.52	1072	1630	0.197
Scandium	Sc	21	44.956	2.99	1539	2730	0.569
Seaborgium	Sg	106	263.118	—	—	—	—
Selenium	Se	34	78.96	4.79	221	685	0.318
Silicon	Si	14	28.086	2.33	1412	2680	0.712
Silver	Ag	47	107.870	10.49	960.8	2210	0.234
Sodium	Na	11	22.9898	0.9712	97.85	892	1.23
Strontium	Sr	38	87.62	2.54	768	1380	0.737
Sulfur	S	16	32.064	2.07	119.0	444.6	0.707
Tantalum	Ta	73	180.948	16.6	3014	5425	0.138
Technetium	Tc	43	(99)	11.46	2200	—	0.209
Tellurium	Te	52	127.60	6.24	449.5	990	0.201
Terbium	Tb	65	158.924	8.229	1357	2530	0.180
Thallium	Tl	81	204.37	11.85	304	1457	0.130
Thorium	Th	90	(232)	11.72	1755	(3850)	0.117
Thulium	Tm	69	168.934	9.32	1545	1720	0.159
Tin	Sn	50	118.69	7.2984	231.868	2270	0.226
Titanium	Ti	22	47.90	4.54	1670	3260	0.523
Tungsten	W	74	183.85	19.3	3380	5930	0.134
Un-named	Uun	110	(269)	—	—	—	—
Un-named	Uuu	111	(272)	—	—	—	—
Un-named	Uub	112	(264)	—	—	—	—
Un-named	Uut	113	—	—	—	—	—
Un-named	Unq	114	(285)	—	—	—	—
Un-named	Uup	115	—	—	—	—	—
Un-named	Uuh	116	(289)	—	—	—	—
Un-named	Uus	117	—	—	—	—	—
Un-named	Uuo	118	(293)	—	—	—	—
Uranium	U	92	(238)	18.95	1132	3818	0.117
Vanadium	V	23	50.942	6.11	1902	3400	0.490
Xenon	Xe	54	131.30	5.495×10^{-3}	−111.79	−108	0.159
Ytterbium	Yb	70	173.04	6.965	824	1530	0.155
Yttrium	Y	39	88.905	4.469	1526	3030	0.297
Zinc	Zn	30	65.37	7.133	419.58	906	0.389
Zirconium	Zr	40	91.22	6.506	1852	3580	0.276

The values in parentheses in the column of molar masses are the mass numbers of the longest-lived isotopes of those elements that are radioactive. Melting points and boiling points in parentheses are uncertain.

The data for gases are valid only when these are in their usual molecular state, such as H_2, He, O_2, Ne, etc. The specific heats of the gases are the values at constant pressure.

Source: Adapted from J. Emsley, *The Elements,* 3rd ed., 1998, Clarendon Press, Oxford. See also www.webelements.com for the latest values and newest elements.

APPENDIX G
Periodic Table of the Elements

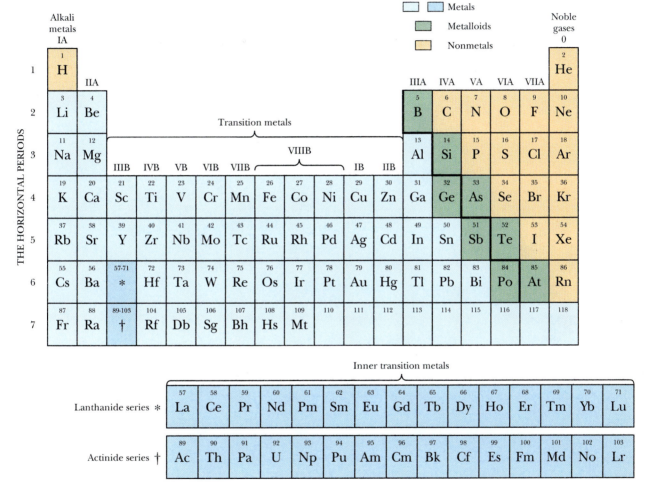

The names for elements 104 through 109 (Rutherfordium, Dubnium, Seaborgium, Bohrium, Hassium, and Meitnerium, respectively) were adopted by the International Union of Pure and Applied Chemistry (IUPAC) in 1997. Elements 110, 111, 112, 114, 116, and 118 have been discovered but, as of 2000, have not yet been named. See www.webelements.com for the latest information and newest elements.

ANSWERS
to Checkpoints and Odd-Numbered Questions, Exercises, and Problems

CHAPTER 13

CP **1.** c, e, f **2.** directly below the rod (torque on the apple due to $\vec{F}_g$, about the suspension, is zero) **3.** (a) no; (b) at site of $\vec{F}_1$, perpendicular to plane of figure; (c) 45 N **4.** (a) at C (to eliminate forces there from a torque equation); (b) plus; (c) minus; (d) equal **5.** d **6.** (a) equal; (b) B; (c) B **Q** **1.** (a) yes; (b) yes; (c) yes; (d) no **3.** a and c (forces and torques balance) **5.** $m_2 = 12$ kg, $m_3 = 3$ kg, $m_4 = 1$ kg **7.** (a) 15 N (the key is the pulley with the 10 N piñata); (b) 10 N **9.** A, then tie of B and C **EP** **1.** (a) 2; (b) 7 **3.** (a) $(-27\hat{i} + 2\hat{j})$ N; (b) 176° counterclockwise from $+x$ direction **5.** 7920 N **7.** (a) $(mg/L)\sqrt{L^2 + r^2}$; (b) mgr/L **9.** (a) 1160 N, down; (b) 1740 N, up; (c) left; (d) right **11.** 74 g **13.** (a) 280 N; (b) 880 N, 71° above the horizontal **15.** (a) 8010 N; (b) 3.65 kN; (c) 5.66 kN **17.** 71.7 N **19.** (a) 5.0 N; (b) 30 N; (c) 1.3 m **21.** $mg\dfrac{\sqrt{2rh - h^2}}{r - h}$ **23.** (a) 192 N; (b) 96.1 N; (c) 55.5 N **25.** (a) 6630 N; (b) 5740 N; (c) 5960 N **27.** 2.20 m **29.** 0.34 **31.** (a) 211 N; (b) 534 N; (c) 320 N **33.** (a) 445 N; (b) 0.50; (c) 315 N **35.** (a) slides at 31°; (b) tips at 34° **37.** (a) 6.5×10^6 N/m²; (b) 1.1×10^{-5} m **39.** (a) 867 N; (b) 143 N; (c) 0.165 **41.** (a) 51°; (b) $0.64Mg$ **N1.** (a) 30.0°; (b) 51.0 kg; (c) 10.2 kg **N3.** 340 N **N5.** 76 N **N7.** (a) 270 N; (b) 72 N; (c) 19° **N9.** 3.4 m **N11.** (a) 200 N; (b) 360 N; (c) 0.35 **N13.** 0.19 m from the left end **N15.** (a) 1380 N; (b) 180 N **N17.** 56.0 mJ **N19.** 44 N **N21.** (a) 3.9 m/s²; (b) 2000 N on each rear wheel, 3500 N on each front wheel; (c) 790 N on each rear wheel, 1410 N on each front wheel

CHAPTER 14

CP **1.** all tie **2.** (a) 1, tie of 2 and 4, then 3; (b) line d **3.** negative y direction **4.** (a) increase; (b) negative **5.** (a) 2; (b) 1 **6.** (a) path 1 (decreased E (more negative) gives decreased a); (b) less (decreased a gives decreased T) **Q** **1.** (a) between, closer to less massive particle; (b) no; (c) no (other than infinity) **3.** $3GM^2/d^2$, leftward **5.** (a) 1 and 2 tie, then 3 and 4 tie; (b) 1, 2, 3, 4 **7.** $U_i/4$ **9.** (a) all tie; (b) all tie **11.** (a)–(d) zero **EP** **1.** 19 m **3.** 29 pN **5.** 1/2 **7.** 2.60×10^5 km **9.** 0.017 N, toward the 300 kg sphere **11.** 3.2×10^{-7} N **13.** $\dfrac{GmM}{d^2}\left[1 - \dfrac{1}{8(1 - R/2d)^2}\right]$ **15.** 2.6×10^6 m **17.** (b) 1.9 h **21.** 4.7×10^{24} kg **23.** (a) $(3.0 \times 10^{-7}$ N/kg)m; (b) $(3.3 \times 10^{-7}$ N/kg)m; (c) $(6.7 \times 10^{-7}$ N/kg · m)mr **25.** (a) 9.83 m/s²; (b) 9.84 m/s²; (c) 9.79 m/s² **27.** (a) -1.3×10^{-4} J; (b) less; (c) positive; (d) negative **29.** (a) 0.74;

(b) 3.7 m/s²; (c) 5.0 km/s **31.** (a) 5.0×10^{-11} J; (b) -5.0×10^{-11} J **35.** (a) 1700 m/s; (b) 250 km; (c) 1400 m/s **37.** (a) 82 km/s; (b) 1.8×10^4 km/s **39.** 2.5×10^4 km **41.** 6.5×10^{23} kg **43.** 5×10^{10} **45.** (a) 7.82 km/s; (b) 87.5 min **47.** (a) 6640 km; (b) 0.0136 **49.** (a) 1.9×10^{13} m; (b) $3.5R_P$ **53.** 0.71 y **55.** $\sqrt{GM/L}$ **57.** (a) 2.8 y; (b) 1.0×10^{-4} **61.** (a) no; (b) same; (c) yes **63.** (a) 7.5 km/s; (b) 97 min; (c) 410 km; (d) 7.7 km/s; (e) 92 min; (f) 3.2×10^{-3} N; (g) no; (h) yes, if the satellite–Earth system is considered isolated **N1.** (a) 1.0×10^3 kg; (b) 1.6 km/s **N3.** 7.2×10^{-9} N **N5.** (a) $\frac{1}{2}$; (b) $\frac{1}{2}$; (c) B, by 1.1×10^8 J **N7.** on the x axis, at $x = -5.00d$ **N9.** $(-1.88d, -3.90d, 0.489d)$ **N11.** 1.1% **N13.** $-Gm(M_E/R + M_m/r)$ **N15.** (a) 8.83×10^{-11} J; (b) 1.88×10^{-5} m/s; (c) 5.69×10^{-9} J; (d) 7.54×10^{-5} m/s; (e) The center-of-mass frame is an inertial frame, and in it the principle of conservation of energy may be written as in Chapter 8; the reference frame attached to sphere A is a noninertial frame because of the acceleration, and the principle cannot be written as in Chapter 8. The answer to (b) is correct. **N17.** 9.24×10^{-5} rad/s **N19.** (a) $0.414R$; (b) $0.5R$ **N21.** (a) $GMmx(x^2 + R^2)^{-3/2}$; (b) $v = [2GM(R^{-1} - (R^2 + x^2)^{-1/2})]^{1/2}$

CHAPTER 15

CP **1.** all tie **2.** (a) all tie (the gravitational force on the penguin is the same); (b) $0.95\rho_0$, ρ_0, $1.1\rho_0$ **3.** 13 cm³/s, outward **4.** (a) all tie; (b) 1, then 2 and 3 tie, 4 (wider means slower); (c) 4, 3, 2, 1 (wider and lower mean more pressure) **Q** **1.** e, then b and d tie, then a and c tie **3.** (a) 2; (b) 1, less; 3, equal; 4, greater **5.** (a) moves downward; (b) moves downward **7.** all tie **9.** (a) downward; (b) downward; (c) same **EP** **1.** 1.1×10^5 Pa or 1.1 atm **3.** 2.9×10^4 N **5.** 0.074 **7.** (b) 26 kN **9.** 5.4×10^4 Pa **11.** (a) 5.3×10^6 N; (b) 2.8×10^5 N; (c) 7.4×10^5 N; (d) no **13.** 7.2×10^5 N **15.** $\frac{1}{4}\rho gA(h_2 - h_1)^2$ **17.** 1.7 km **19.** (a) $\rho gWD^2/2$; (b) $\rho gWD^3/6$; (c) $D/3$ **21.** (a) 7.9 km; (b) 16 km **23.** 4.4 mm **25.** (a) 2.04×10^{-2} m³; (b) 1570 N **27.** (a) 670 kg/m³; (b) 740 kg/m³ **29.** (a) 1.2 kg; (b) 1300 kg/m³ **31.** 57.3 cm **33.** 0.126 m³ **35.** (a) 45 m²; (b) car should be over center of slab if slab is to be level **37.** (a) 9.4 N; (b) 1.6 N **39.** 8.1 m/s **41.** 66 W **43.** (a) 2.5 m/s; (b) 2.6×10^5 Pa **45.** (a) 3.9 m/s; (b) 88 kPa **47.** (a) 1.6×10^{-3} m³/s; (b) 0.90 m **49.** 116 m/s **51.** (a) 6.4 m³; (b) 5.4 m/s; (c) 9.8×10^4 Pa **53.** (a) 74 N; (b) 150 m³ **55.** (b) 2.0×10^{-2} m³/s **57.** (b) 63.3 m/s **59.** (a) 180 kN; (b) 81 kN; (c) 20 kN; (d) 0; (e) 78 kPa; (f) no

61. (a) 0.050; (b) 0.41; (c) no; (d) Lay back on the surface, slowly pull your legs free, and then roll over to the shore. **N1.** (a) 0.25 m²; (b) 6.13 m³/s   **N3.** 8.80×10^{-2} N   **N5.** 5.11×10^{-7} kg   **N7.** 4.00 cm   **N9.** 17 cm   **N11.** 4.69×10^5 N   **N13.** 9.7 mm   **N15.** 1.5 g/cm³   **N19.** 43 cm/s   **N21.** 0.64 atm   **N23.** -1.1×10^3 Pa   **N25.** 44.2 g

CHAPTER 16

CP   **1.** (sketch x versus t) (a) $-x_m$; (b) $+x_m$; (c) 0   **2.** a (F must have form of Eq. 6-10)   **3.** (a) 5 J; (b) 2 J; (c) 5 J   **4.** all tie (in Eq. 16-29, m is included in I)   **5.** 1, 2, 3 (the ratio m/b matters; k does not)   **Q**   **1.** c   **3.** (a) 2; (b) positive; (c) between 0 and $+x_m$ **5.** (a) toward $-x_m$; (b) toward $+x_m$; (c) between $-x_m$ and 0; (d) between $-x_m$ and 0; (e) decreasing; (f) increasing   **7.** (a) π rad; (b) π rad; (c) $\pi/2$ rad   **9.** (a) varies; (b) varies; (c) $x = \pm x_m$; (d) more likely   **11.** b (infinite period; does not oscillate), c, a   **13.** one system: $k = 1500$ N/m, $m = 500$ kg; other system: $k = 1200$ N/m, $m = 400$ kg; the same ratio $k/m = 3$ gives resonance for both systems   **EP**   **1.** (a) 0.50 s; (b) 2.0 Hz; (c) 18 cm   **3.** (a) 0.500 s; (b) 2.00 Hz; (c) 12.6 rad/s; (d) 79.0 N/m; (e) 4.40 m/s; (f) 27.6 N   **5.** $f > 500$ Hz   **7.** (a) 6.28×10^5 rad/s; (b) 1.59 mm   **9.** (a) 1.0 mm; (b) 0.75 m/s; (c) 570 m/s²   **11.** (a) 1.29×10^5 N/m; (b) 2.68 Hz   **13.** 7.2 m/s   **15.** 2.08 h   **17.** 3.1 cm   **19.** (a) 5.58 Hz; (b) 0.325 kg; (c) 0.400 m   **21.** (a) 2.2 Hz; (b) 56 cm/s; (c) 0.10 kg; (d) 20.0 cm below y_i **23.** (a) 0.183A; (b) same direction   **29.** (a) $(n + 1)k/n$; (b) $(n + 1)k$; (c) $\sqrt{(n + 1)/n}\,f$; (d) $\sqrt{n + 1}\,f$ **31.** 37 mJ   **33.** (a) 2.25 Hz; (b) 125 J; (c) 250 J; (d) 86.6 cm   **35.** (a) 130 N/m; (b) 0.62 s; (c) 1.6 Hz; (d) 5.0 cm; (e) 0.51 m/s   **37.** (a) $\frac{3}{4}$; (b) $\frac{1}{4}$; (c) $x_m/\sqrt{2}$ **39.** (a) 16.7 cm; (b) 1.23%   **41.** (a) 39.5 rad/s; (b) 34.2 rad/s; (c) 124 rad/s²   **43.** 99 cm   **45.** 5.6 cm   **47.** (a) $2\pi\sqrt{\dfrac{L^2 + 12d^2}{12gd}}$; (b) increases for $d < L/\sqrt{12}$, decreases for $d > L/\sqrt{12}$; (c) increases; (d) no change   **49.** (a) 0.205 kg·m²; (b) 47.7 cm; (c) 1.50 s   **53.** $2\pi\sqrt{m/3k}$ **55.** (a) 0.35 Hz; (b) 0.39 Hz; (c) 0   **57.** (b) smaller   **59.** 0.39   **61.** (a) 14.3 s; (b) 5.27   **63.** (a) $F_m/b\omega$; (b) F_m/b **N1.** $+1.91$ rad (or -4.37 rad)   **N3.** $+1.82$ rad (or -4.46 rad)   **N5.** the pendulums with lengths of 0.80 m and 1.2 m   **N7.** 833 N/m   **N9.** no, it turns back at $x = 12$ cm   **N11.** (a) 0.50 m; (b) 0.939 mJ   **N13.** (a) 11 m/s; (b) 1.7×10^3 m/s²   **N15.** (a) 0.015; (b) no   **N17.** 1.53 m   **N19.** (a) $y_m = 0.008$ m, $T = 0.18$ s, $\omega = 35$ rad/s; (b) $y_m = 0.07$ m, $T = 0.48$ s, $\omega = 13$ rad/s; (c) $y_m = 0.03$ m, $T = 0.31$ s, $\omega = 20$ rad/s   **N21.** (a) 2.0 s; (b) 18.5 N·m/rad   **N23.** 0.93 s   **N25.** (a) 1.72 ms; (b) 11.2 ms   **N27.** (a) 228 g, 477 g; (c) 8.0 cm

CHAPTER 17

CP   **1.** a, 2; b, 3; c, 1 (compare with phase in Eq. 17-2, then see Eq. 17-5)   **2.** (a) 2, 3, 1 (see Eq. 17-12); (b) 3, then 1 and 2 tie (find amplitude of dy/dt)   **3.** (a) same (independent of f); (b) decrease ($\lambda = v/f$); (c) increase; (d) increase   **4.** (a) increase; (b) increase; (c) increase   **5.** 0.20 and 0.80 tie, then 0.60, 0.45   **6.** (a) 1; (b) 3; (c) 2   **7.** (a) 75 Hz; (b) 525 Hz   **Q**   **1.** 7d **3.** (a) $\pi/2$ rad and 0.25 wavelength; (b) π rad and 0.5 wavelength; (c) $3\pi/2$ rad and 0.75 wavelength; (d) 2π rad and 1.0

wavelength; (e) $3T/4$; (f) $T/2$ **5.** (a) 4; (b) 4; (c) 3   **7.** a and d tie, then b and c tie   **9.** d   **11.** (a) decrease; (b) disappears   **EP**   **1.** (a) 3.49 m^{-1}; (b) 31.5 m/s   **3.** (a) 0.68 s; (b) 1.47 Hz; (c) 2.06 m/s   **7.** (a) $y(x, t) = 2.0 \sin 2\pi(0.10x - 400t)$, with x and y in cm and t in s; (b) 50 m/s; (c) 40 m/s   **9.** (a) 11.7 cm; (b) π rad   **11.** 129 m/s   **13.** (a) 15 m/s; (b) 0.036 N   **15.** $y(x, t) = 0.12 \sin(141x + 628t)$, with y in mm, x in m, and t in s   **17.** (a) $2\pi y_m/\lambda$; (b) no   **19.** (a) 5.0 cm; (b) 40 cm; (c) 12 m/s; (d) 0.033 s; (e) 9.4 m/s; (f) $5.0 \sin(16x + 190t + 0.93)$, with x in m, y in cm, and t in s   **21.** 2.63 m from the end of the wire from which the later pulse originates   **25.** (a) 3.77 m/s; (b) 12.3 N; (c) zero; (d) 46.3 W; (e) zero; (f) zero; (g) ± 0.50 cm   **27.** 1.4y_m **29.** 5.0 cm   **31.** (a) 0.83y_1; (b) 37°   **33.** (a) 140 m/s; (b) 60 cm; (c) 240 Hz   **35.** (a) 82.0 m/s; (b) 16.8 m; (c) 4.88 Hz   **37.** 7.91 Hz, 15.8 Hz, 23.7 Hz   **39.** (a) 105 Hz; (b) 158 m/s   **41.** (a) 0.25 cm; (b) 120 cm/s; (c) 3.0 cm; (d) zero   **43.** (a) 50 Hz; (b) $y = 0.50 \sin[\pi(x \pm 100t)]$, with x in m, y in cm, and t in s   **45.** (a) 1.3 m; (b) $y = 0.002 \sin(9.4x) \cos(3800t)$, with x and y in m and t in s   **47.** (a) 2.0 Hz; (b) 200 cm; (c) 400 cm/s; (d) 50 cm, 150 cm, 250 cm, etc.; (e) 0, 100 cm, 200 cm, etc.   **51.** (a) 323 Hz; (b) eight   **N1.** $y(x, t) = (3.0$ mm$)$ $\sin[(16$ m$^{-1})x - (240$ s$^{-1})t]$ **N3.** π rad $- 0.3398$ rad ≈ 2.8 rad, or $-(\pi$ rad $+ 0.3398$ rad$) \approx -3.5$ rad   **N5.** π rad $- 0.2527$ rad ≈ 2.9 rad, or $-(\pi$ rad $+ 0.2527$ rad$) \approx -3.4$ rad   **N7.** 4.24 m/s   **N9.** 1.1 ms   **N11.** (a) 0.16 m; (b) 240 N; (c) $y(x, t) = 0.16$ $\sin(\pi x/2) \sin(10\pi t)$, with y and x in meters and t in seconds   **N13.** (a) 3.29 mm; (b) 1.55 rad; (c) 1.55 rad   **N17.** (a) 10 W; (b) 20 W; (c) 40 W; (d) 26 W; (e) 0   **N19.** 0.845 g/m   **N21.** 260 Hz

CHAPTER 18

CP   **1.** beginning to decrease (example: mentally move the curves of Fig. 18-7 rightward past the point at $x = 42$ m)   **2.** (a) 0, fully constructive; (b) 4λ, fully constructive   **3.** (a) 1 and 2 tie, then 3 (see Eq. 18-28); (b) 3, then 1 and 2 tie (see Eq. 18-26)   **4.** second (see Eqs. 18-39 and 18-41)   **5.** loosen   **6.** a, greater; b, less; c, can't tell; d, can't tell; e, greater; f, less   **7.** (measure speeds relative to the air) (a) 222 m/s; (b) 222 m/s   **Q**   **1.** pulse along path 2   **3.** (a) 2.0 wavelengths; (b) 1.5 wavelengths; (c) fully constructive, fully destructive   **5.** (a) exactly out of phase; (b) exactly out of phase   **7.** (a) one; (b) nine   **9.** (a) increase; (b) decrease   **11.** all odd harmonics   **13.** d, e, b, c, a **EP**   **1.** divide the time by 3   **3.** (a) 79 m, 41 m; (b) 89 m   **5.** 1900 km   **7.** 40.7 m   **9.** (a) 0.0762 mm; (b) 0.333 mm   **11.** (a) 1.50 Pa; (b) 158 Hz; (c) 2.22 m; (d) 350 m/s   **13.** (a) $343(1 + 2m)$ Hz, with m being an integer from 0 to 28; (b) $686m$ Hz, with m being an integer from 1 to 29   **15.** (a) 143 Hz, 429 Hz, 715 Hz; (b) 286 Hz, 572 Hz, 858 Hz   **17.** 15.0 mW   **19.** 36.8 nm   **21.** (a) 1000; (b) 32   **23.** (a) 59.7; (b) 2.81×10^{-4} **25.** (b) 5.76×10^{-17} J/m³   **27.** (b) length²   **29.** (a) 5200 Hz; (b) amplitude$_{SAD}$/amplitude$_{SBD}$ = 2   **31.** (a) 57.2 cm; (b) 42.9 cm   **33.** (a) 405 m/s; (b) 596 N; (c) 44.0 cm; (d) 37.3 cm   **35.** (a) 1129, 1506, and 1882 Hz   **37.** 12.4 m   **39.** (a) node; (b) 22 s   **41.** 45.3 N   **43.** 387 Hz   **45.** 0.02   **47.** 17.5 kHz   **49.** (a) 526 Hz; (b) 555 Hz   **51.** (a) 1.02 kHz; (b) 1.04 kHz   **53.** 155 Hz   **55.** (a) 485.8 Hz; (b) 500.0 Hz; (c) 486.2 Hz; (d) 500.0 Hz   **57.** (a) 598 Hz;

(b) 608 Hz; (c) 589 Hz **59.** (a) 42°; (b) 11 s **N1.** (a) 572 Hz; (b) 1140 Hz **N3.** (a) $s(x, t) = (6.1 \text{ nm}) \cos[(9.2 \text{ m}^{-1})x - (3.1 \times 10^3 \text{ s}^{-1})t]$; (b) $s(x, t) = (5.9 \text{ nm}) \cos[(9.8 \text{ m}^{-1})x - (3.1 \times 10^3 \text{ s}^{-1})t]$ **N5.** $0.236v$ **N7.** (a) 5 dB (same as at any plotted distance r); (b) 3.2 **N9.** (a) 21 nm; (b) 35 cm; (c) 24 nm; (d) 35 cm **N11.** (a) 14; (b) 14 **N13.** (a) 0.34 nW; (b) 0.68 nW; (c) 1.4 nW; (d) 0.88 nW; (e) 0 **N15.** (a) $L(V - v)/Vv$; (b) 364 m **N17.** 39.3 Hz and 118 Hz **N19.** 20 kHz **N21.** (a) 88 mW/m²; (b) $A_4 = 0.75A_3$ **N23.** no; assuming a point source, the power is only 18 W **N25.** second, sixth, and tenth **N27.** If only the length is uncertain, it must be known to within 10^{-3} cm. If only the time is imprecise, the uncertainty must be no more than one part in 6000. **N31.** 400 Hz

CHAPTER 19

CP **1.** (a) all tie; (b) 50°X, 50°Y, 50°W **2.** (a) 2 and 3 tie, then 1, then 4; (b) 3, 2, then 1 and 4 tie (from Eqs. 19-9 and 19-10, assume that change in area is proportional to initial area) **3.** A (see Eq. 19-14) **4.** c and e (maximize area enclosed by a clockwise cycle) **5.** (a) all tie (ΔE_{int} depends on i and f, not on path); (b) 4, 3, 2, 1 (compare areas under curves); (c) 4, 3, 2, 1 (see Eq. 19-26) **6.** (a) zero (closed cycle); (b) negative (W_{net} is negative; see Eq. 19-26) **7.** b and d tie, then a, c (P_{cond} identical; see Eq. 19-32) **Q** **1.** 25 S°, 25 U°, 25 R° **3.** A and B tie, then C, D **5.** (a) both clockwise; (b) both clockwise **7.** c, a, b **9.** sphere, hemisphere, cube **11.** (a) at freezing point; (b) no liquid freezes; (c) ice partly melts **EP** **1.** 0.05 kPa, nitrogen **3.** 348 K **5.** (a) $-40°$; (b) 575°; (c) Celsius and Kelvin cannot give the same reading **7.** (a) Dimensions are inverse time. **9.** $-92.1°$X **11.** 960 μm **13.** 2.731 cm **15.** 29 cm³ **17.** 0.26 cm³ **19.** 360°C **23.** 0.68 s/h, fast **25.** 7.5 cm **27.** (a) 523 J/kg·K; (b) 26.2 J/mol·K; (c) 0.600 mole **29.** 42.7 kJ **31.** 1.9 times as great **33.** (a) 33.9 Btu; (b) 172 F° **35.** 160 s **37.** 2.8 days **39.** 742 kJ **41.** 82 cal **43.** 33 g **45.** (a) 0°C; (b) 2.5°C **47.** 8.72 g **49.** A: 120 J, B: 75 J, C: 30 J **51.** -30 J **53.** (a) 6.0 cal; (b) -43 cal; (c) 40 cal; (d) 18 cal, 18 cal **55.** (a) 0.13 m; (b) 2.3 km **57.** 1660 J/s **59.** (a) 16 J/s; (b) 0.048 g/s **61.** 0.50 min **63.** (a) 17 kW/m²; (b) 18 W/m² **65.** 0.40 cm/h **67.** (a) 90 W; (b) 230 W; (c) 330 W **N1.** $-92.1°$X **N5.** 333 J **N7.** 220 m/s **N9.** (a) 415 g; (b) 0.031\$US **N11.** C is proportional to $T^{-2/3}$ **N13.** 9.1 s; the clock is running slowly **N15.** (b) 71 cm; (c) 41 cm **N17.** 0.27 mm **N19.** 66°C **N21.** 21.3°C **N23.** 4.83×10^{-2} cm³ **N25.** 10.5°C

CHAPTER 20

CP **1.** all but c **2.** (a) all tie; (b) 3, 2, 1 **3.** gas A **4.** 5 (greatest change in T), then tie of 1, 2, 3, and 4 **5.** 1, 2, 3 ($Q_3 = 0$, Q_2 goes into work W_2, but Q_1 goes into greater work W_1 and increases gas temperature) **Q** **1.** increased but less than doubled **3.** tie of a and c, then b, then d **5.** 1–4 **7.** 20 J **9.** (a) 3; (b) 1; (c) 4; (d) 2; (e) yes **11.** (a) 1, 2, 3, 4; (b) 1, 2, 3

EP **1.** 0.933 kg **3.** 6560 **5.** (a) 5.47×10^{-8} mol; (b) 3.29×10^{16} **7.** (a) 0.0388 mol; (b) 220°C **9.** (a) 106; (b) 0.892 m³ **11.** $A(T_2 - T_1) - B(T_2^2 - T_1^2)$ **13.** 5600 J **15.** 100 cm³ **17.** 2.0×10^5 Pa **19.** 180 m/s **21.** 9.53×10^6 m/s **23.** 1.9 kPa **25.** 3.3×10^{-20} J **27.** (a) 6.75×10^{-20} J; (b) 10.7 **31.** (a) 6×10^9 km **33.** 15 cm **35.** (a) 3.27×10^{10}; (b) 172 m **37.** (a) 6.5 km/s; (b) 7.1 km/s **39.** (a) 1.0×10^4 K; (b) 1.6×10^5 K; (c) 440 K, 7000 K; (d) hydrogen, no; oxygen, yes **41.** (a) 7.0 km/s; (b) 2.0×10^{-8} cm; (c) 3.5×10^{10} collisions/s **43.** (a) $\frac{2}{3}v_0$; (b) $N/3$; (c) $122v_0$; (d) $1.31v_0$ **45.** $RT \ln(V_f/V_i)$ **47.** $(n_1C_1 + n_2C_2 + n_3C_3)/(n_1 + n_2 + n_3)$ **49.** (a) 6.6×10^{-26} kg; (b) 40 g/mol **51.** 8000 J **53.** (a) 6980 J; (b) 4990 J; (c) 1990 J; (d) 2990 J **55.** (a) 14 atm; (b) 620 K **59.** 1.40 **61.** (a) In joules, in the order Q, ΔE_{int}, W: $1 \rightarrow 2$: 3740, 3740, 0; $2 \rightarrow 3$: 0, -1810, 1810; $3 \rightarrow 1$: -3220, -1930, -1290; Cycle: 520, 0, 520; (b) $V_2 = 0.0246$ m³, $p_2 = 2.00$ atm, $V_3 = 0.0373$ m³, $p_3 = 1.00$ atm **N1.** 1.9×10^4 K **N3.** (a) $(p_f V_f - p_i V_i)/(1 - \gamma)$ **N5.** (b) 125 J; (c) to the gas **N7.** 9.2×10^{-6} **N9.** 349 K **N11.** (a) 1.44×10^3 m/s; (b) 5.78×10^{-4}; (c) 71%; (d) 2.03×10^3 m/s; (e) 4.09×10^{-4}; (f) increase; (g) decrease **N13.** 3.6 GHz **N15.** 7.73×10^3 K **N17.** tablespoon **N19.** (a) diatomic; (b) 446 K; (c) 8.10 **N21.** (b) work done by environment: 7.72×10^4 J, energy absorbed as heat: 5.46×10^4 J; (c) 5.17 J/mol·K; (d) work done by environment: 4.32×10^4 J, energy absorbed as heat: 8.86×10^4 J, molar specific heat: 8.38 J/mol·K **N23.** (a) 35.73; (b) decreased; (c) increased

CHAPTER 21

CP **1.** a, b, c **2.** smaller (Q is smaller) **3.** c, b, a **4.** a, d, c, b **5.** b **Q** **1.** unchanged **3.** b, a, c, d **5.** equal **7.** (a) same; (b) increase; (c) decrease **9.** (a) same; (b) increase; (c) decrease **11.** (a) 0; (b) 0.25; (c) 0.50 **EP** **1.** 14.4 J/K **3.** (a) 9220 J; (b) 23.0 J/K; (c) 0 **5.** (a) 5.79×10^4 J; (b) 173 J/K **7.** (a) 14.6 J/K; (b) 30.2 J/K **9.** (a) 57.0°C; (b) -22.1 J/K; (c) $+24.9$ J/K; (d) $+2.8$ J/K **13.** (a) 320 K; (b) 0; (c) $+1.72$ J/K **15.** $+0.75$ J/K **17.** (a) -943 J/K; (b) $+943$ J/K; (c) yes **19.** (a) $3p_0V_0$; (b) $\Delta E_{int} = 6RT_0$, $\Delta S = \frac{3}{2}R \ln 2$; (c) both are zero **21.** (a) 31%; (b) 16 kJ **23.** (a) 23.6%; (b) 1.49×10^4 J **25.** 266 K and 341 K **27.** (a) 1470 J; (b) 554 J; (c) 918 J; (d) 62.4% **29.** (a) 2270 J; (b) 14 800 J; (c) 15.4%; (d) 75.0%, greater **31.** (a) 78%; (b) 81 kg/s **33.** (a) $T_2 = 3T_1$, $T_3 = 3T_1/4^{\gamma-1}$, $T_4 = T_1/4^{\gamma-1}$, $p_2 = 3p_1$, $p_3 = 3p_1/4^\gamma$, $p_4 = p_1/4^\gamma$; (b) $1 - 4^{1-\gamma}$ **35.** 21 J **37.** 440 W **39.** 0.25 hp **41.** $[1 - (T_2/T_1)]/[1 - (T_4/T_3)]$ **45.** (a) $W = N!/(n_1! \, n_2! \, n_3!)$; (b) $[(N/2)! \, (N/2)!]/[(N/3)! \, (N/3)! \, (N/3)!]$; (c) 4.2×10^{16} **N1.** (a) 4.45 J/K; (b) no **N3.** 0.0368 J/K **N11.** (a) 4.67 kJ/s; (b) 4.17 kJ/s **N17.** (a) 66.5°C; (b) 14.6 J/K; (c) 11.0 J/K; (d) -21.2 J/K; (e) 4.40 J/K **N19.** (a) 3.75; (b) 710 J **N21.** 0.141 J/K·s **N23.** (a) 1; (b) 6; (c) 0; (d) 2.47×10^{-23} J/K

PHOTO CREDITS

INDEX

Figures are noted by page numbers in *italics;* tables are indicated by t following the page number.